Directory of Sport Science

5th Edition

—

A Journey Through Time – The Changing Face of ICSSPE

ICSSPE

CIEPSS

International Council of Sport Science and Physical Education (ICSSPE)

Human Kinetics
Distribution Partner

Dedicated to Professor Dr. Gudrun Doll-Tepper, ICSSPE President

Directory of Sport Science

5th Edition

—

A Journey Through Time – The Changing Face of ICSSPE

Jan Borms

ICSSPE

Distribution Partner: Human Kinetics

British Library Cataloguing in Publication Data
A catalogue record for this book is available from the British Library

Directory of Sport Science, 5ᵗʰ Edition
Jan Borms (ed.)
ISSN: 1729-3227
ISBN-10: 0-7360-8736-2
ISBN-13: 978-0-7360-8736-0

© 2008 by ICSSPE
Managing Editor: Katrin Koenen
Language Editor: Tamie Needham
Layout and Typesetting: Astrid Lange and Anke Thomas
Cover Design: Astrid Lange
Graphics: Jörg Mailliet

Printed in the United States of America   10   9   8   7   6   5   4   3   2   1

Distribution Partner:
Human Kinetics
Web site: www.HumanKinetics.com

United States: Human Kinetics
P.O. Box 5076
Champaign, IL 61825-5076
800-747-4457
e-mail: humank@hkusa.com

Canada: Human Kinetics
475 Devonshire Road Unit 100
Windsor, ON N8Y 2L5
800-465-7301 (in Canada only)
e-mail: info@hkcanada.com

Europe: Human Kinetics
107 Bradford Road
Stanningley
Leeds LS28 6AT
United Kingdom
+44 (0) 113 255 5665
e-mail: hk@hkeurope.com

Australia: Human Kinetics
57A Price Avenue
Lower Mitcham
South Australia 5062
08 8372 0999
e-mail: info@hkaustralia.com

New Zealand: Human Kinetics
Division of Sports Distributors NZ Ltd.
P.O. Box 300 226 Albany
North Shore City
Auckland
0064 9 448 1207
e-mail: info@humankinetics.co.nz

Directory of Sport Science, 5ᵗʰ Edition

A Journey Through Time – The Changing Face of ICSSPE

Part I:
A Journey Through Time – The Changing Face of ICSSPE

Part II:
Directory of Sport Science, 5ᵗʰ Edition

Sport Scientific Disciplines

Thematic Areas of Sport Science

Appendix

Foreword

Comte Dr. Jacques Rogge
President of the International Olympic Committee

Happy 50th anniversary to the International Council of Sport Science and Physical Education (ICSSPE)! Fifty years of existence and presence on the international sports science stage is an achievement that deserves our recognition and congratulations.

To mark this unique occasion, the ICSSPE is publishing a special edition of the "Directory of Sport Science", highlighting over 20 sports science disciplines and thematic areas, and providing an overview of the past 10 years of the ICSSPE.

As I am myself from the medical profession, I know the work carried out over the years by the ICSSPE for the development and promotion of a healthy lifestyle based on regular physical activity – one of the principles of the Olympic Movement.

As you know, sports science is a fundamental part of our knowledge, and has developed particularly in recent decades as our society has become more interested in physical activity, nutrition and health promotion, as well as in socio-cultural aspects and psychology.

It is thanks to the close collaboration between the IOC and sports organisations such as the ICSSPE that sciences applied to sport and physical education occupy an important place at the service of the athlete, whether amateur or elite.

Once again, a happy anniversary to the ICSSPE family! And please accept my best wishes for all your undertakings to promote sports science in the sports world.

Foreword

Wilfried Lemke
Special Adviser to the United Nations Secretary-General
on Sport for Development and Peace
United Nations Under-Secretary-General

I congratulate the International Council of Sport Science and Physical Education (ICSSPE) on its 50th anniversary.

For decades, ICSSPE's fundamental goal has been to support scientific research in sport, physical education and physical activity. In its global efforts, ICSSPE has always strived to provide access to physical education and sport to all, irrespective of their social, cultural and economic background, age, sex or religion.

Since the creation of the mandate of the Special Adviser to the United Nations Secretary-General on Sport for Development and Peace in 2001, the United Nations has placed a strong emphasis on sport as a means to promote education, health, development and peace. In many of the projects implemented during the International Year of Sport and Physical Education 2005, ICSSPE was an important partner; since the International Year, ICCSPE has continued to successfully advocate for sport as an instrument in post-disaster interventions and peacebuilding activities.

ICSSPE's endeavours in assisting members and partners to resort to sport in achieving the Millennium Development Goals, as well as ICSSPE's projects relating to sport and development issues continue to be very much in line with my mandate as the Special Adviser.

It gives me great pleasure to help honour ICSSPE, a dedicated partner of the United Nations, on its 50th anniversary and I look forward to a continued synergy between our two organisations.

Preface

Prof. em. Dr. Jan Borms
Editor

The year 2008 marks the 50th anniversary of ICSSPE, the International Council of Sport Science and Physical Education. The Editorial Board of ICSSPE, meeting in March 2007 in Pretoria, proposed a special publication in conjunction with this memorable event in combination with the 5th Edition of its Directory of Sport Science. The concept of this publication was developed at ICSSPE's Headquarters in Berlin and accepted at the Editorial Board meeting in Warsaw that same year. I was approached in September to become the editor of the volume and author of the historical chapter of the past decade of the Council's activities. Once all arrangements were settled, not much time remained since one of the challenges of this undertaking was to have this publication released in August 2008, on the occasion of the first International Convention on Science, Education and Medicine in Sport in Guangzhou, China.

"Sport Science and Physical Education: A Journey Through Time", the name conceived for this volume, reflects the activities and developments of Sport Science and Physical Education in the world within the framework of ICSSPE's goals and mission.

Part 1 of this book features a historical chapter about ICSSPE. It was more than a challenge to pick up the history of ICSSPE from the time where Steve Bailey had left it (1996). Steve is the author of the magnum opus "Science in the Service of Physical Education and Sport", a complete history of the first forty years of ICSPE/ICSSPE (1958-1996). The chapter, named "A Journey Through Time, The Changing Face of ICSSPE" continues the story of development of the world's foremost international organisation representing sport, sport science and physical education from 1996 until the present day. We did not specifically follow a chronological model but rather preferred to research specific themes that were landmark in ICSSPE's recent history. Within these themes, however, we followed a chronological line. I delved into my personal archives and those of the central administration, especially the several hundreds of pages of meeting minutes of ICSSPE's Boards to have objective witnesses. My lasting impression is that of an immense amount of work and service performed by motivated and committed people. To increase the readability of this text and also to make it accessible for a large readership, I tried to merge anecdotal with factual information.

Part 2, the most comprehensive part in this book, contains the Directory of Sport Science edition No 5. The first edition released in 1998, then named "Vade Mecum", included eleven scientific disciplines and evolved into the Directory of Sport Science. The current edition covers sixteen scientific disciplines ranging from Adapted Physical Activity to Sport Pedagogy and introduces also the development of nine thematic areas of sport science for example Women and Sport, HEPA, Physical Education, Sport and Development, Sport and Human Rights, Doping in Sport, and Sport Governance.

The objective is to disseminate up-to-date, structured, and easy-to-use information about the different sport science disciplines and thematic areas, including many links to networks and websites. The contributors to this Directory are recognised scholars who have all been involved in their respective disciplines.

Publishing a book like this is a major undertaking in which many persons are involved. I would therefore like to acknowledge all guest authors for their willingness to write or update the featured articles on the scientific disciplines and the thematic areas of sport science. We are also thankful for the reviewers whose expertise, professional and scientific dedication helped significantly in the accomplishment of this delicate task. I would also like to thank Gudrun Doll-Tepper, Christophe Mailliet and Detlef Dumon for interviews to clarify sometimes conflicting information.

I would like to extend a special "thank you" to two persons: Katrin Koenen, the current Publications and Scientific Affairs Manager, was the trusted coordinator of this project, and furthermore, a pleasant person to work with. Tamie Devine, with whom I worked in my capacity as chair of the Editorial Board from 2003 unil the end of 2004 when we both retired from ICSSPE, was a critical but outstanding reviewer of the historical chapter. I was very pleased that she enjoyed so much to be once more involved in an ICSSPE project and the feelings were reciprocal.

Last but not least, it was a pleasure to 'return' to ICSSPE, although it may have distracted me from my leisure time activities as an emeritus professor! It was exciting to know that this book was dedicated to Gudrun Doll-Tepper in recognition of her twelve years of ICSSPE's Presidency and especially that this had to be kept as a secret until the truth was revealed the day of the presentation of this book.

Jan Borms

Affligem, June 2008

Contact

Prof. Dr. em. Jan Borms
Blakmeers 31
1790 Affligem
Belgium
Phone: +32 (0)53 667108
Email: jborms@skynet.be

Part I:
A Journey Through Time –
The Changing Face of ICSSPE

Jan Borms

The International Council of Sport and Physical Education (ICSPE) was created in 1958 through the will and commitment of a group, which included Ernst Jokl, Carl Diem, Fritz Duras and William Jones, who saw the need for a world-wide representative group for the growing fields of sport science. So dynamic was activity in the field that a name change occurred in 1983 to the International Council of Sport Science and Physical Education (ICSSPE).

Instrumental in the creation, development and elevation of ICSSPE to become the global force that it is today in advocacy, service and knowledge dissemination in the various fields of sport, sport science and physical education, have been the well regarded Presidents. These individuals, with their unique skill sets and vision, worked collaboratively with ICSSPE members to place the Council in the well-respected stead that it enjoys today.

Fritz Duras served from 1956 to 1960 as Honorary President of the Provisional Committee. Sir Phillip Noel-Baker served from 1960 to 1976, Sir Roger Bannister from 1976 to 1982, August Kirsch from 1983 to 1990, Paavo Komi from 1991 to 1996 and Prof. Dr Gudrun Doll-Tepper from 1996 to 2008.

The Presidents were each ably served by Secretary Generals during their terms. The Secretary Generals (later Executive Directors) were:

R. Williams Jones   1960-1968
Julien Falize   1968-1976
John Coghlan   1976-1982
Werner Sonnenschein   1983-1990

Sonja Boelaert-Suominen 1991-1992
André-Noël Chaker   1992-1996
Christophe Mailliet   1996-2006
Detlef Dumon   2006-present

A complete history of the first forty years of ICSPE/ICSSPE (1958-1998) has been detailed in Steve Bailey's magnum opus "Science in the Service of Physical Education and Sport"1. This period of time essentially saw important relationships and links forged with bodies including the United Nations Educational, Scientific and Cultural Organisation (UNESCO), the World Health Organisation (WHO), the International Olympic Committee (IOC) and other international organisations that were instituted during the same time frame. Key activities and events with which the Council was involved included: establishment of the Research Committee, which supported congresses and scientific meetings around the world; the Declaration of Sport, credited with bringing attention to fair play and ethics in sport; creation of the Pierre de Coubertin International Fair Play Trophies; the First International Conference of Ministers and Senior Officials Responsible for Physical Education and Sport (MINEPS); signing of a Cooperation Agreement between ICSSPE and IOC; support to developing countries; and coordination of activities in the Sport for All field.

The following chapter picks up the history of ICSSPE from 1996 and continues the story of development of the world's foremost international organisation representing sport, sport science and physical education.

1996, A New Administration for a New President

In 1990, Gudrun Doll-Tepper, professor at the Freie Universität in Berlin, Germany, became a formal member of ICSSPE when she was delegated to the General Assembly as a representative of the International Federation of Adapted Physical Activity. Prior to this, during the late 1980s, she had followed the activities of ICSSPE and had received patronage for the International Symposium on Adapted Physical Activity held in Berlin, June, 1989. She became a member of the Executive Board and from 1990 to the present, has never missed a meeting, even hosting it in Berlin in 1992. In 1995, when the then ICSSPE President, Paavo Komi, decided not to complete his term of office, he asked Prof. Dr. Doll-Tepper, on behalf of the President's Committee, if she would consider nominating to succeed him [2]. Prof. Dr. John Kane, from England was another candidate for the position but withdrew prior to the elections in Dallas at the 1996 Pre-Olympic Congress. At the 1996 General Assembly, Doll-Tepper was well prepared and impressed the audience when explaining a ten-point plan to be implemented if she was elected as the next ICSSPE President. The ten-point plan was to: (1) develop closer links within the ICSSPE family; (2) revisit the current structures within ICSSPE; (3) strengthen cooperation with the International Olympic Committee (IOC), the United Nations Educational, Scientific and Cultural Organisation (UNESCO) and the World Health Organisation (WHO); (4) develop new partnerships with international organisations in sport, sport science and physical education; (5) cooperate more closely and make scientific knowledge of sport and practical experiences available, especially to those in developing countries; (6) prepare thoroughly for the Pre-Olympic Scientific Congress in Australia in 2000; (7) improve services to members by publication of newsletters, journals and books; (8) use new technology to make more information available at a faster speed; (9) emphasise the cultural aspects of physical education and sport; and (10) become an even more powerful voice in the world, an advocate for the values of physical activity and sport.

Prof. Dr. Gudrun Doll-Tepper was unanimously elected and became ICSSPE's fifth President and the first woman to hold the position.

The new President felt that she had inherited many good things from the previous Komi administration and the transition from one administration to the other was made seamless by the former Secretary-General André-Noël Chaker. Gudrun Doll-Tepper believed it to be absolutely crucial that such a large, international organisation as ICSSPE should have professional staff at its headquarters, who were available on a daily basis. This became one of the first, important issues that she had to convince the Berlin government of, that is, the need for full-time staff and consequent financial assistance to run the office efficiently. The first staff member to join the office was a former student of Doll-Tepper's, Ms. Deena Scoretz from Canada, who would become the coordinator of publications until 2000. The search for a Secretary General took somewhat more time and probably Gudrun's greatest luck at the beginning of her administration was to have 'discovered' Mr. Christophe Mailliet. He was someone from outside the physical education world, thus without preconceived opinions and expectations, had communication and diplomatic talents, excellent linguistic skills, was intelligent and able to quickly understand the problems and what was expected of the position. Gudrun Doll-Tepper describes him still today as "the right man, in the right place, at the right time".

Before she had put forward her candidacy, Doll-Tepper had to explain to the governing mayor of Berlin what her plans were and what was expected from the government in terms of financial support. The situation at that time (1995/1996) was such that the International Paralympic Committee (IPC), presently headquartered in Belgium, was also looking for a permanent site. Doll-Tepper visited the Governing Mayor, at that time Mr. Diepgen, and presented to him the two possibilities, that were, of either the IPC or ICSSPE having headquarters in Berlin. After the Paralympics of Seoul and Barcelona, the IPC was much more well-known than ICSSPE, so in a way the governing mayor was more sympathetic to the IPC.

When Berlin lost its candidature for hosting the Olympic and Paralympic Games of 2000, and because Bonn was another candidate city for hosting the IPC, the city administration decided not to stand in the way of another city hosting the IPC. It then became clear that the Berlin government would support ICSSPE. The most important steps in establishing the present-day ICSSPE Executive Office, from Gudrun Doll-Tepper's perspective today, was to secure the continuous, structural, long-term financial support from the Berlin and Federal governments.

The first year of the Berlin administration was both very active and successful. The President represented ICSSPE at several international congresses in different parts of the world and had tried to attend as many world-wide events as possible. This resulted in increased visibility of ICSSPE and in new partnerships and levels of co-operation.

One of the very first initiatives in the Doll-Tepper administration (1997) was to bring together ICSSPE members with partner organisations such as UNESCO. The lead officer at UNESCO was Arthur Gillette who had a great interest in ancient, traditional sports and games. It was believed to be a good idea to work together with UNESCO to demonstrate ICSSPE's efforts at collaboration. The outcome was a very interesting multidisciplinary conference in June 1997, with many well-known speakers. The event was also supported by Doll-Tepper's own university, which was something she had always wanted.

A New Strategic Structure for ICSSPE

While attending the Paralympics in Atlanta, 1996, Doll-Tepper experienced a strategic planning meeting facilitated by a company called Group Solutions Inc. from Atlanta, USA, that used the latest information and computer technology. Back in Germany, she convinced Executive Board members and her team in Berlin that it would be a good idea to arrange a similar computer-based session to identify ICSSPE's structure and tasks in the future. The IOC was successfully approached for financial assistance and the event was organised in Lausanne, Switzerland. During a two-day meeting in March 1998, all leading ICSSPE officers, as well as representatives from external organisations such as UNESCO, the IOC and the WHO, attended and were given the opportunity to express their opinions on a large number of questions and position statements, all related to ICSSPE[3]. Delegates shared views on what ICSSPE stood for, what made it a unique organisation with an important and central place in the world of sport science and physical education, and what its strength, weakness, opportunities and threats were, a kind of SWOT analysis. Participants shared ideas on how they saw the organisation and identified areas that could form part of a working plan so that short-, mid- and long-term goals and activities could be determined. It was not only a successful session in terms of the information gleaned, but also an interesting one using computer technology to answer and vote on posed questions with collective answers displayed on a screen but with votes and comments kept anonymous. The outcomes were condensed into key competencies and values for ICSSPE. On the second day, prioritising and voting began.

This meeting helped the new administration, in many respects, to plan a strategy for the future, which has been followed ever since. The model of the 'new' ICSSPE that came out of the Lausanne sessions is displayed in Fig. 1. Many of the new ideas had to be incorporated into new Statutes and Bye-laws at the next Board meeting (held at the end of that same year).

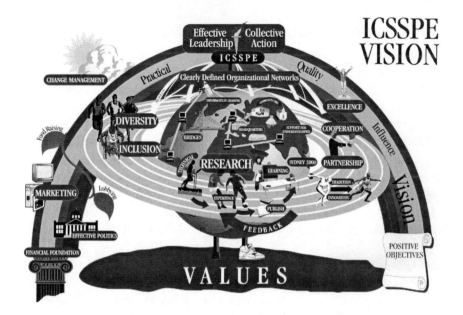

Figure 1: ICSSPE's Vision

The Associations' Board, a New Concept in the New Structure

One of the issues noted at the strategic meeting in Lausanne was a sense that the structure of the Board at that time did not give enough voice and influence to the different associations that were members of ICSSPE. This problem was resolved by establishing an "Associations' Board". A structure similar to the Associations' Board is that of the General Association of International Sport Federations (GAISF), in which all international presidents and organisations come together and identify what should be done within the organisation. In the case of ICSSPE, this meant that the Associations' Board would determine what should be done in addition to what the Editorial Board, Executive Board, President's Committee and the General Assembly were doing.

The 60th Executive Board meeting[4], organised by Victor Matsudo and his team in Sao Paulo, Brazil, in October 1999, was the first that included the Associations'

Board and was held under the new Statutes and Bye-Laws which had come into force after the General Assembly and Executive Board meetings in Barcelona, 1998. The Associations' Board had to be formally established and the roles of the organs of the International Council of Sport Science and Physical Education (the Council) defined so that they complemented each other in their work.

At its session of October 12, 1999, the member associations present nominated Prof. Dr. Karen DePauw from the International Federation for Adapted Physical Activity as interim Speaker until the position could be ratified at the 19th General Assembly, which was to be held in Brisbane, 2000. Other representatives nominated were Prof. Dr. Ronald Feingold from the Association Internationale des Ecoles Supérieures d'Education Physique (AIESEP) and Dr. Mari-Kristin Sisjord, representing the International Sociology of Sport Association (ISSA).

At the same session, the Associations' Board established criteria for membership, with potential members identified as: a) full ICSSPE members; b) international in scope, i.e. cover more than one continent; c) active in sport science, physical education and/or physical activity, i.e. have their primary interest in one or more of these fields; d) have an open membership; and e) be formally constituted. It was decided that existing members of ICSSPE should be the first informed of these decisions.

At the General Assembly in Brisbane, 2000[5], the Speaker presented the findings of the Associations' Board's deliberations from its 2nd meeting, held on September 7, 2000 in Brisbane. Karen DePauw referred to the ICSSPE Statutes, which defined the establishment and implementation of a Working Programme as well as a Research Agenda of topics that formed part of the Working Programme (see Table 1), and which needed to be addressed by ICSSPE members. Such a programme meant to provide a framework for collective action, uniting the very diverse ICSSPE membership. A common plan of action, based on common interests and shared values, was aimed at strengthening commitment and clarifying the roles and contributions of ICSSPE member organisations in unified action.

Table 1: ICSSPE's Working Programme in three sections: service, science and advocacy

Service	Science	Advocacy
• Develop service in languages other than English • Further the theory/practice links in ICSSPE's work and publications • Develop knowledge transfer systems • Strengthen links and communication among members • Play a pro-active role in capacity building and developing new connections • Develop the brokering and dissemination of information • Try to reach its multiple constituencies, particularly students, more effectively • Develop sending of materials to developing countries • Develop the concept of the Pre-Olympic Scientific Congresses as a service to ICSSPE	• Assure follow-up to the World Summit on Physical Education (International Committee for Sport Pedagogy (ICSP)) • Monitoring the implementation of MINEPS III • Develop position statements • Develop connections to: IOC, UN, World Bank, UNESCO, WHO, World Anti Doping Agency, International Federations, corporations, sponsors, others	• Prepare a research agenda (see below) and search for possibilities of implementation • Develop inter-disciplinary links within ICSSPE and other disciplines • Continuously develop the concept of the Pre-Olympic Scientific Congresses • Continue and/or initiate the collection of baseline data on physical education and physical activity at national level

The proposed ICSSPE research topics were:

• Core knowledge for physical education and sport;
• Technology, physical education and sport;
• Professional development;

- Creation/promotion of safe environments in physical education and sport;
- Sustainability: healthful, environmentally sensitive active living;
- Role of physical education and sport in development;
- Risk prevention;
- Sport and business;
- Sport information;
- Education in the Olympic Movement;
- Equity, social justice and inclusiveness in physical education and sport;
- Physical education and sport infrastructures.

This Working Programme and Agenda was submitted to the General Assembly for discussion and approval at its next meeting.

At the meetings in Beijing, June, 2001[6], the Associations' Board presented three overarching themes to focus on, with all themes being presented against a background of globalisation (see also Figure 2):

1. Ethics and Professionalisation;
2. Quality Physical Education (follow-up to World Summit on Physical Education, see part 5 of this chapter); and
3. Healthy living across the life-span and the human development/human performance continuum.

All ICSSPE Boards, joined in one meeting, discussed the outcomes of the first strategic planning meeting as well as the future of the organisation. From these discussions, the following model emerged.

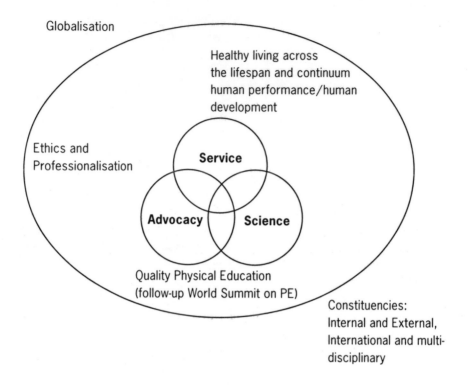

Figure 2: Science, Service, Advocacy

At the Associations' Boards meeting in Thessaloniki, 2004, Prof. Dr. Michael Mc-
Namee (UK) was unanimously elected to become the next Speaker for a 2 year period. Ron Feingold and Susi-Käthi Jost were both re-elected as Representatives of
the Associations' Board to the Executive Board and Karen DePauw was honoured
by the ICSSPE President for her outstanding work as the inaugural Speaker of the
Associations' Board[7].

Between 2004 and 2006, the Associations' Board developed a matrix to assess
its progress toward the Working Programme. At the meeting of the Board in Berlin,
20068 the Speaker asked the associations to complete a matrix with appropriate
content and return it to the Executive Office for collation. No feedback was every
received so no formal report prepared.

What ICSSPE had been trying to realise in the years following the remarkable 1998 Lausanne meeting was to link the work of the associations with that of the other Boards and committees. In the first years, joint sessions were organised to ensure that the information flow was going into both directions, from Executive Board to the Associations' Board, and vice-versa.

In light of the establishment of the Associations' Board and its likely overall profile, and considering current developments in the biophysical/biomedical field, the President's Committee deemed it important to try to attract more organisations from the natural sciences area. A challenge had always been to maintain a good balance between the various scientific disciplines and the social-educational and physical education-oriented groups. A well-balanced Associations' Board would indeed have a significant impact on the potential for quality research and publications.

There was little doubt about the growing weight, influence and effect of the Associations' Board, reflecting the fact that it has become the "think-tank" of ICSSPE.

The Gender Equity Plan

Under the Working Programme of advocacy, Executive Board member Dr. Anita White presented at the Executive Board meeting in Sao Paulo, Brazil, 1999[4] a draft proposal for ICSSPE's Gender Equity Plan, which she had prepared with input from several individuals. As background, she explained that ICSSPE was one of the first organisations to endorse the Brighton Declaration (delivered in 1994 at the inaugural International Conference on Women and Sport, to provide principles that should guide action intended to increase the involvement of women in sport at all levels and in all functions and roles). She explained that the main difficulty encountered in preparing this draft proposal was due to the structure of ICSSPE as an umbrella organisation with autonomous and independent member organisations of various kinds. During discussion at the Executive Board meeting, a number of comments and suggestions were made. It was agreed that member organisations should be encouraged to respond to this Gender Equity Plan and reflect on the percentage of women in positions of authority and the real input of women in their organisations, including contribution to the organisations' publications. Although ICSSPE's influence on its member organisations is thought to be limited, it could set the example by reaching such a political decision as to adopt a Gender Equity Plan. It was recommended that guidelines be prepared for the organisation of Pre-Olympic Congresses and other ICSSPE supported events to take into account the Gender Equity Plan. Likewise, it was agreed that congress organisers be requested to consider these guidelines before patronage from ICSSPE was provided. The ICSSPE Gender Equity Plan was adopted in principle under the condition that changes be made in order to take into account the recommendations of the Boards. Anita White incorporated these comments into the document, which was approved in 1999 by the Executive Board, and integrated as an annex to the Statutes and Bye-laws. It has since been effectively applied on several occasions, such as in the context of Pre-Olympic Congresses, or when deciding which organisations should receive patronage and/or financial support from ICSSPE.

ICSSPE Membership

Membership of ICSSPE falls into four categories: A, B, C and D, depending on the status of the member. The work of the Council occurred, amongst other strategies, through special committees and working groups made up of ICSSPE members. Examples include the former "Sport and Mass Media Committee" and "International Working Group on Ergometry". These, and other committees and working groups, were the backbone of the Council. They were formed by individuals sharing similar interests with the aim of establishing coordinated and international structures. Several of them became international societies and remained under ICSSPE's umbrella, while others became independent societies and left ICSSPE.

As early as 1996, ISAK (the International Society for the Advancement of Kinanthropometry, formerly a working group of the Council), questioned the need for them to be a member of the larger ICSSPE family. Although the group later decided to become a full member in 1999, they agreed to review their membership after three years. The International Research Group on Biochemistry of Exercise, also an ex-officio member of the Council, decided to withdraw its membership of ICSSPE when they had to pay an annual membership fee.

The status of yet another ICSSPE Working Group (the "World Commission on Biomechanics of Sport") was unclear given their plans to change its name to "World Commission of Science and Sport". The general opinion of the Executive Board was that the intended change did not reflect the work of this Committee properly.

The new name could generate identification problems, since sport sciences were a much broader field than the one covered by the Working Group. The new name could be misleading and it was suggested that a name be found which reflected the status and the function of this Working Group properly. In the end, the proposed name remained "World Commission of Science and Sports" (WCSS) and the group maintained close co-operation with the Associations' Board.

These three examples prompted the President's Committee, while meeting in Orlando, 2000, to discuss an important question: *"What do I gain from being a member of ICSSPE?"*[9]. In light of the composition of the Associations' Board and the internal discussions that had taken place in the three (former) ICSSPE working groups about the perceived benefits of being a member, it became clear that for organisations active in the biomedical/natural sciences area, being a member of ICSSPE (and/or the Associations' Board) was not a too attractive option.

It was decided to organise a symposium at the Pre-Olympic Congress in Brisbane on "The Role of a Multidisciplinary Organisation in an Age of Specialisation"[10].

Three questions formed the basis of ICSSPE's symposium at the Pre-Olympic Congress:

• What is the role of a multidisciplinary organisation in an age of specialisation?
• Does success in the 21st century require stronger global connections?
• Do we need collective thinking?

In order to bring these issues to the table, two panellists representing ICSSPE opened the panel discussion - Gudrun Doll-Tepper and Karen DePauw. ICSSPE believed that it was not alone in facing these questions and challenges and therefore invited other international organisations, as well as member organisations and working groups, to join the Symposium panel. The following speakers addressed the three questions above and the current issues facing their respective organisations: Prof. Dr. K.M. Chan, Vice-President, International Federation of Sports Medicine (FIMS); Prof. Dr. Mike Marfell-Jones, President ISAK; and Prof. Dr. Tom Reilly, Chair of WCSS. Prof. Dr. em. Jan Borms, Chair of the ICSSPE Editorial Board, moderated the session, drawing from his more than 30 years of experience within ICSSPE and its predecessor, the International Council of Sport and Physical Education (ICSPE).

ICSSPE's President began with a brief historical overview of the Council's development. She and DePauw then presented three arguments supporting a multidisciplinary organisation in an age of specialisation: 1) the *holistic approach*, emphasising that current tasks and challenges with regard to sport can best be met with a collective and multidisciplinary approach; 2) a unified voice through interdisciplinary co-operation and collaboration enables a greater impact for sport and physical activity internationally in all sectors of society – health, education, community development, etc (the *advocacy approach*); and 3) the multidisciplinary network within a multidisciplinary organisation increases the accessibility of scientific knowledge by means of publications, congresses (the Pre-Olympic Congresses were excellent examples) and joint projects. Additionally, contacts to other international bodies, such as UNESCO, IOC, WHO, FIMS are available (*Contacts and Partnerships approach*).

They concluded by citing the following statement from Steve Bailey:

> "ICSSPE can be at once an emporium, a bazaar, a stimulator of ideas and a promoter of co- operation on many levels. The ability of ICSSPE to act as a forum for interchange and multi-disciplinary stimulation is what makes it crucial" (Bailey, 1996, p 312).

Each of the other three organisations (FIMS, ISAK and WCSS) briefly presented the current status of their membership and the challenges they faced with regard to international collaboration.

Despite the differing circumstances of the three organisations, several common themes arose and discussion with the symposium audience ensued.

All participants agreed that working together was necessary, as long as it was possible for autonomous organisations and sport science disciplines to maintain their individual identity. Several benefits of global and multidisciplinary connections included: sharing of knowledge, reducing of confusion, holistic approaches to issues, brokerage functions, credibility and power in numbers. Factors identified that blocked effective collaboration included:

- Definitions – It is difficult to talk as one voice when definitions for sport/exercise science/physical education, etc. are not universally agreed upon.

- Perceived "rivalries" and a lack of recognition between the so-called "hard" and "soft" sciences can lead to stereotypes and/or prejudices and block successful communication and co-operation.

In conclusion, it was felt important to realise that multidisciplinary organisations, such as ICSSPE and FIMS, were only as strong as their members. Naturally, each organisation required specific, individual networks to develop their field of work optimally, however there was strong consensus that it was better to move together, rather than in opposition or in competition. Moreover, all organisations need to change with the times and adapt to emerging job opportunities. The success of the Brisbane 2000 Pre-Olympic Congress is testament to the enormous potential within the sport science community as a whole and is an excellent example of multidisciplinary collaboration.

This symposium and the interaction it encouraged reinforced the importance of constructive dialogue in determining a common ground for future co-operation.

World Summits on Physical Education

Beginning in 1998, President Doll-Tepper and Vice-President Prof. Dr. Margaret Talbot successfully presented the idea of organising a World Summit on physical education to the President's Committee. The impetus to invite leaders and practitioners to a World Summit gained momentum as ICSSPE recognised the need for a platform to share the ground-breaking results of the world-wide audit on the State and Status of School Physical Education[11]. This study, financially supported by the IOC, set the context for the Summit, which was hosted in Berlin from November 3-5, 1999.

A total of 250 representatives from governments (including five ministers), non-governmental and inter-governmental organisations, universities, schools and research centres attended the action-oriented conference and the issue of quality physical education was a recurrent theme throughout the discussions. Participants from over 80 nations attended, but despite the diversity, there was unanimous agreement on two fundamental issues:

- Physical Education is a right for all children and a fundamental component of their development and education; and

- Strategies and actions are needed to ensure that quality physical education is implemented and supported world-wide.

All of the eight keynote presentations had built a convincing case for physical education centred on these two issues, using evidence from a range of scientific research and practical examples. Proceedings were published afterwards with extensive reference lists appending each paper, leading readers to more detailed literature on physical education.

Gudrun Doll-Tepper and Christophe Mailliet still consider the organisation of this Summit as one of the landmarks of the ICSSPE era in Berlin. The success of this Summit was the result of the dedication of many individuals and particularly the financial contribution of a large number of international, federal, local and private organisations, which was testament to the President's skills as a fundraiser. Members of the International Committee of Sport Pedagogy (ICSP) played an important guiding role in the planning process for the Summit and in the creation of its scientific programme. In particular, Vice-President Margaret Talbot was recognised as a driving force.

In the closing plenary session, the "Berlin Agenda for Action (Part One)" was adopted by all Summit delegates. This 'Call for Action' was aimed at governments and reinforced the importance of physical education in the development and life-long education process. ICSSPE had also committed itself to coordinating and publishing the "Berlin Agenda (Part Two)" for Physical Educators themselves to use in their own countries. Delegates agreed that the physical education profession itself must address the urgent need to improve the quality of physical education delivered in schools.

In summary, the World Summit succeeded in presenting the broad spectrum of scientific evidence surrounding the benefits of physical education and in raising the issues facing its practical implementation.

MINEPS

MINEPS is the name given to the event hosted by UNESCO, where all ministers and senior officials responsible for physical education and sport meet at an international level. One of the outcomes of MINEPS I, held in Paris, 1976, was a document on physical education and sport, published in 1978. This document, the International Charter of Physical Education and Sport, is still used by many people. MINEPS II was held in Moscow, November, 1988 and ICSSPE was pleased to learn that UNESCO planned MINEPS III at the end of November/early December in 1999 in Punta Del Este, Uruguay. This was soon after the Berlin Summit so Doll-Tepper and Mailliet believed this to be a unique opportunity to attend and present the results of the World Summit on Physical Education and the Berlin Agenda for Action, which was translated into different languages to ensure wide spread dissemination. The important outcome was that in the final congress document of MINEPS III, the Berlin Agenda and the Berlin Summit were integrated and thus endorsed in a way, by ministers and senior officials from 62 United Nations member States, three observer states and 38 intergovernmental or non-governmental organisations.

Undoubtedly, documents such as the Berlin Agenda for Action and the MINEPS III Recommendations were important levers. They have been circulated widely, raising awareness in the media and general public, and the Berlin Agenda for Action has been endorsed by many national and international organisations. These documents, however, could not make a difference on their own but were important first steps in building momentum for action at the national and local levels.

The 2nd World Summit on Physical Education was successfully held in Magglingen, Switzerland, December 2-3, 2005 under the leadership of ICSSPE and the Swiss Federal Office of Sport (BASPO), with Executive Board member Walter Mengisen as key organiser. It took place under the patronage of UNESCO, GAISF, the WHO and IOC and was also supported by the United Nations Special Adviser on Sport for Development and Peace, Mr. Adolf Ogi.

While the first World Summit on Physical Education stood at the beginning of a broad international debate on the current state of physical education world-wide, the participants at the 2nd World Summit reviewed the status of physical education, following presentations and analysis of both the positive and negative developments since 1999. A reflection on the intrinsic role of physical education in education and its potential to contribute to the achievement of the United Nations' Millennium Development Goals, based on current research results, took place during the plenary sessions and workshops. Other important topics were also discussed, such as: the development of quality standards and benchmarks for physical education which are based on scientific evidence and contribute particularly to personal and social development; the development of effective and modular strategies to secure and further develop physical education as an essential component of education; and the integration of high-quality and culturally sensitive physical education policies world-wide into education and sports policies.

More than 150 decision-makers and researchers from the areas of science, politics, sports and education representing nearly 40 countries attended this 2nd World Summit, making it a truly international forum for exchange and debate. Participants were members and partners of ICSSPE, relevant/responsible ministries, specialised organisations, media representatives, sport federation officers, UN agencies and non-government organisations.

In order to prepare for the Summit, a Steering Committee was established with Prof. Dr. Karen P. DePauw (ICSSPE), Prof. Dr. Maria Dinold (Chair, ICSP), Prof. Dr. Gudrun Doll-Tepper (ICSSPE President), Mr. Christophe Mailliet (ICSSPE Executive Director), Mr. Walter Mengisen (BASPO) and Margaret Talbot. An Expert Committee, which was actively involved both in the preparations for the 2nd World Summit on Physical Education, as well as in the Summit itself, was composed of: Prof. Dr. Chantal Amade Escot (France), Prof. Dr. Lateef O. Amusa (Nigeria), Prof. Dr. Richard Bailey (United Kingdom), Prof. Dr. Wolf-Dietrich Brettschneider (Germany), Prof. Dr. Ming-Kai Chin (Hong Kong, China), Prof. Dr. Karel J. van Deventer (South Africa), Prof. Dr. Ronald S. Feingold (United States of America), Prof. Dr. Kenneth Hardman (United Kingdom), Prof. Dr. Uwe Pühse (Switzerland) and Prof. Dr. Manoel José Gomes Tubino (Brazil).

At the end of the 2nd World Summit, participants adopted the Magglingen Commitment for Physical Education. This document highlighted the need for implementation of recommendations and declarations made at previous conferences and meetings, but also provided new directions for the future of physical education. This document would come to served as the basis for the future strategy of ICSSPE, its partners and members to further develop physical education world-wide, and its distribution was encouraged by any available means.

Old and New Partnerships

IOC

The relationship between the IOC and ICSSPE dates back to October 11, 1984 when the respective Presidents, Samaranch and Kirsch, signed an agreement outlining collaboration of the two organisations in the field of sports science[12]. The way was then paved for ICSSPE to function as a 'recognised organisation' of the IOC. The IOC Executive Board had granted this status to ICSSPE in November, 1983, thereby making it the second sport science organisation (after The Fédération Internationale de Médecine du Sport (FIMS)/International Federation of Sports Medicine) to be thus acknowledged.

The relationship with the IOC developed to its strongest through the personal attention of President Paavo Komi during his term of office between 1991 and 1996. One of the important items on the agenda of ICSSPE's new president, Gudrun Doll-Tepper was to visit the IOC. She and the Executive Director, Christophe Mailliet had two meetings with the IOC, in October 1997 and January 1998. One of the outcomes was that the IOC invited the Executive Board to come to Lausanne in 1998 to hold the strategic meeting, which they had supported financially. Two new scientific projects proposed by ICSSPE were also accepted and fully funded by the IOC. These two projects were on the State and Status of School Physical Education and the History of the Paralympic Movement. IOC President Samaranch also confirmed that the Pre-Olympic Congress in Brisbane, 7-13 September 2000, would receive the same support as the Congress in Dallas. He also requested ICSSPE's support and input into preparations for the upcoming IOC International Conference on Education and Olympism. This represented a great opportunity and responsibility for ICSSPE and a new step in the cooperation between the two organisations.

In 1983, Prof. Dr. Edelfrid Buggel, then ICSSPE's Treasurer, made a private suggestion to IOC President Samaranch to institute the "Sport Science Award of the IOC". This developed into a specific award coordinated by ICSSPE on behalf of the IOC President and was first presented in 1989. This award carried with it a purse of USD 5,000, which increased to USD 7,500 a year later (Bailey, 1996, p.214)[1]. In 1990, only 9 rather low profile nomination papers were submitted for the award and the jury recommended that no award should be made. The Council began to look at the nature of the award's regulations with a suggested new format of selection involving the monitoring of peer-reviewed journals by the award jury.

A few years later at the President's Committee meeting in Brisbane, 1999 and thus under Doll-Tepper's presidency, the system for the Sport Science Award of the IOC President was found to be outdated and not reliable. After discussion, the President's Committee decided to propose a new procedure to the IOC. Vice President Prof. Tony Parker and Christophe Mailliet prepared a draft immediately following the meeting, with the main aspects of the new procedure outlining that: the award would be granted annually both for biomedical and social sciences; the prize would be increased to USD 10,000 to be split between the two categories; and would recognise the continuous work of a team or an author rather than just one paper.

In July 1999, the IOC President accepted the new procedures. At the Executive Board meeting in Sao Paulo, 1999 it was decided that due to this change, there would be no award made for the year 1998 and that the deadline for the 1999 award would be extended[4].

The recipients of the 1999 Award were the HERITAGE Family Study Research Group (biomedical sciences) and Allen Guttmann (social sciences). The recipients of the 2002 Award were Roland Renson (social sciences) and the Copenhagen Muscle Research Centre (biomedical sciences).

At the Executive Board meeting in Edmonton, 2005, the Board decided that it was important to continue to have a joint award with the IOC, since it would further the relationship between the two organisations[13]. The IOC wanted to have a joint, single award that would be granted with or through ICSSPE, however there were other awards given by ICSSPE members (e.g. AIESEP[14]) in cooperation with the IOC. There was a misconception on the part of the IOC, as far as ICSSPE's structure and influence on its members was concerned. The Executive Director reported that it had become increasingly difficult to market the award and obtain nominations of the superior quality that was desired. The criteria were felt as being too high level and potentially elitist and the protocols too cumbersome. This lead to the conclusion that the criteria and procedures would have to be amended and simplified if the award was to retain significance. It was also concluded that the award policy of ICSSPE should be re-thought, in light of the expressed priorities and Working Programme. For example, it would seem that the Philip Noel Baker Research Award would reward services to ICSSPE and the community while other awards could focus on science and advocacy and should recompense people who contributed to a broad agenda. As a conclusion, the Board expressed its wish to review the award policy of ICSSPE and requested the Vice-President 'Science' to prepare a corresponding proposal for the upcoming meeting, which would also look into the involvement of the IOC in the procedure.

Later in the same year (Executive Board meeting, Berlin, 2006), Tony Parker, presented several options regarding the future of awards being bestowed by ICSSPE. He reported that most particularly the IOC Sport Science Award had come to a standstill due to operational difficulties and little public recognition8. A number of alternatives were presented, from which it was agreed that a new award should recognise both young researchers and colleagues in developing countries. Proposals have been made to the IOC and negotiations on an award have, at time of publication, just been taken up again.

In 2007, Gudrun Doll-Tepper was elected, in her capacity of ICSSPE President, as member of the Education Committee of the Post-Graduate Research Grant Programme of the IOC. Since 2008, she has been a member of the IOC Women and Sport Commission, as well as being a member of the Culture and Education Consultative Group for the Youth Olympic Games.

FIMS

The collaboration with FIMS dates back to the 1970s and has been enhanced since then largely due to good personal relationships and the hard work of ICSSPE's representatives, Prof. Dr. Marcel Hebbelinck and Prof. Dr. Jim Skinner and several other Board members. Tony Parker became the new liaison between the Council and FIMS in 1997 and was also an elected FIMS Board member. FIMS was facing similar challenges as ICSSPE in terms of restructuring and redefining the organisation and wanted to open its membership. Possible areas for collaboration were identified, including the preparation of position statements, travelling fellowships and cooperation with the FIMS Education Commission. The FIMS Executive

Committee was standing for re-election at their World Congress in Orlando (USA) in May-June, 1998, which was organised by Prof. Dr. Skinner. Prof. Dr. Kai-Ming Chan (Hong Kong) was elected as new President of FIMS and became the successor to Prof. Dr. Eduardo de Rose (Brazil).

At the 2002 World Congress on Sports Medicine, which took place in June in Budapest, there were two joint ICSSPE-FIMS sessions: a symposium on obesity and a workshop on regional development. Both were well attended and plans were made to continue this type of cooperation in the future.

In 2004, ICSSPE offered a symposium on "Sports with Special Needs" at the XXVIII FIMS World Congress in Oman being the first symposium ever on "Sport for Persons with a Disability" at a FIMS Congress.

IPC

During the Berlin administration, links between the Council and the IPC were strengthened and one of the most remarkable achievements was the inclusion of the IPC in future Pre-Olympic Scientific Congresses (see Pre-Olympic Congress and the New Style Conference in this chapter).

Besides many other collaborations, the History of the Paralympic Movement was a major book project directed by ICSSPE with funding from the IOC. It was initiated in 1997/98 and completed by Steve Bailey in 2007. Other publications that have been published in collaboration with the IPC were a Bulletin (September 2005) featuring Sport for People with Disabilities and Perspectives Volume 7: Sport for Persons with a Disability (2007).

A fruitful exchange occurred and support was given by the IPC for the Sport and Development and Sport in Post Disaster Intervention seminars coordinated by ICSSPE in 2006.

UN

In 2001, Mr. Adolf Ogi, former President of Switzerland, was appointed as Special Adviser on Sport for Development and Peace to the United Nations Secretary

General. Mr Ogi was, amongst other duties, responsible for the organisation of the 1st Magglingen Conference on Sport and Development, which took place in February 2003 in Magglingen, Switzerland. The key result, the Magglingen Declaration, affirmed the delegates' commitment to sport and development, stating that sport is a human right and an ideal learning ground for life's essential skills. President Gudrun Doll-Tepper featured as a key-note speaker on the impressive list of invited presenters.

The ICSSPE Office has also been in regular contact with the Swiss Academy for Development (SAD), which works closely with the UN Task Force in this area. SAD is advised by a scientific board, composed of recognised specialists from within the scientific communities. President Doll-Tepper is one of the members of this scientific board.

The year 2005 was proclaimed by the UN General Assembly as the United Nations International Year of Sport and Physical Education (IYSPE). This event was seen as a once in a life-time opportunity to demonstrate the role and importance of ICSSPE as a leading organisation in the area of sport, sport science and physical education, to reflect on the global governance of sport and to develop strategies for collective efforts. It was believed that the focus should be on designing action strategies for the future, based on the available knowledge, with the overall aim of poverty alleviation. To this end, ICSSPE prepared a position statement based on the official UN documents for the International Year, and co-organised the Second World Summit on Physical Education to celebrate and support the International Year.

Also in the context of the IYSPE 2005, ICSSPE organised an International Forum on Sports and Development - Economy, Culture, Ethics and hosted a seminar on Rehabilitation through Adapted Physical Activity and Sport following the devastating tsunami in South East Asia in December, 2004. Supported by the UN and in cooperation with the German Federal Ministry of the Interior, the seminar was held in Bangkok, Thailand in October, 2005. It was closely followed by a second seminar on Sport and Reconstruction in the Tsunami Region held in Berlin, Germany, in January, 2006.

As a result of the interest generated at these seminars, ICSSPE developed a comprehensive training seminar with the intention of providing disaster workers with the practical skills and knowledge required to deliver sport and physical activity programmes in disaster settings. Supported by the German Red Cross, the first edition of this international training seminar was held in November, 2007 in Rheinsberg, Germany, and a participant handbook was produced to provide guidance in applying the theories in a disaster relief situation. A second international training seminar is planned for November, 2008, again in Rheinsberg, Germany.

WADA

At the Executive Board meeting in Sao Paulo, Brazil in 1999, it was found indispensable that sport science and physical education be included in the governing bodies of the World Anti-Doping Agency (WADA). It was also felt that ICSSPE's role should be in the educational part of anti-doping activities, e.g. in developing attitudinal changes among athletes, including athletes with disabilities[4].

In the Winter of 2001, a meeting took place in Berlin between WADA Secretary General, Harri Syväsälmi and ICSSPE. Based on the results of this meeting, an agreement of cooperation between WADA and ICSSPE was elaborated and ICSSPE was requested to comment on the first draft of the World Anti-Doping Code. This was circulated to the Association's and the Executive Board for comments and input.

With the financial support of UNESCO, a multi-lingual brochure on anti-doping prevention, that was targeted specifically at young athletes in developing countries, was developed in 1999 by Prof. Dr. Colin Higgs and colleagues. This leaflet was called Champions don't cheat and was distributed widely by ICSSPE and UNESCO to several organisations and individuals.

At the Executive Board meeting in Berlin, 2006, the Speaker of the Associations' Board, Mike McNamee proposed that ICSSPE should have a public position on the issue of doping. "Sport Science is necessary to improve performance" he said "but at the same time, sport science disciplines want to, and must express their position on performance enhancement. ICSSPE's targets are broader than the narrower discussion of WADA. Science should serve sport in an ethical way, and the contribution of ICSSPE should be defined. Prevention is one of the crucial areas that has been forgotten in a long time, and should play a central role in ICSSPE's work. ICSSPE definitely is against drugs, which is a classic multidisciplinary issue. ICSSPE should not focus exclusively on performance enhancement, but also health and ethics are important".

Besides statements that were given at the 1st IOC Anti-Doping Conference and at UNESCO's Round-Table, ICSSPE raised doping issues, in particular on prevention and education on many occasions and hosted two symposia in cooperation with International Sport Lawyer Association (ISLA), one of them on Doping and Corruption in Sport in May, 2007.

The Board agreed that ICSSPE had already provided critical and constructive perspectives on the work of WADA and on the UNESCO Anti-Doping Convention and decided that it would develop a broader position statement. When developing this statement, a reflection on its appropriateness, the nature of the expert knowledge needed, and its quality should form the basis of the work undertaken.

World Sport Forum (Lausanne, 1998)

One of the new relationships of ICSSPE in the 1990s was generated through its involvement in the planning of the 1st World Sport Forum held in St. Moritz, Switzerland, March 8-11, 1998, which had been organised by the World Federation of the Sporting Goods Industry (WFGSGI). Karen DePauw, John Andrews, Victor Matsudo, Margaret Talbot and President Doll-Tepper took part in the Forum either as speakers, moderators or rapporteurs. Later, the forum organisers decided not to repeat this event.

GAISF

Contacts were established with the General Assembly of International Sport Federations (GAISF) during its Congress and General Assembly in October, 1997 in Duisburg, Germany. Many of the international federations, which were members of GAISF, had scientific committees and the need to contact these was identified by ICSSPE as a means to strengthen the cooperation.

The ICSSPE President attended almost every GAISF's General Assembly and spoke at the 2002 meeting in Colorado Springs on women and leadership. A decision was taken to implement a new event, named SportAccord, organised by GAISF, ASOIF (Association of Summer Olympic International Federations) and AWOIF (Association of Winter Olympic International Federations). The first SportAccord took place in Madrid in 2002, the second in Lausanne in 2003 and through the direct connections and support of ICSSPE, Berlin bid successfully for 2005. Gudrun Doll-Tepper also represented ICSSPE at the events in 2007 in Beijing and in 2008 in Athens.

Dimensions of Performance – Seminar in cooperation with ISTAF

Since 2005, in cooperation with the organiser of the ISTAF, one of the International Amateur Athletics Federation's (IAAF) Golden League athletics meetings, ICSSPE has organised an annual one-day seminar in Berlin, Germany, focusing on themes relevant to high performance sport. This event has provided ICSSPE with a closer link to sport federations. In 2005, the symposium focused on Altitude Training: Myths and Methods. This was followed in 2006 with Talent Identification and Development: A Multidisciplinary Approach. In 2007, the symposium involved stakeholders in education, as its theme was Learning to Win: Elite Schools of Sport. Most recently, in 2008, ICSSPE has continued to strengthen the relationship between sport federations, sport scientists and educational institutions with Degrees of Success: University Education and Elite Sport.

WHO

ICSSPE has always been in contact with the World Health Organisation (WHO). That has led to a cooperation in 2004, when WHO commissioned ICSSPE to prepare a series of technical papers to assist in the preparation of resources for young people, in the context of World Move for Health Day. ICSSPE members, experts in each of the topics covered, prepared the following documentation:

- Girls' Participation in Physical Activities and Sports benefits, patterns, influences and ways forward - Richard Bailey, Ian Wellard and Harriet Dismore
- Physical Activity and its Impact on Health Behaviour among Youth - Prof. Dr. Denise Jones-Palm and Prof. Dr. Jürgen Palm
- Young People With Disability in Physical Education, Physical Activity and Sport, in and out of Schools - Prof. Dr. Claudine Sherrill

During the Executive Board meeting[13] in Edmonton, 2005, it was decided that the Council should have a more active role in the Move for Health initiatives of WHO and that Agita Mundo should be asked for recommendations. Agita Mundo, Spanish for 'move the world' was created by Dr. Victor Matsudo from Sao Paulo, Brazil, with the mandate to encourage physical activity participation across the globe. Matsudo went on to become Regional Coordinator for Latin America on ICSSPE's Executive Board in 2000. It was also decided that ICSSPE should promote a broad understanding of health and be aware of political and economic considerations and their importance in health policy, to maximise its impact. ICSSPE should be auditing international declarations for health-related issues and the Declaration of Hammamet (UN Conference Sport and Health) should be sent to all members. Another item of importance was the Anti-Doping Convention of UNESCO, which ICSSPE should support and promote. Concerning the topic of professionalisation, it was agreed that the current debate should focus on the conflicting concepts of market competence and civic science, which were affecting the profession. Generally speaking, a regular update on the items defined in the Working Programme (3 main strands: Healthy living across the lifespan / Quality physical education / Ethics and professionalisation) should be included on the agenda of the Executive Board in the future.

Pre-Olympic Congresses and the 'New Style' Conference

Pre-Olympic Scientific Congress, Brisbane, 2000

Tony Parker was appointed in 1997 by the conference organisers, Sports Medicine Australia (SMA), as Chair of the Organising Committee of the Pre-Olympic Congress in Brisbane, Australia, 7-13 September, 2000. The logo of the Conference, as well as the term "Pre-Olympic" had been accepted by the IOC President who also had accepted the invitation to open the Congress. The IOC would also provide (at least) the same financial support as for the previous Congress in Dallas. Christophe Mailliet's visit to Brisbane in December, 1997, assisted with the setting up and structuring of the Conference and in getting acquainted with the organisers both at the local and national levels.

SMA encountered a number of difficulties, which resulted in significant changes in the organisational framework of the Congress. The Australian Council of Health,

Physical Education and Recreation (ACHPER) had become the third partner in the organisation. However, the situation in terms of budget and sponsoring opportunities was difficult because of the IOC's and the Sydney Organising Committee of the Olympic Games' (SOCOG's) brand protection policies in this field.

Other difficulties encountered during the months in the lead up to the event were due, amongst others, to serious competition with the IOC Congress. Unlike many other cases, there was no substantial financial support from official bodies for the Congress and it was clear that the time had come to re-think the whole concept of the Pre-Olympic Congress.

As was the case at previous Congresses, there were a number of critical remarks. The International Sociology of Sport Association (ISSA) expressed the position that there had been a lack of inclusion and that their premises and statements were not taken into account. There had also been communication problems, along with criticism concerning the choice of themes (only one strand for socio-cultural issues out of six), the selection of keynote speakers (gender equity), the allocation of time and space, the fact that no funds were available for the ICSSPE Associations, and that the congress fee of AUD 500 was too high.

The Pre-Olympic Congress opened with an impressive group of speakers including the Honourable Terry Mackenroth, Queensland Minister for Communication and Planning and Minister for Sport, Dr. Shane Conway, President of Sports Medicine Australia and His Excellency Dr. Pál Schmitt, Member and Chief Protocol of the International Olympic Committee, followed by a spectacle of colour, movement and music at the impressive Brisbane Convention and Exhibition Centre. This ceremony was preceded by practitioner workshops and was followed by four intense days of a multidisciplinary scientific programme and opportunities for networking and socialising. The main scientific programme provided delegates with a diverse and extensive range of sessions, well balanced across the different disciplines represented within the ICSSPE membership and also offering opportunities for addressing topical issues in a multidisciplinary forum. The quality of the programme was enhanced by excellent presentations from a large number of keynote and invited speakers representing the various discipline areas. These speakers also provided leadership in a broad range of topical symposia and workshops, which were well received by delegates and often provoked media and community interest. Keynote speakers were also well supported by free communications sessions with 740 papers presented in platform or poster format in addition to 38 symposia and workshops. The response to the programme from delegates was very positive and many indicated their satisfaction and appreciation of the opportunity to attend sessions and hear papers outside of their normal area of interest. In other words, they were confirming a major goal of ICSSPE conferences, which was to encourage and provide opportunities for cross-disciplinary interactions and understanding. It was generally agreed that the venue and facilities at the Brisbane Convention and Exhibition Centre were outstanding and enhanced opportunities for delegates to move easily between rooms during the programme, which often involved 10 concurrent sessions.

Over 1300 delegates attended the Congress and the preceding workshops, with international delegates comprising about 60% of the cohort – a tribute to the widespread promotion, assisted by the many groups and organisations involved.

An award system was incorporated into the Congress with plaques presented at the closing ceremony in recognition of scientific excellence. The awards were presented for best paper presentations in the various discipline categories following assessment by expert panels. An additional award was presented for best paper on a topic related to women in sport and this was a reflection of the significant number of presentations and symposia associated with women in sport and exercise[15].

Tony Parker, who did a remarkable work as Congress Chair of the 2000 Pre-Olympic Congress, presented a detailed report to the Boards. Foci of the Congress were on developing countries and practitioner involvement. He said that bringing the various organisations under the umbrella of the Congress and national and international satellite symposia was an issue. He also referred to problems in negotiating with the organising committee of the Games (SOCOG) and the IOC Medical Commission. For example, the organisers of the 2000 Pre-Olympic Congress were not allowed to approach sponsors. He advised that ICSSPE must assure that both its Congress and the IOC Congress would be treated equitably in the future. He also thanked the ICSSPE Executive Office for its support in organising the stay of young scholars from developing countries through an Olympic Solidarity grant. Finally, he also thanked Amanda Costin (now Smyth) for her invaluable support in the organisations of the Congress.

In Beijing, 2001 the debriefing of the 2000 Pre-Olympic Congress showed a positive balance for the organisation as documented in the Independent Audit Report to Australian Sports Medicine Federation Ltd. prepared by Ernst & Young[6].

The 2004 Pre-Olympic Congress Thessaloniki, August 2004

In the years preceding the 2004 Pre-Olympic Congress, the Council's boards were regularly informed on the organising and scientific committees' progress by Profs. V. Klissouras and S. Kellis (Chair and Secretary of the scientific committee) who acted as the main liaisons. The President's Committee also met twice in Thessaloniki (2001 and 2002) to solve a number of problems.

The 2004 Pre-Olympic Congress was the largest ever held in terms of parallel sessions (12 parallel sessions in each time slot) and the Congress venue was the most suitable choice for this "feat", considering the fact that the Aristotle University was the most convenient place in Thessaloniki, in terms of its location in the heart of the city[16]. The Congress was attended by 1253 people from approximately 70 countries with presentations from Europe, America, Asia, Africa and Australia. Of interest to note was the attendance of people coming especially from Asian countries, about 36% of the total number of participants, as well as from developing countries, which numbered 45% of total participants. The organisers were extremely pleased to welcome the delegation of the Chinese Olympic Committee, as well as approximately 200 Chinese, many of who were visiting a European country for the first time.

As it was the case in previous Congresses, several member associations were holding their annual meetings while also contributing to the Congress scientific programme.

There were a total of 1013 (440 oral and 573 poster) free presentations during the Congress. 5 keynote speakers were hosted, presenting their lecture on an important topic of international interest during the day's plenary session and one presenting the Ernst Jokl Lecture before the Congress closing ceremony. This Pre-Olympic Congress was characterised by the great variety of sessions. There were 4 interdisciplinary round tables, 11 disciplinary current issues sessions, 3 Socratic debates (a session presented for the first time at an ICSSPE Pre-Olympic Congress), 13 workshops, 45 symposia, 3 colloquia, 2 tutorial sessions and 2 special events sessions. The total number of presentations in those various Congress sessions reached 230.

The scientific programme was characterised by more pedagogy and sociology sessions, making up 35% of the total number of sessions. Approximately ten percent of the Congress sessions were dedicated to Adapted Physical Education issues. The President of the International Paralympic Committee (IPC) Mr. Phil Craven was one of the distinguished invited guests, who contributed with his presence as a Congress delegate, speaker and chairperson.

A scientific event such as the 2004 Pre-Olympic Congress would not have been possible without the support and assistance of several public, private and international organisations/associations and of the 90 volunteers. These people, mostly students of the various departments of physical education and sport science at Australian universities actually displayed the spirit of Greek hospitality and good will, manifesting one of the country's most important cultural traits.

The 2004 Pre-Olympic Congress gave delegates the opportunity to meet people from all over the world, develop new scientific and personal contacts, learn and acquire an experience to remember. This was most likely the greatest satisfaction that Congress organisers (organising and scientific committee) aimed for and achieved after four long years of preparatory work.

At the Executive Board meeting in Berlin, 2006, the Executive Director reported on the financial deficit of the 2004 Pre-Olympic Congress in Thessaloniki, which was due to the unfulfilled promise of financial support by the Municipality of Thessaloniki. The Associations' Board Speaker, Mike McNamee, reminded the Board about the difficulties in organising the Congress and thanked the Executive Director and the ICSSPE President for their efforts. He also expressed thanks to Prof. Dr. Spiros Kellis for bridging with local teams and having done such great work in organising the Congress.

The road towards a joint Congress in 2008 with ICSSPE, IOC, IPC and FIMS

Concerns about the organisation of one of ICSSPE's most important events, the Pre-Olympic Congress, arose in the middle of 1988, prior to the Cheonan Congress in Korea. The IOC Medical Commission had announced that the First IOC World Congress on Sport Sciences would be held in 1989 in Colorado Springs. From that time, six more IOC congresses were organised over the following 14 years with ICSSPE hosting four Pre-Olympic Congresses between 1988 and 2004. In spite of fierce competition between the two organisations, in terms of financing and recruiting top speakers, the IOC, and particularly President Samaranch, remained an important sponsor for the ICSSPE Congress. With the installation, however, of a new IOC President, Dr. Jacques Rogge, in 2001 and the passing away of Prince de Mérode (long-standing Chairman of the IOC Medical Commission), it was obvious that this odd situation of two international sport science congresses needed to be clarified. The IOC announced that it would not continue to support ICSSPE's Pre-Olympic Congress in its current form, however it was supportive of a joint initiative put forward by ICSSPE, FIMS and the IPC. There were several rounds of discussion, most notably concerning the name, which needed to reflect the identity of the involved partners. All groups had previously hosted their own congresses but decided, with the exception of FIMS, to cancel theirs in favour of a joint event. The Associations' Board of ICSSPE agreed that the model used at the 2004 Thessaloniki Pre-Olympic Congress should be followed, that is, an event with a strong multi-disciplinary balance and reflecting a wide range of topics and disciplines. ICSSPE would act as the permanent secretariat for the new joint Congress.

All parties involved agreed with the new concept of a joint initiative, with the inaugural event being the 2008 International Convention on Science, Education and Medicine in Sport (or 2008 ICSEMIS) in Guangzhou, China. Mr Lauri Tarasti acted as the legal advisor for ICSSPE in preparation of the memorandum of agreement, which the IOC President Rogge signed on 30 August, 2006.

Research Projects

Some research projects, initiated under Paavo Komi's presidency, had not yet been finalised when the new President Doll-Tepper commenced her administration in Berlin in 1997. One was led by Roland Naul entitled Sporting Lifestyle, Motor Performance and Olympic Ideals of Youth in Europe, the other, led by Kari Fasting, named The Experience of Sport in the Lives of Women in Different European Countries. Support for a further two projects was reported to the Brisbane Executive

Board meeting[17] in February, 1996 for a study on genetic limitations of sporting performance (Vasilis Klissouras) and a study on the implications of the media coverage of sport (James Halloran). All projects were finalised in 1998-1999, with some finishing sooner than others and perhaps not in due time as a result of unforeseen circumstances. Gudrun Doll-Tepper informed the President's Committee in Brisbane in 1999 of problems encountered during the projects, including no acknowledgement of the Council's role in securing the grants to conduct them, concerns with quality standards and quality control of the reports and problems with a disrespectful foreword in one of the reports. The need to define new procedures for presenting requests, reviewing final reports and increasing reliability was stressed. It was suggested that ICSSPE should define itself as a broker between ideas, priorities and sponsors. The IOC had asked for a meeting at the soonest possible date for the completed research reports to be handed over. Before this could occur, a number of actions were taken. In the course of the meeting in Brisbane, and after having been requested by the President's Committee to do so, Dr. Keith Gilbert, a member of the Executive Board, reviewed the report by James Halloran and found it to be inappropriate both in content, structure and format. After having considered this, the President's Committee decided that the reports of Halloran, Fasting and Klissouras had to first be (blind) reviewed before being submitted to the IOC in July, 1999. This experience highlighted to the Council that a new research policy was required to define clearer rules to ensure the quality of future projects and increase ICSSPE's credibility and accountability

Two new ICSSPE sponsored projects were accepted by the IOC in the late 1990s. The first project, developed by Dr. Ken Hardman on behalf of the International Committee of Sport Pedagogy, was the State and Status of School Physical Education. This was funded to USD 46,500 in 1999.

The second project was a publication by Steve Bailey entitled Athlete First - A History of the Paralympic Movement, which received wholehearted support from the International Paralympic Committee and its President at the time, Prof. Dr. Robert Steadward. This project was funded at USD 19,750 by the IOC. The production of this book took a long time due to, amongst other things, difficulties encountered between the IPC and the author. The IPC requested corrections that were not only factual but also had to do with the interpretation of specific events. ICSSPE defended the position of the author and his entitlement to form an opinion on the historical events that took place, at the same time recognising that factual errors, if any, needed to be addressed and corrected. Eventually, the IPC and Bailey reached an understanding on both the contentious sections and the title, which was also under discussion, and the book was finally, after ten years, published at the end of 2007 by Wiley.

As a result of a meeting between Doll-Tepper and Mailliet with the President of the IOC in 1999, ICSSPE was asked to develop proposals for three new projects focussing on: the contribution of sport science and physical education to Agenda 21 of the Olympic Movement; the role of sport in education; and the development of sport science and physical education in developing countries[18].

The first project didn't progress further than preliminary discussions and since no individual or organisation was identified to take the lead, no comprehensive proposal was developed.

It was decided that Ron Feingold, Ken Hardman, Margaret Talbot, Jean-Francis Gréhaigne and Karen DePauw would prepare a proposal on he role of sport in education for submission to the IOC. This project took a while to commence but eventually took the form of the SpinEd (The Role of Physical Education and Sport in Education) project, headed by Richard Bailey. Much data of this project was used for and presented at the 2nd World Summit on Physical Education in Magglingen, 2005.

The final SpinEd report was presented to the IOC and MINEPS IV at the end of 2004. There was universal acceptance of the outcome of the research, however not agreement on what physical education actually is. SpinEd proved to be a starting point for discussions. The review of literature was made available through the project's website, http://spined.cant.ac.uk, and a relaunch as well as Japanese mirror site was planned. The Athens Declaration of MINEPS IV included a substantial amount of language from SpinEd and there was a positive response from participants at this meeting.

The third project topic identified by the IOC, the development of sport science and physical education in developing countries, was to be planned by Regional Coordinator Victor Matsudo but was never considered as the IOC decided to fund only one project proposal.

The Editorial Board, Flagship of ICSSPE

The Beginning

At the General Assembly in Quebec, 1976, agreement was reached "that the Council needed a Bulletin or a journal in order to communicate well with the membership"[19]. The editors of the Bulletin that was created were Professors F. Lotz, P. McIntosh, M. Hebbelinck and E. Buggel, who reaffirmed in London (Executive Board meeting December, 1976) that the German Democratic Republic would take on the expense of printing the Bulletin. The inside cover of Bulletin number 10 (some years later in 1984), contains the same names as those from the first issue, with John Coghlan, former Secretary-General and Werner Sonnenschein, Secretary-General at that time as additions. Gunther Rotter was also mentioned as contributing, under the heading of "Editing". In 1985, Peter McIntosh who, until then had acted unofficially as the chair of the group of editors, resigned. For issue 12 (June, 1988) a chair (John Coghlan) and an Editorial Board (EB) are mentioned for the first time. Aarne Koskela (Finland) and Heidi Kunath (Germany) joined this Board.

Issue 14 in 1991 was the last Bulletin printed in digest format. In the same year, Herbert Haag and Henry Järvinen (from the ICSSPE Office) joined the EB.

In June, 1992 issue 15 was printed in Finland and in a journal format. As Paavo Komi had come into office with the wish to raise the publication's profile, there was encouragement for continued development. A significant step was made with the employment of a part-time publicity officer at the Jyväskylä ICSSPE office. Throughout the Komi administration, the EB regularly produced the Bulletin, in addition to Sport Science Review, Sport Science Studies, Technical Studies, the Calendar and brief statements as to the current status of each sport science discipline, which later evolved to become the Vade Mecum.

New Developments for the Editorial Board

When Gudrun Doll-Tepper unveiled her plans for the new administration, she made it very clear that she considered the activities of the Editorial Board to be of the utmost importance to the Council. She was convinced that the EB reflected what was happening in the world of sports science and physical education through its publications. The fact that the status of the EB was upgraded in the Statutes, which were changed in Barcelona, 1998, (see Finances, Marketing and Statutes Changes in this section: EB at the same level as the other Boards and its Chair sitting on the President's Committee) reflected the respect of the President and that of the General Assembly. The first full time staff member working alongside the Executive Director, Ms. Deena Scoretz, became ICSSPE's first Publications Manager. She and Christophe Mailliet attended all EB meetings in an ex-officio capacity and were instrumental in the smooth running of its work. Gudrun Doll-Tepper remembers that "it was always a pleasure to attend the Editorial Board meetings because its members always came up with original ideas but also implemented what was being suggested or decided in the other Boards and the President's Committee. The Editorial Board was also visionary in understanding what are the topics, what are the issues, but also what kind of technology shall be used to spread the news and share the information. The Bulletin was always a publication that we could take with us, we could not only inform our members but also show it to others and tell them what we are doing. And yes, the EB could be seen as the Council's flagship as they produce something that can be seen"[20].

Berlin, Germany and Budapest, Hungary

In the first year of the Berlin administration (1997), two EB meetings were organised, with the first taking place in Berlin. John Coghlan was still chairperson and Steve Bailey, Aarni Koskela and Guido Schilling had joined Marcel Hebbelinck, Herbert Haag and Albert Remans (ICSSPE's Deputy Secretary General at UNESCO).

The second EB meeting was organised in Budapest and hosted by András Mónus[21]. Prof. Jan Borms, who had left ICSSPE since the dissolution in 1988 of the Research Committee (he was its Secretary-General 1975-1988), joined the Board. This meeting also marked the definite departure of Marcel Hebbelinck as he stepped down from his position after many dedicated years of service to the Board (since 1976) and to ICSSPE (since 1964).

Ongoing publications were reviewed at these meetings, including the Sport Science Studies, Technical Studies and IOC Research Studies. The first edition of the Vade Mecum included eleven scientific disciplines and evolved into the Directory of Sport Science in 1998.

A new proposal was presented to the EB called Women and Girls and Sport: Best Practices. Plans for this had been discussed with Mr. Gilette from UNESCO, at the Berlin meeting, and John Coghlan had been in contact with Anita White (ICSSPE's contact for an upcoming congress on Women and Sport in Namibia) who had recommended Dr. Darlene Kluka to be the liaison between the Editorial Board, the International Association of Physical Education and Sport for Girls and Women (IAPESGW) and the congress in Namibia. Dr. Kluka, then not familiar with ICSSPE, was pleased to take on this role and be more involved in the work of the EB. Later, she was invited to the Board meeting and became a much appreciated and hard working member and then Chairperson in 2004.

Steve Bailey, having accepted the position of Managing Editor of the EB, presented a comprehensive description of the aims, structure and potential themes of a new publication series for the Board's review and comments. The aim of the new publication was to: replace the Sport Science Review; emphasise ICSSPE's umbrella function; encourage interdisciplinary dialogue; promote the exchange of high-level research in sport science and physical education; promote the benefits of ICSSPE membership and generate income surplus. He considered a single theme for each issue (two per year), with disciplinary papers and one commentary paper, which would guide theory into practice and suggest future actions needed in that area (theme). The Board was in agreement with these fundamental ideas. After a constructive discussion regarding practical and other aspects of the journal, it was agreed that the title of the new publication should be Perspectives - The Multidisciplinary Journal of Sport Science and Physical Education.

The two first volumes (Competition in School Sport, 2000 and Physical Activity and Ageing, 2001) were edited by Steve Bailey with EB members forming a complete Editorial Board for the series. Fifteen invited experts from the Executive Board and member organisations formed the Advisory Review Board, chaired by Prof. M. Hebbelinck. It was the start of a new venture for ICSSPE.

Since the first two volumes, seven more have been published: Volume 3 The Business of Sport (Kluka and Schilling, 2001); Volume 4 Sport and Information Technology (Ghent, Kluka and Jones, 2002); Volume 5 Aspects of Sport Governance (Kluka, Stier and Schilling, 2005); Volume 6 Health Enhancing Physical Activity (Oja and Borms, 2004); Volume 7 Sport for Persons with a Disability (Higgs and Vanlandewijck, 2007); Volume 8 Children, Obesity and Exercise (Hills, King and Byrne, 2007) and Volume 9 Talent Identification and Development – The Search for Sporting Excellence (Fisher and Bailey, 2008).

Present at the meeting where Perspectives was discussed was Hans Jürgen Meyer of Meyer & Meyer Verlag (M&M). Mr. Meyer explained his intention of focusing on the sport science market and his interest in working with ICSSPE as a long-term partner. No commitments were made with Meyer, however, as the Berlin office was negotiating with at least three other publishing companies at the time.

In addition to supporting the commencement of the Perspective series, Deena Scoretz worked enthusiastically on the Bulletin and her work paid off. The publication developed a new look, style and range of content.

Lausanne, Switzerland

The following year, the EB meeting was held in Lausanne where the other Boards were also meeting to discuss the restructuring of the Council (see part 2). At this EB meeting, Aarni Koskela stepped down from his position after years of commitment to the group's activities.

In Budapest, during a candid and constructive discussion, members had already voiced their opinions regarding the next EB chairperson and new EB members. Steve Bailey had then suggested that a rotating chair may be worthy of consideration. Ultimately, it was the President's decision as to who would take over as chair of the EB. Doll-Tepper agreed with the EB's Rotation Policy concerning its Chairperson and proposed Dr. Guido Schilling, who accepted her invitation to take on the first term as Chairperson with the rotation keeping him in place for two years.

In the period between the Budapest and the Lausanne meetings, the Executive Office had come to an agreement with Meyer & Meyer Verlag to produce the Perspectives series.

Volume 9 of Sport Science Studies, Research in Sport Management: Implications for Sport Administrators (Edited by Daniel Soucie) was completed and printed in March, 1998.

The Annotated Bibliography of Traditional Play and Games in Africa, edited by Jeroen Scheerder and Roland Renson was another in-house publication, produced in both hard copy and on diskette. This was published as part of the Technical Studies series.

The IOC funded a research study by R. Naul, M. Pieron et al. called Sporting Lifestyle, Motor Performance, and Olympic Ideals of Youth in Europe, which was successfully completed in July, 1998.

The Sharing Good Practice series, initiated in Budapest with Volume 1 Women and Girls in Sport was expanded with two more volumes (Volume 2 - Physical Education Practice and Volume 3 - Coaching Education/Development). These were led by Darlene Kluka, who had been invited to the EB meeting as a guest.

Thanks to new co-operation with a graphic design agency in Berlin, much progress was made during this time in developing ICSSPE's website.

A new initiative, the Share the Knowledge Programme, was presented to the EB as a cooperation with Human Kinetics Publishing (HKP). HKP had books and some journals available that were not fit for sale or were overstocked, all of which could be donated to institutions in developing countries that were in need of resources. Since this time, the ICSSPE office has coordinated the project, following strict criteria as to which institutes will receive this assistance.

Barcelona, Spain

The EB meeting in Barcelona, 1998 was John Coghlan's last and the President rightly thanked him for his role as Chair of the Board for the first two years of the Berlin administration22. Coghlan had been involved in the publication business of the Council since 1984 and before that had made himself a name as Secretary General in the Bannister administration.

A long discussion took place on the Perspectives Series. The stress, due to time pressures in the signed contract with Meyer & Meyer Verlag, gave rise to a discussion as to whether the Council needed to produce another scientific journal series. It was agreed that multidisciplinary information could also be presented in the form of a monograph and the Perspectives publications would be produced on a more flexible time-line (at least one per year). One Perspectives Monograph per year was considered sufficient given that the Executive Office staff proposed to expand the Bulletin to include more content-rich articles addressing theme topics. The intention of the Executive Office was to produce a magazine that would be of interest to individuals inside and outside ICSSPE's membership. It had continually received good feedback about the Bulletin so plans were underway to expand the publication and increase frequency of production to 3 times per year. The Bulletin was seen as having great potential to reach a large audience and to reinforce the Council's umbrella function by including more multidisciplinary content.

Finally, the first edition of ICSSPE's Vade Mecum was completed and distributed to members in May, 1998 with the 11 disciplines agreed upon at the previous EB meeting. This first edition was a start and could certainly be improved on. Now that the structure was in place, the Executive Office would begin to work on a second, more comprehensive Vade Mecum. In general, the Office received good feedback from members, however many members noted that several (important) disciplines were not represented.

Magglingen, Switzerland

Guido Schilling hosted the next EB meeting in Magglingen in May, 1999. This was also his first meeting as the new chair. He had previously gained enough experience to provide a warm welcome to his guests at the Federal School of Sports and it was no surprise that the beautiful setting had a positive effect on the spirit of the sessions.

UNESCO had agreed to grant US$5 000 in support of the publication, Women, Sport and Physical Activity: Sharing Good Practice. There was a large demand for the book and discussions continued with UNESCO regarding their role in its distribution.

Translations into French and Spanish were not yet confirmed, but efforts would be made to ensure the publication was as useful as possible for a large market – particularly in developing countries.

It was decided to focus on best practices in physical education in Volume 2 of Sharing Good Practice, linking the data collected at the 1st World Summit on Physical Education, 1999. Volume 3 would focus on best practices in coaching with data in the collection linked to the Summer Olympic Games in Sydney, 2000. Plans continued for the expansion of the Bulletin, including a feature section with contributions from various disciplines on a specific theme, and new printing, distribution (with a publisher, rather than in-house) and marketing agreements with the intention to produce a more professional publication that was also attractive to non-ICSSPE members. Nevertheless, it was difficult to maintain a well-rounded and balanced selection of articles (i.e. including social and biomedical sciences). Communication with members and Executive Board members working in the biomedical sciences had proved difficult and they did not seem to see the relevance of submitting their work/research/news to the Bulletin, a non-scientific, non-refereed journal.

The Newsletter was now being produced as often as possible, sent by e-mail and also available on the website.

Orlando, USA

Due to tragic family circumstances, Guido Schilling sent his apologies for the meetings in Orlando, 2000 and advised that he could not complete his term as EB chair. Jan Borms agreed to complete the term, which ended as planned at the end of 2000. The revised Statutes stated that the Chair was to be elected by the Board for a 4 year period, which would have run from January 2001 to December 2004. Jan Borms was elected as the next EB chair and lead his first EB meeting in Orlando. To date, only a very vague handout from 1982 had been prepared in relation to the EB membership. No concrete criteria existed, nor did a job description for Board members. Additionally no term of office had been set, so it was during the Orlando meeting that the terms of office of members was discussed, as well as the appointment of new members to the Board. Issues raised included: What criteria should be considered when selecting new members? Should a term of office be set for Board membership? Is the 4 year term (in effect for Executive Board members) also appropriate? What are the specific duties of Board members? How can involvement in the development of publications be increased most effectively?

Brisbane, Australia

Before the EB meeting started in Brisbane (September, 2000), the ICSSPE President officially congratulated Jan Borms on behalf of the Council and thanked him for his contributions to date and his willingness to continue as Editorial Board Chair.

Following on from discussions during the Orlando meetings, Board members unanimously voted in favour of establishing a four year term of office. With the resignation of the Finnish member Aarni Koskela, the Editorial Board's membership was now reduced to 5 members, whereas the Statutes stated that the Board may consist of a maximum of 8. It was agreed that current members needed to stagger their terms of office to assist in continuity of the publications programme and work of the EB. The remaining discussion turned to new membership and the Board was reminded of the candidates proposed. Dr. Darlene Kluka briefly introduced the candidates and members had the opportunity to review their curriculum vitae's.

Candidates were:

Biomedical - Dr. Pekka Oja, Finland; Dr. Jonathan Reeser, USA; Dr. Russell Pate, USA. Social Sciences - Dr. Denise Jones, South Africa; Dr. William Stier, Jr., USA.

The key issues in the discussion on new candidates were regional representation, representation of the full range of sport science disciplines, experience in publications/editing, being a member of an ICSSPE member organisation, availability to attend meetings, areas of specialisation that complemented other areas already represented on the Board (i.e. the Editorial Board should include experts from a range of biomedical and social sciences), a mix of experience in the field of expertise and/or academic standing and PhD being desirable, but not required.

Taking these criteria into consideration and the feedback received from the individual candidates, it was decided that Dr. Pekka Oja from the UKK Institute and Dr.

Denise Jones from IAPESGW be invited to join the Board.
The other two candidates Dr. Jonathan Reeser and Dr. William Stier, Jr. were invited to become advisors to the Editorial Board in light of their respective organisations not yet being members of ICSSPE as openings for full Board membership were foreseen for 2002.

The duties of Board members were described as follows:

Vision: responsibility for Working Programme of ICSSPE's Editorial Board. This point included establishing a vision and direction for ICSSPE's publications, attendance at a minimum of one meeting per year and regular contact with the Executive Office.

Advice: the Council relies on Board members to stay up to date on the relevant publications in their own fields of expertise so that they can effectively guide publication plans. Specifically, EB members should be able to suggest themes and authors who could fit the needs/fill gaps of the publications planned and offer feedback on the reception of ICSSPE publications in their respective networks. Additionally, EB members' input into the Council's marketing strategies is very important.

Practical Involvement: in addition to the points mentioned above, the following specific contributions were suggested: commitment to write one article per year for the ICSSPE Bulletin or to provide one article from a colleague; provide publicity for ICSSPE and the Council's publications (at events, in networks or other publications); promote membership and the Council's publications; solicit suggestions for new publications (from ICSSPE membership and beyond); willingness to edit and review contributions submitted for publication within ICSSPE's publication programme.

The meeting also provided an opportunity to review the problems with the marketing and sales of ICSSPE's publications. Deena Scoretz reported to the Board the current situation regarding marketing, sales and subscriptions, and summarised the most recent meeting with Meyer & Meyer Verlag. Unfortunately, M&M did not feel that the ICSSPE Bulletin was marketable outside of ICSSPE's membership. The low number of subscribers to date illustrated the need to make some changes to the way in which the ICSSPE Bulletin was promoted. At the most recent meeting in Berlin between the publisher and ICSSPE, the Executive Office staff stressed that the Bulletin was no longer an internal 'newsletter' and began brainstorming with Hans Jürgen Meyer on ways in which visibility and awareness of the publication could be increased. In preparation for the development of a new strategy, a reader's survey was included in Bulletin Number 30.

During the meeting with M&M, the inherent challenge in ICSSPE's membership structure (i.e. only one ICSSPE Bulletin is sent to each organisation) was identified and it was agreed that the individual members of ICSSPE member organisations would be an excellent target market. Editorial Board members agreed that sport science institutes and libraries should be contacted world-wide and encouraged to subscribe. Board members noted that a higher subscription price should be established for libraries.

While several members could justify and explain the choice of Vade Mecum as the publication's title, it was agreed to change the title for the third edition to The ICSSPE Directory of Sport Science, with Vade Mecum as a subtitle. This more descriptive title was hoped to increase sales and distribution. It was agreed that the 3rd edition should be produced in the same format as the 2nd with, if possible, additional sport science disciplines included. Margaret Talbot had suggested that Sport Feminism be included.

Unfortunately, the Brisbane meeting was also the last meeting of Deena Scoretz who had made a remarkable contribution as Publications Manager. She returned to her homeland of Canada in January, 2001 but not before she had acquainted her successor with ICSSPE business. Her successor was Amanda Costin who had been Tony's Parker's "right hand" in the preparation of the Pre-Olympic Congress 2000.

Magglingen, Switzerland

Dr. Guido Schilling was no longer a member of the EB but he was again the gracious host of the 2001 meeting in Magglingen and attended all sessions. Since finishing with the EB, he had become an honorary member of ICSSPE. Bill Stier joined the Board as its seventh member and Gretchen Ghent as the eighth. Jonathon Reeser (USA) and Andrew Hills became advisory members.

The Board discussed the current inclusion of a research section into the Bulletin. This was a key issue as the Bulletin was not a peer reviewed publication so care was needed when publishing research as inclusion of only one research finding could be misleading for readers not familiar with the area. Board members agreed that the name of the section should be changed to "sport science". The Board also decided that congress reports, book reviews and position statements would be included as a valuable addition.

The English edition of the proceedings of the World Summit on Physical Education were printed in early 2001 and were later followed by German, Japanese, Chinese and Polish editions.

Budapest, Hungary; Manchester, UK; and Marshfield, USA

The discussion regarding transfer of the Bulletin to an online format was initiated at the 2001 EB meeting in Budapest, again excellently organised by András Mónus. As expected, discussion centred around the question of how to make it available to members only and to convince some Board members who showed resistance to the new style and format, namely electronic versus hard copy. The rationale behind the option of going online was not just a matter of following new trends but it was also a more cost effective choice because production of the Bulletin, including the mailing, was very expensive and the intention was to reach more readers by sharing the login code among the members of all ICSSPE member organisations.

In Manchester, 2002, the Executive Office presented documentation regarding the advancement of the Bulletin as an on-line source. The Board agreed that this would be a great initiative and they also liked the idea of continuing with a printed version of the Bulletin so that it could be used as a promotional tool and also be sent to members in developing countries. It was further suggested, that there may not be a need to outlay extensive printing costs if some type of inexpensive corporate folder was developed where information could be inserted professionally. It was further suggested that the folder could have a CD holder so that the papers could briefly identify what the articles were about and then the CD be included with full text.

The Board further suggested that they would like an evaluation undertaken to determine if these actions resulted in increased activity on the ICSSPE website or increased attention and response to the Bulletin.

Looking back on the issue, it is surprising that there was actually little resistance from the larger ICSSPE family with this change in production style of the Bulletin. At the meeting in Marshfield, Wisconsin, 2003 (well organised by Jonathon Reeser), it was announced that the first online Bulletin had been completed successfully and that a CD version had also been created for those members who had no access to the internet. The Board decided that with the large number of publications planned for 2004, especially within the Perspectives series of ICSSPE, the other very valuable proposals and information that had been received could instead be used as an addendum to the Directory of Sport Science. This implied a great deal of work to develop and it was agreed that a taskforce would be needed to direct this action[23].

The publication, Physical Education: Deconstruction and Reconstruction: Issues and Directions, was completed in late May, 2003, and was distributed to all ICSSPE members. The Board recognised Ken Hardman's effort to finally bring this publication to fruition after a long process of work.

Although the topic of criteria and direction of ICSSPE's Publication Strategy was not on the initial EB agenda, discussion surrounding this item took place at many times during the meeting. The final decisions, plus the developed diagrammatic representation of ICSSPE's publication strategy are presented below (see Figure 3).

The strategy was developed based upon the strategic model (science, service, advocacy) created by the Associations' Board as the areas which direct ICSSPE's activities. The EB reiterated that publications should be multidisciplinary, regionally and gender balanced, and cover a topical, current issue related to sport science or physical education. The Perspectives series fitted best within the desired objectives and it was thus agreed that it becomes ICSSPE's signature series.

It was further agreed that the Sport Science Series did not fit well within the desired publication plan and thus it would be phased out and the planned volumes would instead be included in the Perspectives series, adapted to the different format.

Gretchen Ghent had undertaken a very valuable task to ensure that all ICSSPE publications were listed on catalogues of the Library of Congress, the British Library, Sport Discus and World Cat.

In addition, ICSSPE's Newsletter, now called ICSSPE News, had become a monthly publication.

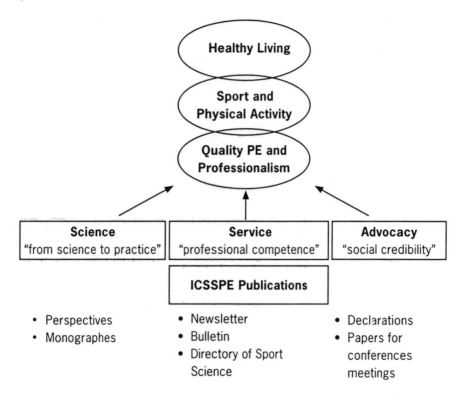

Figure 3: Model of Publication Strategy as based upon ICSSPE's Strategic Model.

Unfortunately, this was also Amanda Costin's last meeting. Amanda, like Deena Scoretz, made a very substantial and professional contribution to the work of the Editorial Board. She introduced her successor to the Board members, Tamie Devine, also an Australian, who had been invited as a guest to the meeting and was going to take over the remaining 18 months of Amanda's contract.

Thessaloniki, Greece

In Thessaloniki, August, 2004 Christophe Mailliet, Executive Director, informed the Board that Tamie Devine, Publications and Scientific Affairs Manager of ICSSPE, would be leaving the ICSSPE Office at the end of 2004. He thanked her, also on behalf of the entire EB, for her hard work and collegiality and for the privilege of having been able to work with her over the previous 18 months.

It was also the last meeting for Jan Borms who had decided not to extend his mandate for another four year term. Later, the President thanked him at the General Assembly and he received Honorary Member status "in recognition of outstanding services in the interest and for the benefit of the Council".

Prof. Dr. Darlene Kluka was unanimously elected as new Chair of the Editorial Board for the period 2004-2008 and it is very likely that with her talents, energy and leadership she will keep the EB evolving in the good direction, as it had in the preceding years. Unfortunately, due to personal circumstances and problems with staffing in the Berlin headquarters, the work of the EB was somewhat hindered in the year following the Thessaloniki meetings. The EB met again in 2006 (Berlin) and in 2007 (Pretoria and Warsaw). Pekka Oja, Herbert Haag (a member with a long standing history in the EB), Bill Stier, Jonathon Reeser and Denise Jones finished their terms during this time.

It was a sign of vitality, dynamics and commitment that new members were ready to replace the former EB members. The current and nominated members of the Board are: Prof. Dr. Darlene Kluka, USA, whose expertise lies in the fields of Motor Behaviour and Women and Sport; Gretchen Ghent, Canada, Sport Information; Prof. Dr. Andrew Hills, Australia, Health Issues, Nutrition, Obesity; Prof. Dr. Kari Keskinen, Finland, Sport Biology; Prof. Dr. Anneliese Goslin, South Africa, Sport Management; Prof. Dr. Richard Bailey, United Kingdom, Sports Pedagogy; Prof. Dr. Alexander Woll, Germany, Research Methodology and Health Education.

Ex-officio members are Vice President – Programmes and Services, Colin Higgs; Executive Director, Mr. Detlef Dumon and Ms. Katrin Koenen, Publication Manager. Advisors to the Board are Mr. Jens Sejer Andersen; Prof. Dr. Jinxia Dong; Prof. Dr. Abel Toriola; Prof. Herbert Haag; and Prof. Dr. Rosa López de D'Amico.

Katrin Koenen became the new Publication and Scientific Affairs Manager in June, 2006. She had previously worked for ICSSPE on a project basis, assisting with the Rehabilitation through Adapted Physical Activity and Sport for Children and Youth Affected by the Tsunami in Southeast Asia conference held in Bangkok, 2005, and its follow up event in January 2006 in Berlin.

As a former student of Gudrun Doll-Tepper, Koenen graduated with an Elementary School Teacher degree and had been working as Public Relations Consultant and Communication Manager for different sport events including the IAAF Golden League ISTAF meeting in Berlin and at Paralympic Games 1996, 2000, 2004 and 2006. In 2007, Christophe Mailliet left his Executive Director role at ICSSPE after conducting the organisation's business very successfully for ten years. He took a new position as Executive Manager of Streetfootballworld, also based in Berlin. Detlef Dumon was appointed as the new Executive Director after serving as ICSSPE Communications and Finances Manager for the previous six years. This internal change allowed for a smooth transition and for ICSSPE to continue working directly with all projects. Detlef Dumon graduated with a degree in Journalism and Linguistics in Berlin and had worked as a lecturer in the Czech Republic for five years. After returning to Berlin he worked as an Office Manager and Public Relations Consultant for two communication agencies. In one of these roles, he had been responsible for media relations for the 1999 World Summit on Physical Education in Berlin.

Finances, Marketing and Statutes Changes

Secretary-General Christophe Mailliet had been in place for approximately one year when he attended his second Executive Board Meeting in Lausanne, March, 1998. This was the site where ideas about restructuring of ICSSPE were agreed upon. Many of these new ideas, however, had still to be streamlined and moulded in new Statutes and Bye-laws for the next Board meeting at the end of that same year. Today, Christophe Mailliet still remembers vividly the meetings of the Executive Board and General Assembly in Barcelona, November 1998 because they were the first real test of his capacity as a manager as he was very much involved in guiding this important item on the agenda.

The changes to Statutes and Bye-laws was a direct result of the session in

Lausanne, where there was the general will for structural changes to make the Council a more effective organisation on the one hand but also an organisation that would have more impact at the scientific and political levels. It was also intended that ICSSPE be an organisation that would really be able to speak with one voice representing the whole field in the general context of the sport movement. Mailliet recalls that "the Executive Board was a strange mix of individuals who were elected to be on that Board and people who were representing the so-called ex-officio members of the Executive Board, which were in fact organisations. So there was a strange construction and mixture of tasks, which was not really effective because the elected individuals had to look after the management of the organisation. Basically that was their function and then the people representing organisations had to represent in a way the political and scientific interests of their particular field and so therefore they were sort of blocking each other and there was never really much output from the Executive Board meetings. At least that was my feeling"[24].

The basic concept and rationale behind the proposed statutes' changes was to separate the management aspects of the Council from the political and content aspects, which would be dealt with by the Associations' Board. The latter became the "think tank" of ICSSPE, generating new ideas where one would try to find common positions on specific topics or issues. Formation of the Associations' Board was a major structural change that would have to be integrated into new statutes and structures. Another proposed important structural change was to raise the status of the Editorial Board, which was until then, basically a working committee without a direct voice in the political life of the organisation. Thus, the Executive Board would be reduced in size and put into a more managerial function, the Associations' Board would reunite the previous ex-officio members, the (international) organisations would deal with specific aspects of sports science or physical education, and the Editorial Board status would be raised so that it was on par with the Executive and Associations' Boards. As a consequence, the composition of the President's Committee was changed to include the Chairs of the three Boards, along with the President, the Vice-Presidents and the Executive Director (without vote). A first draft of new Statutes was developed by a working group composed of Anthony Parker, Karen DePauw and Christophe Mailliet. The final documents for Barcelona were the product of a comprehensive consultation process, which took place over several months. The immense work of rewriting and streamlining the Statutes, more or less completely, and drafting amendments preceded a vote by members. Mailliet, who was charged with this task with the help of Board members, thoroughly enjoyed this work and the consultative sessions, which he believes were a milestone in ICSSPE's existence. There were heated discussions during the meetings with a juridical construction needed to make it all happen. The Council had become so large because of the steady increase in membership, that the General Assembly would never have been able to make decisions because it was unlikely to have enough organisations present to reach the required quorum. The new Statutes and Bye-Laws were accepted and came into force on 1 February, 1999.

Looking back, the changes proved to be successful because very shortly afterwards, one of the first actions of the Associations' Board was to organise the first World Summit on Physical Education, which was held the year after. For Mailliet, 1998 and 1999 were undoubtedly the years that paved the way for almost everything that followed. Another highlight of this period of time was that Mailliet's work identified him as one who could definitely follow in the footsteps of his illustrious predecessors John Coghlan, Werner Sonnenschein and André Noel Chaker.

As a side note, it is worth mentioning that in Barcelona the name of Secretary General of the Council had been changed to Executive Director. The new name did not give necessarily more status or authority but it was felt that it better reflected the role because the function was more of a manager-type position whereas Secretary General was more a structural position.

The financial security of the Council had always been a prime concern of all administrations and it was no surprise that from the onset of her mandate, President Doll-Tepper secured the existence and the future of her headquarters through her network of contacts with the federal and Berlin governments and private enterprises.

In spite of a fairly optimistic future when the Berlin administration took off, Doll-Tepper put strategies for improving ICSSPE's financial situation on the agenda.

Even with support from Germany, ICSSPE was dependent on subventions such as the IOC's, which could be cancelled at any moment. Therefore, the Executive

Board, meeting in Sao Paulo, 1999, was convinced that an adequate and cautious investment policy should be developed using existing assets of ICSSPE as a base, thereby keeping a reserve which would allow for the running of the Executive Office for 3 months in case of unforeseen events. It also found that the membership fees should reflect the actual cost of membership service. Indeed, calculation of the cost of servicing members made it clear that membership fees were not adequate anymore in relation to the development of membership service and ICSSPE activities in recent years[4]. The Board decided to work towards generating more income through publication sales, to provide better service for national institutions, to propose to the upcoming General Assembly to raise the fees by 10%, and to propose to integrate an inflation rate of 5% in the annual membership fee.

Various possibilities of membership development were discussed by the President's Committee at the Sao Paulo meeting, including corporate membership and individual membership. It was found that ICSSPE should think of putting in place a fundraising and/or financial advisory committee constituted of representatives of companies interested in ICSSPE's work. It was emphasised that it was crucial for ICSSPE to identify priorities and projects, since this would simplify the search for new sources of funding. Regarding individual membership, it was felt that it could constitute a potential threat to the member organisations. It could also result in a loss of profile for ICSSPE as an umbrella organisation. On the other hand, joint membership in ICSSPE and one of its member organisations could be considered and could possibly be an incentive to join.

The Future

The year 2008 will mark the end of the presidency of Gudrun Doll-Tepper. The fifth President continued to grow and maintain ICSSPE's influence, like her famous predecessors had done before' Noel-Baker, Bannister, Kirsch and Komi. Thanks to her charisma, enthusiasm, positive outlook and creativity, she realised the vision she had from the beginning of her administration, which she was able to share with many other capable and enthusiastic supporters. She worked hard for and succeeded in making ICSSPE a real umbrella organisation, respected in the world by partners like the IOC, UNESCO, IPC, FIMS, WHO and others.

Her legendary vitality and high energy levels brought her to visit all corners of the globe, making presentations at numerous congresses and business meetings, re-establishing old connections and making new ones. She definitely put her stamp on the organisation but she admits that this could not have been possible without the capable people in the different Boards, Committees and ad hoc working groups of ICSSPE. The Berlin administration was characterised by the structural changes in the Council that were realised and streamlined in Statutes and Bye-Laws and which made the organisation more workable. Organising the World Summits on Physical Education and the attention to quality physical education were just some of the many initiatives that took place in the period 1997-2008.

The headquarters were staffed by full- or part-time professionals who had very good working and personal relationships between themselves and particularly with Gudrun Doll-Tepper, whom they described as always open for critique and new ideas, who was never negative, never bossy, who never had a personal conflict and who was admired for her leadership.

At the end of this presidency, there seems to again be financial concern for the future, as was the case at the end of each of the former administrations. Possible strategies for coping with these inevitable changing situations have to be pro-posed and discussed. Corporate funding should be sought along with assistance from the IOC for more substantial support, since ICSSPE is the only organisation of its kind in this field. Since the independence of ICSSPE is crucial, a fundraising process whereby corporate sponsors are identified should be put in place as soon as possible.

At one of its meetings[25] the Executive Board agreed in principle to pursue, as a strategic goal, the separation of the ICSSPE administration from the President. The current situation, whereby the administration is based in the home city of the President, is per se unstable and represents a disadvantage for those potential candidates who wish to run for the position, but cannot create similar conditions. Therefore, the establishment of permanent headquarters and continuous employ-ment of professional staff will be needed, and should be pursued as another stra-tegic goal.

The widening economic gap between scientists of different regions in the world is another area that the new administration will need to address. The issue is somewhat similar to those addressed by ICSSPE around gender and the situation of developing countries, so may require policy to be implemented. The 'new' disciplines within ICSSPE such as sports law, sport economics and management and sport policy should also be further developed. The concepts of Ethics and Professionalisation have been mentioned in many reports and during one of the meetings of the Associations' Board, it was stated that the interface between medicine, ethics and sport is an area that needs to be tackled.

Also looking to the future, the 2012 International Convention on Science, Education and Medicine in Sport will tentatively be jointly hosted with IOC, IPC and FIMS. Preliminary contacts to the London Organising Committee and the British Olympic Association have already been established by Margaret Talbot, with further discussion planned for the northern Autumn, 2008.

Elections for the new President are to be held in Guangzhou, China, August, 2008. As ICSSPE prepares to enter a new era, it's members and the international sport, sport science and physical education communities can be assured that the professionalism and commitment of the organisation will continue as before.

References and Footnotes

1. Bailey Steve, *Science in the Service of Physical Education and Sport*, John Wiley & Sons, Chicester, 1996
2. Kirsch (1982-1988) was re-elected in 1988 for a term of office that would last until 1994 but ended his term in 1992. Komi took over from him for final 2 years (until 1994) and was then elected in 1994. This term was through to 2000 but Komi finished in 1996. Gudrun Doll-Tepper was elected in 1996 but in the meantime the statutes had changed the President's term from 6 to 4 years, meaning that Doll-Tepper's term should end in 2000. Legal advice made it clear that the years of service provided by Doll-Tepper during the non-completed term of Komi did not count as years of elected presidency so when she ran officially in 2000 and was elected, this commenced her first term of president. She was elected for a second and final term in 2004, to service until 2008.
3. Agenda and Provisional Minutes of the Executive Board Meeting, Lausanne, 1998
4. Agenda and Provisional Minutes of the Executive Board meeting, Sao Paulo,

1999
5. Agenda and Provisional Minutes of the General Assembly, Brisbane, 2000
6. Agenda and Provisional Minutes of the Associations' and Executive Board meetings, Beijing, 2001
7. Agenda and Provisional Minutes of the Associations' and Executive Board meetings, Thessaloniki, 2004
8. Agenda and Provisional Minutes of the 8th Associations' Board Meeting and the 67th Executive Board Meeting, Berlin, 2006
9. Agenda and Provisional Minutes of the President's Committee meeting, Orlando, 2000
10. Scoretz D., *The Role of a Multidisciplinary Organisation in an Age of Specialisation*, Bulletin, 31, February 2001, 22-23
11. Hardman K. & Marshall J., *World-wide Survey of the State and Status of School Physical Education*, University of Manchester, 2000
12. Bulletin 10, March 1985, 23
13. Agenda and Provisional Minutes of the 7th Associations' Board Meeting and the 66th Executive Board Meeting, Edmonton, 2005
14. *AIESEP wished to retain its own prize, which goes back to President Samaranch. Furthermore, it was a prize specifically to reward excellence in physical education, which makes it different in nature to other existing awards.*
15. *Congress 2000 – a Multidisciplinary Achievement*, Bulletin 31, February 2001, 16-18
16. 2004 Pre-Olympic Congress, Bulletin 43, January 2005, http://www.icsspe.org/index.php?m=15
17. Agenda and Provisional Minutes of the Executive Board meeting, Brisbane, 1996
18. Minutes of the IOC/ICSSPE meeting, Lausanne, July 29 1999
19. Bulletin, 1, 1977
20. Interview with Gudrun Doll-Tepper, January 2008
21. Minutes of the Editorial Board Meeting, Budapest, 1997
22. Minutes of the Editorial Board Meeting, Barcelona, 1998
23. Minutes of the Editorial Board meeting, Marshfield, Wisconsin, June 2003
24. Interview with Christophe Mailliet, January 2008
25. Agenda and Provisional Minutes of the 5th Associations' Board Meeting and the 64th Executive Board Meeting, Pretoria, 2003

Part II
Directory of Sport Science, 5th Edition

Preface

Three primary objectives of the International Council of Sport Science and Physical Education (ICSSPE) are: 1) to encourage international co-operation in the field of sport science; 2) to facilitate differentiation in sport science whilst promoting the integration of various branches; and 3) to make scientific knowledge of sport and physical education available. The Directory of Sport Science is developed in line with these three objectives. The intention of ICSSPE's Editorial Board is to produce a publication listing fundamental information about the wide range of sport science and physical education disciplines and thematic areas.

The Directory is a 'first stop' for information on the diverse fields that comprise sport science and physical education. The information included directs readers to additional information sources. Although many of the authors are active in ICSSPE member organisations, the Directory of Sport Science is not meant to be a 'Who's Who' in sport science, but rather a 'What's What', that is, a description of what components make up the 'whole' we refer to as sport science.

To make the Directory as easy-to-use as possible, authors have prepared their information according to the following outline (clarified in Introductory Comments):

Part 1. General Information - including the discipline's Historical Developments, Function, Body of Knowledge, Methodology, Relationship to Practice and Future Directions. This section also contains a reference list of any information that was directly referenced to in this section.

Part 2. Information Sources - this section lists relevant Journals, Reference Books, Encyclopaedias, and Book Series (all alphabetically). Conference/Workshop Proceedings are listed with the most recent publications first. Additionally, relevant Data Banks and Internet Sources (Internet addresses or listservs) are included.

Part 3. Organisational Network – describes the organisations working in each respective sport science field at the International Level, Regional Level and National Level. When Specialised Centres exist, these are also included. Finally, due to the increase in the number of opportunities available for international studies, Specialised International Degree Programmes are also listed where applicable.

Part 4. Appendix Material – includes Terminology, Position Statement(s), Varia and Free Statement(s) when appropriate.

The contact information of the authors is included in each chapter. We encourage readers to contact the authors directly for additional information.

It is our hope that this publication will deepen the common understanding of each facet of sport science, thereby strengthening the field as a whole. Undoubtedly, additional fields, not included in this Directory, are active in sport science and we anticipate that the Directory will grow in the years to come.

Please do not hesitate to contact the ICSSPE Executive Office with additional suggestions for future editions. The information contained in the Directory will be fully updated every two years. This 2008 Edition is the fifth of the Directory.

International Council of Sport Science and Physical Education (ICSSPE/CIEPSS)

Introductory Comments

Herbert Haag

The International Council of Sport Science and Physical Education (ICSSPE) is an 'organisation of organisations' that forms a network encompassing the fields of sport science and physical education. The mission of ICSSPE is to represent sport science and physical education world-wide by disseminating research results, implementing relevant projects and encouraging links between disciplines and members. The Directory of Sport Science is a publication reflecting ICSSPE´s mission.

To better understand the material included in the Directory of Sport Science, with the aim that knowledge acquired from this book may furthermore enhance professional work within the academic fields and thematic areas of sport science, these introductory comments outline the following:

1. Sport Science - Terminology
2. Sport Science - Historical Dimension
3. Directory of Sport Science - Content Structure
4. Directory of Sport Science - Formal Structure.

1. Sport Science - Terminology

In order to understand the basis upon which the Directory of Sport Science has been constructed, it is of utmost importance to clarify concepts and constructs which underline and formulate sport science and also constitute sport science as an academic discipline. The terminology presented here forms the foundation of the Directory of Sport Science content; the terms are linked into three groups:

a. Sport, Exercise, Physical Activity and Movement;
b. Physical Education as a Form of Education;
c. Sport Science as an Academic Discipline.

The following definitions come from a variety of sources, including dictionaries (Anshel, 1991; Beyer, 1992; Haag and Haag, 2003; Röthig and Prohl, 2003). A clear understanding of the underlying terminology is very beneficial to the understanding and further development of an academic discipline.

1.1. Sport, Exercise, Physical Activity and Movement

Sport

"A precise definition of sport is impossible due to the great variety of meaning in colloquial language. Everything that is understood under the term sport is less determined by the scientific analyses of its boundaries than by daily usage and historically developed and transmitted ties to social, economic, political and judicial structures. The conceptual understanding of sport is always subject to historical processes and it cannot be established once and for all. Sport is practised in certain types of sport (in clubs, sport associations); as athletics (amateur sport; performance sport; top sport; professional sport); in the school as school sport; predominantly for reasons of health as in recreational sport, sport for compensation, "Trimm" action (physical fitness) for the fun of playing sport for all, mass sport; leisure time sport, vacation sport; within specific institutions as industrial sport; college sport; sport in the armed forces; police sport; prison sport; in groups with typical characteristics as women's sport; sport for children and youth; sport for the disabled; sport for the aged; within denominational associations (church and sport); in the form of artistic movement arrangements as dance, social dance, ballet, jazz dance; rhythmic gymnastics, etc. In contrast to the activities of daily life and work, sports activity gains its special characteristics by the change in meaning of the content of action; in other words, sport is based on the modification of contexts of reality in terms of ritualisation and symbolism of realistic behaviour. Sport actions (predominantly motor actions) are to a certain extent "freed" activities, which transcend the purposeful decisions of the workaday world. This is not to say that they lack purpose, but they are not totally bound by traditional considerations of utility. Fishing, which is primarily done to catch fish, for example, becomes sport (e.g. competitive casting) when the goal lies in the perfection and precision of the movement process. To this end, procedures and rules are set which have to be interpreted as ritualistic actions when compared with the original action. Fishing is modified into so-called "dry-fishing", emphasising the (athletic) alienation and modification of the initial intention. Motor activities and social interaction are usually characteristics of sport. Criteria such as performance, competition, rules, ideal types of experiences and organisation are phenomena used to characterise sport, but have different significance in different situations. The designation of non-motor activities as sport is predominantly done on traditional grounds (e.g. chess)" (Beyer, 1992, pp 574-575).

Exercise

"The term Exercise refers to:

1. An operation for processing learning contents such as motor skills through repeated execution, if necessary under different conditions (e.g. in the form of repeated attempts to do the skill). With each repetition, the internal organisation of the process changes, as well as the nature and the way in which the attempt is executed. Exercise can be classified within the context of a sequence of instructional steps or learning stages according to the principles of the psychology of learning. In connection with the stabilisation of movement, exercise causes, according to cybernetic theory, a shift in the regulation of movement from external to internal loop controls (automation), while at the same time the anticipation is focused on new events. Depending on the learning contents and the type of learning (reactive conditioning, learning through trial and error, learning as the result of the reinforcing effect of exercise (law of effect according to Thorndike; breaking of habits; tracking behaviour, etc.)), there are specific rules for the use of exercise in education (distributed practice, massed practice, massive exercise).
2. The repeated execution of relatively simple, automated skills for the improvement of physical abilities, if necessary under more difficult conditions.
3. The objectified form or the result of an exercise process. All compounds of the term exercise relate to this sense of term" (Beyer, 1992, pp 709-710).

Physical Activity

"Totality of all motor activities that are realised in motor learning, exercising (training) and in sport competition. Most likely based on the knowledge about the central role of the body in the life of the human being as well as on an original play and performance drive of the human being, physical activity has been known since the earliest developmental phases of the human being, as utilitarian, work and combat movements such as walking, running, jumping, hopping, carrying, pushing, pulling, hanging, supporting, swinging, rolling, throwing, catching, kicking, balancing, etc. The predominant functions of physical activity are the securing of existence and health (e.g. dietetics), life interpretation in culture (e.g., in Greek Antiquity; some traces of this can still be found in modern Olympic ceremonies) and realisation of a certain educational ideal (e.g., with the Philanthropist). Irrespective of the various cultures, physical exercises have been the main content of all historic sport movements (calisthenics, gymnastics, play and sport movement) and follow goals from the areas of health, recreation, performance, forming of personality, etc." (Haag and Haag, 2002, pp 363-364).

Movement

"From a physical viewpoint the change in location of an object (a mass) over time in relation to another object viewed as a resting entity or in relation to a reference system defined by other objects; movement is, therefore, always relative, an absolute movement does not exist. Human movement or rather the human being's urge to move is part of his nature and represents a central human behaviour. Human movement can be regarded as a change in location (change of the body's position in space), change in position (changing the position of the body segments in relation to each other) and change in velocity (change of the movement state by means of applying force). Sport is inconceivable without movement" (Haag and Haag, 2002, p 315).

1.2. Physical Education as a Form of Education

The educational dimension of sport, exercise, physical activity and movement can be described by the term "physical education", which historically has been used for denoting the academic field. In some parts of the world, the term physical education still continues to describe the field of sport science. This meaning is, however, rather limited. In reality it stands for an educational dimension, for a process and an area of educating human beings.

Physical Education

"Originally (in reform pedagogy) the complete education of the human being starting from the body. The term originates from reform pedagogy, which intended to make a contribution towards the total education of the human being on the basis of human physicality by means of movement, sport and play. After 1945, physical education was understood as holistic education following the fundamental principles of nature, proximity to life, spontaneity, social binding, value orientation and popularity. Today, physical education manifests itself predominantly in movement education, health education, play education and education towards a sensible utilisation of leisure time. It also stands for the pedagogic training of motor characteristics, skills and abilities within the framework of physical culture, which is predominantly used in the former socialist nations. In many English-speaking countries, the term physical education is still used to refer to sport instruction in schools and to the field of physical education teacher training and sport science in general" (Haag and Haag, 2002, p. 363).

1.3. Sport Science as an Academic Discipline

This academic discipline has been labelled in many ways throughout the world, using terms such as kinesiology, physical activity science, gymnology, sportology, kinanthropology, kinetics and movement sciences. In the Directory of Sport Science, the term sport science is used as it is gaining world-wide acceptance. However, an important prerequisite would be that science is understood as a comprehensive term, representing natural, social-behavioral, political-economical and cultural sciences.

Sport science can be described in the following way: "Sport science represents a system of scientific research, teaching and practice to which knowledge from other disciplines is integrated. It is the purpose and function of sport science to investigate questions, which have been identified on a scientific basis. Finally, solutions have to be applied, in order to explain, control, and if necessary, change the practice of sport. Sport science is mainly visible through its body of knowledge, which is the result of the scientific endeavour in regard to sport. Sport science is a relatively young science. It is an example of a so-called theme, integration, interdisciplinary and cross disciplinary scientific field in contrast to the long established sciences like philosophy, medicine, law and mathematics which can be called discipline-oriented sciences" (Haag, 1994, p. 15-16).

It is important to note that other definitions of these terms are possible and can be found by referring to relevant dictionaries (see References).

2. Sport Science - Historical Dimension

The scientific treatment of exercise, physical activity and movement can be traced back to ancient times. These basic actions of human beings have been analysed over time from philosophical, educational, historical, social and medical perspectives; however, 'sport science' is not considered as an academic discipline with a long history. The development of sport science must also consider the differentiation and specification within science throughout the ages as more and more academic disciplines have been conceptualised.

The term "concept" refers to the sum of aspects, which should be analysed in order to understand a discipline or thematic area. There are six aspects covered for each discipline and thematic area:

1. *Historical Development*
2. *Function*
3. *Methodology*
4. *Body of Knowledge*
5. *Relationship to Practice*
6. *Future Perspectives*

These six aspects represent a relatively comprehensive perspective of sport science; they also serve as a structure for the Directory of Sport Science, including sport scientific disciplines and thematic areas of sport science; both represent sport science as an academic discipline.

3. Directory of Sport Science - Content Structure

The content structure of the Directory of Sport Science is important, however using a given structure as an analytical model may induce patterns that are too fixed, provide a structure that often does not necessarily transmit the intended idea, create overlaps between certain parts of the structure since sharp borders do not exist and finally, a process orientation may change the structure faster than previously anticipated. Consequently, this structure should be considered as one option to the present understanding of sport science. It may change in the future and may therefore require adaptation in subsequent editions of the Directory.

3.1. Disciplines of Sport Science

In the Directory, a distinction between disciplines and thematic areas can be made within the academic fields covered. Disciplines of sport science are characterised as theory fields of sport science and thus "A theory field refers to an applied discipline of a more or less established academic discipline with a relatively old history. The so-called intra-relationship between the applied sub-discipline and the academic "mother" or "relation" science is of importance within the concept of theory fields" (Haag, 1994, p 50).

Under this definition, the following sport scientific disciplines are represented in the Directory of Sport Science (in alphabetical order): Adapted Physical Activity; Biomechanics of Sport; Coaching Sciences; Kinanthropometry; Motor Behaviour; Motor Development, Motor Control and Motor Learning; Philosophy of Sport; Sociology of Sport; Sport and Exercise Physiology; Sport and Exercise Psychology; Sport and Leisure Facilities; Sport History; Sport Information; Sports Law; Sport Management; Sports Medicine; and Sport Pedagogy.

3.2. Thematic Areas of Sport Science

Thematic areas of sport science may be characterised as theme fields of sport science.

"A theme field is considered a scientific unit which is composed of interrelationships between different theory fields related to a certain subject, which might come directly from the field of movement, play and sport or which might be a theme with dimensions that are not only sport-specific. Themes are considered from an interdisciplinary point of view, where a certain number of theory fields, depending on the given theme, integrate their scientific results. These themes represent the world of sport in a very direct way, while the theory fields have a more abstract and academic discipline-oriented character" (Haag, 1994, p50).

Using this definition, the following thematic areas of sport science are represented in the Directory of Sport Science (in alphabetical order): Athletic Training and Sport Therapy; Comparative Physical Education and Sport; Doping in Sport; Health Enhancing Physical Activity; Physical Education; Sport and Development; Sport and Human Rights; Sport Governance; and Women and Sport.

This list of thematic areas in sport science is open to addition as new topics, themes and issues related to sport, exercise, physical exercise (activity) and movement are constantly developing. To be included in the Directory of Sport Science, it is important that the thematic areas have developed to a similar level as those already included in previous editions. Nevertheless, the line of identifying new thematic areas to be included in the Directory of Sport Science has still to be followed, since this is one means to support the necessary process development of sport science, especially as it is a relatively new science within the world of academia. Often, thematic areas also allow for and even require integrated and interdisciplinary strategies in regard to the sport science disciplines in order to generate scientific knowledge related to a given topical area of sport science.

4. Directory of Sport Science: Formal Structure

The structure of the Directory of Sport Science includes the following parts:

1. *General Information*
2. *Information Sources*
3. *Organisational Networks*
4. *Appendix Material*

Ensuring a structure is followed throughout the Directory of Sport Science enables the reader to make comparisons between the academic areas. If a category is not relevant for an area, it is marked not applicable.

1. *General Information*
1.1. *Historical Development*

Aspects of sport science have originated and been conceptualised in order to develop disciplines and thematic areas, including relevant networks.

1.2. Function Terms

Terms like task, aim, objective or purpose also describe what is meant by the function of sport science. This aspect provides a sound justification to answer the question: Why does one engage in sport science? In general, the explanation of the reasons for working within an academic discipline are explicitly formulated and expressed. The function of an academic discipline has to be clear; otherwise there is no theoretical framework available for the academic field.

1.3. Body of Knowledge

The historical development and the function-oriented conceptualisation of an academic discipline are also linked to the issue of body of knowledge. This aspect answers the question: What is the content of sport science? In this context, it is important that the body of knowledge of sport science is represented in the form of disciplines and thematic areas with a so-called dual approach.

1.4. Methodology

The procedures summarised under methodology are those that can be used to generate scientific knowledge; they are a basic aspect of the concept of an academic discipline. Sport science is characterised by many different approaches: it uses description, correlation and experimental research methods. Hermeneutical (word-based-data) and empirical (number-based-data) are used as epistemological approaches to acquire scientific knowledge. The many aspects, which have to be considered when dealing with sport, exercise, physical activity and movement, result in a large range of methodological approaches.

1.5 Relationship to Practice

The relationship to practice (science transfer) is another integral part of a holistic approach to the conceptualisation of sport science. This issue relates to the fundamental question: What is the practical application of sport science research results? The relationship of theory to practice and vice versa is very noteworthy as researchers and practitioners must be willing to co-operate. The slogans, practice-guided theory and theory-guided practice trys to indicate this important mutual relationship. Therefore, research projects should include researchers and practitioners who are investigating specific questions and then realising scientific transfer at the end of the research process.

1.6. Future Perspectives
The world of science is a world of constant evaluation and re-evaluation as scientific knowledge becomes outdated and continuing research validates newly generated scientific knowledge. It is of particular importance to anticipate future perspectives. These perspectives should be formulated in order to continue research in sport science in the right direction.

2.   Information Sources
2.1. Journals
2.2. Reference Books
2.3. Book Series
2.4. Congress/Workshop Proceedings
2.5. Data Banks
2.6. Internet Sources

In the current information age, it is important to include additional information sources. Resources in traditional print as well as information technology based resources like website locations and PDF files are included in section 2.0 for each discipline and thematic area.

3.   Organisational Networks
3.1. International Level
3.2. Regional Level
3.3. National Level
3.4. Specialised Centres
3.5. International Degree Programmes

An outline of organisational networks will assist in directing the reader to the body that is relevant to their needs, whether they be information, education, instruction or simply assistance. Very often, valuable information and data can be obtained from relevant groups which otherwise would not be received.

4. *Appendix Material*
4.1. *Terminology*
4.2. *Position Statement(s)*
4.3. *Varia*
4.4. *Free Statement*

Material outlined under section 4.0 is often not publicly known and difficult to access, particularly position statements. These documents are developed by many experts in the field, are short as well as concise and are in most cases officially accepted by governing boards.

References

Anshel, M. H. (Ed.). (1991). *Dictionary of Sport and Exercise Sciences*. Champaign (Ill.): Human Kinetics.

Beyer, E. (Ed.). (1992). *Dictionary of Sport Science*. Deutsch-English-Französisch. (2. Aufl.). Schorndorf: Hofmann.

Haag, H. (1994). *Theoretical Foundation of Sport Science as a Scientific Discipline*. Schorndorf: Hofmann.

Haag, H. and Haag, G. (Eds.). (2003). *Dictionary Sport, Sport Education, Sport Science*. Kiel: ISS.

Röthig, P. and Prohl, R. (Eds.). (2003). *Sportwissenschaftliches Lexikon*. (7. Aufl.). Schorndorf: Hofmann.

Contact

Prof. em. Dr. Dr. h.c. Herbert Haag
Christian-Albrechts-Universität Kiel
Institut für Sport und Sportwissenschaften
Olshausenstraße 74
24098 Kiel
Germany
Phone: +49 431 880 3772
Fax: +49 431 880 37 73
Email: sportpaed@email.uni-kiel.de

Adapted Physical Activity Science

Claudine Sherrill and Yeshayahu Hutzler

Contact

Dr. Claudine Sherrill
11168 Windjammer Drive
Frisco, Texas 75034
USA
Email: csherrill1@earthlink.net

Dr. Yeshayahu Hutzler
Zinman College of Physical Education and
Sport Sciences at the Wingate Center
Israel Sport Center for the Disabled, Ramat Gan 52575
Netanya 42902
Israel
Email: shayke@wincol.ac.il

1. General Information

1.1. Historical Developments

Adapted physical activity (APA) takes its name from such terms as *adapt, adaptable* and *adaptation* from the Latin *adaptare*, meaning "to fit, to adjust, to make suitable to," which appeared in the French language in the 1600s. In conjunction, social changes underlying the first services for persons considered "unfit, unworthy, poor, crippled, insane, defective, untrainable, or criminal" (Barnes, 1991; Turner, 1995), also contributed. All persons deemed "too different to fit in" were housed together initially, with no distinction between conditions, in residential facilities established from the 1600s onward (Stiker, 1982). Directors were typically physicians who, by the 1840s, were pioneers in physical activity as a means of treatment or rehabilitation (Sherrill and DePauw, 1997). As medical science matured and professions of education, psychology, sociology and physical culture/medical gymnastics emerged, special needs were addressed, institutions specialised by disability developed and knowledge expanded. By the 21st century, most persons with disabilities in developed countries were living at home, schooled or employed at least partially in the mainstream, with opportunities for physical activity, but requiring tremendous help to access them and achieve a healthy active lifestyle. Despite enormous progress, persons with disabilities continue to experience prejudice, discrimination, stigma and oppression in most cultures (Charlton, 1998; Goffman, 1963).

Adapted physical activity (APA) science is research, theory and practice directed toward persons of all ages underserved by the general sport sciences, disadvantaged in resources, or lacking power to access equal physical activity opportunities and rights. APA services and supports are provided in all kinds of settings. Thus research, theory and practice relate to needs and rights in inclusive (Block and Obrusnikova, 2007) as well as separate APA programs.

Social construction of the ICSSPE sport sciences (general and adapted) parallels the division of medical, educational and social sciences into separate knowledge areas and professions. One branch focuses primarily on "able-bodied" and the other on conditions that the World Health Organization first (1980) categorised as impairments, disabilities and handicaps and later (1991), as impairments, activity limitations and participation restrictions. *Disabilities*, however, is the preferred term of persons who cope with activity and participation challenges and thus the word used herein (e.g., *persons with disabilities, disability sports, disability studies*). APA, of course, has changed at different rates across cultures and disabilities. Major reasons for this are: (a) widespread diversity in how disability is defined and fits into various cultures; and (b) insufficient resources to support alternative opportunities and rights (Ingstad and Whyte, 1995).

From the 1950s onward, scholarship on specific conditions (e.g., blindness, intellectual disabilities, physical disabilities) intensified, philosophy changed, families became involved, advocacy began and charitable and government funding expanded. In some countries, disabilities were defined in law that provided funding, assuming that criteria of classification systems were met. Separate organisations were founded also. For example, the first international sports organisation for persons with disabilities (specifically the deaf) was founded in France in 1924. The *disability sport movement*, which began with the rehabilitative use of physical activity with war veterans in the 1940s by British neurosurgeon Sir Ludwig Guttmann, led to establishment of the Paralympic Games for elite athletes. Illustrative of APA research generated by sport was the First International Medical Congress on Sports for the Disabled, held in Norway, in 1980. This congress featured research mainly by physicians who were medical directors of disability sports programs. Today, almost all international competitions for athletes with disabilities hold research congresses (or the equivalent) in conjunction with their Games and continue the tradition of sharing new APA knowledge (Doll-Tepper, Kroner and Sonnenschein, 2001; Sherrill, 1986; Steadward, Wheeler and Watkinson, 2003).

Revitalisation of special education from the 1950s onward, as well as pioneer legislation supporting equality of rights, resources and opportunities *for all*, has contributed tremendously to APA. Law has been a major force in the creation and maintenance of APA jobs in schools, recreation centres, hospitals, rehabilitation facilities, sport organisations, etc. In response, universities have developed new courses, professional curricula, masters and doctoral degree specialisations and standards for education of personnel needed to fill jobs (e.g., Kelly, 2006; Sherrill, 1988; Van Coppenolle et al., 2003; Winnick, 2005).

The history of APA reveals many names, but today two terms dominate. Adapted physical education (APE), first adopted in 1952 in the United States, focuses on school-based services. Adapted physical activity (APA), first introduced in 1973 by Canadian and Belgium founders of the International Federation of Adapted Physical Activity (IFAPA), is an umbrella term encompassing physical activity for persons of all ages in rehabilitation, sport, recreation and physical education. Alternative terms in other languages exist, of course (see sections 2.1 Journals and 2.2 Reference Books, Encyclopedias etc).

The United Nations (UN), from the 1970s onward, influenced APA programs and research. In 1971 and 1975, respectively, the UN General Assembly adopted the Declaration on the Rights of Mentally Retarded Persons and the Declaration on the Rights of Disabled Persons. The UN-declared International Year of Disabled Persons (1981) and the International Decade of the Disabled Person (1983-1992) provided visibility, knowledge and motivation to found advocacy groups and actively work toward enactment and enforcement of laws and policies that supported rights. Advocacy thus became a new research area, with law and social policy integrated into APA science. In 2006, the UN passed Article 30.5 of the Convention on the Rights of Persons with Disabilities, which stated that persons with disabilities should participate "on an equal basis with others in recreational, leisure and sporting activities." This reflects a trend away from needs-based services toward rights-based opportunities.

1.2. Function

The function of APA science is to provide: (a) theoretical and applied knowledge; (b) highly-qualified professionals and practitioners; and (c) research-based practices (services, supports, activities, programs) that focus on physical activity goals, needs, rights and empowerment of persons of all ages with disabilities in physical education, sport, recreation and rehabilitation. Individuals with disabilities and the contextual factors affecting their movement performance, as well as social inclusion in physical activities of choice, are extremely diverse. APA university preparation functions to provide: (a) introductory courses for generalists; (b) advanced degree specialisations for future university faculty, researchers and administrators; and (c) ongoing research. APA education with different functions are pre-service and in-service in universities, continuing education in the field, instruction for parents and the community at large, and *infusion* practices whereby APA specialists provide generalists with knowledge about disabilities adapted to general subject matter and supports needed to integrate persons with disabilities into their classes or programs. *Infusion* aims to broaden the knowledge base and instructional practices of generalists to better cover individual differences associated with disabilities and enhance supportive attitudes.

1.3. Body of Knowledge

APA science is crossdisciplinary, defined as "the integration of knowledge from many disciplines in the creation of a distinct, unique body of APA knowledge that focuses on [interrelationships among] adaptation or change processes, individual differences and physical activity" (Sherrill, 2004, p. 6). Integration of knowledge, in this definition, means a continuous cycle of testing the applicability and trustworthiness of existing theories and practices, combining relevant knowledge in new and different ways, conducting research on new perspectives and advancing new theories and practices specific to APA. Adaptation is "art and science, used by qualified personnel, of assessing and managing person-task-environment variables in physical activity services, supports and interventions designed to meet unique psychomotor needs and achieve desired outcomes" (Sherrill, 2004, p. 7). Individual differences are conceptualised, not as persons, but as all of the interacting person-task-environment variables that interfere with "average and above" performance and self-determined participation. APA is big-muscle activity associated with the sciences and professions of physical education, recreation, sport and rehabilitation.

Numerous theories guide APA practices. Illustrative of these are Adaptation Theory or Metatheory (Kiphard, 1983; Sherrill, 1995, 2004), Individualised Education Program (IEP) Assessment and Service Delivery theory (Kelly, 2006; Sherrill, 2004; Winnick, 2005); Ecological Task Analysis (ETA) theory (Davis and Broadhead, 2007), and Systematic Ecological Modification Approach (SEMA) theory (Hutzler, 2007). Sherrill (1995) presented APA's need for a metaparadigm (or metatheory), defined as the "shared vision that binds members of a specific discipline together" or the "center or core of an identifiable domain of knowledge with many layers, all integrated with one another." Adaptation was recommended as APA's metatheory. Central phenomena in this metatheory might be: (a) ecosystems (individuals interacting with others and with environmental variables) that cope with barriers to success; (b) physical activity; (c) service delivery; and (d) empowerment, including advocacy. Sherrill (2004) has expanded theorisation in each of her textbook editions, asserting that this framework encourages the construction of many APA theories that link aspects of the named phenomena.

Space prohibits description of theories, but two illustrations reveal a unique domain of knowledge. IEP theory is rooted in legislation, from 1973 onward in the USA, that mandates assessment, service delivery, intervention and evaluation be derived from research and best practices associated with special education, neuroscience, pedagogy and related sciences. APE professionals have applied this theory to delineating Adapted Physical Education National Standards (APENS), with specific knowledges and skills to be demonstrated in the following areas:

- human development;
- motor behaviour;
- exercise science;
- measurement and evaluation;
- history and philosophy;
- unique attributes of learners;
- curriculum and development;
- assessment;
- instructional design and planning;
- teaching, consultation and staff development;
- program evaluation;
- continuing education;
- ethics; and
- communication (Kelly, 2006).

Central to APENS are numerous theoretical frameworks underlying the construction of new assessment instruments for target student groups as well as teacher attitude, performance and accountability. The new Systematic Ecological Modification Approach (SEMA) (Hutzler, 2007) builds on ETA theory and addresses: (a) linking task goals to World Health Organization (WHO) terminology; (b) identifying objective and subjective criteria for referencing effectiveness of goal accomplishment; (c) limiting and facilitating factors in accordance with WHO; (d) adapting variables enhancing activity and participation; and (e) encouraging choices for selecting a movement form.

1.4. Methodology

All research methodologies (both quantitative and qualitative) are used, with a trend toward mixing methodologies to assure comprehensive results. Extending and adapting existing theory to include disability is emphasised, as is developing original theories (Reid, 2000). Data collection techniques are systematically adapted to participants' unique movement and communication abilities. Validation of instruments for specific samples and purposes is necessary. APA research is especially complex because of variability of participants and sometimes availability of only small samples. Researchers use both nonparametric and parametric statistics and alternative approaches such as qualitative research, which includes at least 16 specific subtypes. IFAPA's official research journal *Adapted Physical Activity Quarterly (APAQ)* has published work from over 30 countries during the last decade and is indexed in MEDLINE and other sources. Additionally, IFAPA researchers publish in numerous disciplinary journals as well as periodicals devoted to specific types of disabilities. Global trends toward accountability increasingly require that all practices (interventions) be evidence-based or research-grounded, hence prioritising experimental research.

1.5. Relationship to Practice

Practical applications focus on improvement of quality of life for all (not just those with disabilities) though application of APA knowledge at many levels. Inclusiveness is facilitated through the elimination of attitudinal, aspirational, architectural, transportation and communication barriers and occurs concurrently in home, school and community environments. Accessibility to community resources makes everyone's lives better, especially those of underserved minorities (e.g., elderly, disabled, poor). APA teaches persons to become increasingly independent, productive and self-determined in choosing healthy activity when diverse options and settings are assured. APA science aims toward collaboration between specialists and generalists in all aspects of life that will afford underserved persons equal physical activity opportunity and rights.

1.6. Future Perspectives

The 21st century opens opportunities for APA to become a major science, business and education priority. Some reasons include: (a) the inactivity epidemic has increased dramatically worldwide and at a higher rate among persons with disabilities; (b) society is aging with longer lifespans and greater needs for health and fitness services at reasonable costs; (c) society is paying increasing attention to lifelong health-behaviors; (d) APA is grounded in "healthy" motivational factors not always found in therapeutic settings; (e) APA has long experience in training multiple task engagement, now accepted as a high priority in exercise training; and (f) governments and international bodies are increasingly accepting social rights of minorities, including persons with disabilities. APA is prepared to meet changing needs and provide leadership for cooperation among organisations.

References

Barnes, C. (1991). *Disabled people in Britain and discrimination.* London: Hurst & Co.

Block, M.E. and Obrusnikova, I. (2007). Inclusion in physical education: A review of the literature. *Adapted Physical Activity Quarterly, 24,* 103-124.

Charlton, J.I. (1998). *Nothing about us without us: Disability oppression and empowerment.* Berkeley: University of California Press.

Hutzler, Y. (2007a). A systematic ecological model for adapting physical activities: Theoretical foundations and practical examples. *Adapted Physical Activity Quarterly, 24,* 287-304.

Hutzler, Y. and Sherrill, C. (2007). Defining adapted physical activity: International perspectives. *Adapted Physical Activity Quarterly, 24,* 1-20.

Ingstad, B. and Whyte, S.R. (1995). *Disability and culture.* Berkeley, CA: University of California Press.

Reid, G. (2000). Future directions of inquiry in adapted physical activity, *Quest, 52,* 370-382.

Sherrill, C. (1995). Adaptation theory: The essence of our profession and discipline. In I. Morisbak and P.E. Jorgensen (Eds.). *10th International Symposium on Adapted Physical Activity Proceedings* (pp. 31-45). Oslo: BB Grafisk.

Sherrill, C. and DePauw, K.P. (1997). Adapted physical activity and education. In J.D. Massengale and R.A. Swanson (Eds.). *The history of exercise and sport science* (pp. 39-108). Champaign, IL: Human Kinetics.

Turner, B.S. (1995). *Medical power and social knowledge.* London: Sage.

For others, see section 2.2 Reference Books, Encyclopedias etc.

2. Information Sources

2.1. Journals (Illustrative)

Adapta: A Revista Profissional Da Sobama. Sociedade Brasileira de Atividade Motoro Adaptada (SOBAMA).
Adapted Physical Activity Quarterly, Human Kinetics.
European Journal of Adapted Physical Activity, European Union Federation of Adapted Physical Activity.
Motorik, Zeitschrift fur Motopadagogik und Mototherapie. Hofman Verlag.
Palaestra, Challenge Publications, Illinois
Praxis der Psychomotorik. Borgmann Publishing.

2.2. Reference Books, Encyclopedias, etc.

Davis, W.E. and Broadhead, G.D. (Eds.). (2007). *Ecological task analysis: Looking back. Thinking forward.* Champaign, IL: Human Kinetics.

Doll-Tepper, G., Kroner, M. and Sonnenschein,W. (2001). *New horizons in sport for athletes with a disability*: VISTA '99 Conference. Oxford: Meyer & Meyer Sport.

DePauw, K.P. and Gavron, S.J. (2005). *Disability sport* (2nd ed.). Champaign, IL: Human Kinetics.

Goffman, E. (1963). *Stigma: Notes on the management of spoiled identity.* New York: Doubleday.

Kelly, L.E. (Ed.). (2006). *Adapted physical education national standards: National Consortium for Physical Education and Recreation for Individuals with Disabilities* (2nd ed.). Champaign, IL: Human Kinetics.

Kiphard, E.J. (1983). Adapted physical education in Germany. In R.L. Eason, T.L. Smith and F. Caron (Eds.). *Adapted physical activity: From theory to application:*

Proceedings of the 3rd ISAPA (pp. 25-32). Champaign, IL: Human Kinetics.

Mauerberg-deCastro, E. (2005). *Atividade fisica adaptada.* Sao Paulo, Brasil: Tecmedd.

Sherrill, C. (Ed.). (1986). *Sport and disabled athletes.* Champaign, IL: Human Kinetics.

Sherrill, C. (Ed.). (1988). *Leadership training in adapted physical education.* Champaign, IL: Human Kinetics.

Sherrill, C. (2004). *Adapted physical activity, recreation, and sport: Crossdisciplinary and lifespan* (6th ed.). Boston: McGraw-Hill.

Steadward, R.D., Wheeler, G.D. and Watkinson, E.J. (Eds.). (2003). *Adapted physical activity.* Edmonton, Canada: University of Alberta Press.

Stiker, H-J. (1982). *Corps infirmes et societes.* Paris: Aubier Montaigne.

Van Coppenolle, H., Potter, J.C., Van Peteghem, A., Djobova, S. and Wijms, K. (Eds.). (2003). *Inclusion and integration through adapted physical activity.* Leuven, Belgium: THENAPA.

Winnick, J.P. (Ed.). (2005). *Adapted physical education and sport* (4th ed.). Champaign, IL: Human Kinetics.

2.3. Book Series

Several series are available, generally funded by government or private agencies. Illustrative volumes are:

Australian Sport Commission
1995. *Willing and able: An introduction to inclusive practices.*
2001. *Give it a go: Including people with disabilities in sport and physical activity.*

European Union APA Projects. Contact H. Van Coppenolle at
www.kuleuven.ac.be/thenapa/education/index.htm for books and DVDs.

2.4. Congress/Workshop Proceedings

Illustrative volumes of ISAPA proceedings are listed chronologically:

Eason, R., Smith, T. and Caron, F. (Eds.). (1983). *Adapted physical activity: From theory to practice. Proceedings of 3rd ISAPA, New Orleans.* Champaign, IL: Human Kinetics.

Doll-Tepper, G., Dahms, C., Doll, B. and von Selzam, H. (Eds.). (1990). *Adapted physical activity: An interdisciplinary approach. Proceedings of 7th ISAPA.* Berlin: Springer-Verlag.

Yabe, K., Kusano, K. and Nakata, H. (1994). *Adapted physical activity: Proceedings of 9th ISAPA, Yokohoma.* Tokyo: Springer-Verlag.

Morisbak, I. And Jorgensen, P.E. (Eds.). (1995). *Quality of life through adapted physical activity and sport – a lifespan concept. Proceedings of 10th ISAPA.* Oslo and Beitostolen, Norway: ISAPA Organisers.

Dinold, M., Gerber, G. and Reinelt, T. (Eds.). (2003). *Towards a society for all – Through adapted physical activity. Proceedings of 13th ISAPA, Austria.* Vienna: Austrian Federation of Adapted Physical Activity.

Mauerberg-de Castro, E. and Campbell, D. (2007). *Book of proceedings of 16th ISAPA.* mauerber@rc.unesp.br

2.5. Data Banks

Organisations listed under other headings maintain data banks.

2.6. Internet Sources

Adapted Physical Activity Council (APAC), American Alliance for Health, Physical Education, Recreation, and Dance (AAHPERD)
   www.aapar.org

Adapted Physical Activity Quarterly (APAQ), the official journal of the International Federation of Adapted Physical Activity
   www.humankinetics.com

Palaestra, a journal.
   www.palaestra.com

Council for Exceptional Children (CEC) www.cec.sped.org Human Kinetics
   www.humankinetics.com

International Council of Sport Science and Physical Education (ICSSPE)
   www.icsspe.org

International Federation of Adapted Physical Activity
   www.ifapa.biz

International Paralympic Committee
   www.paralympic.org

Special Olympics International, Inc.
www.specialolympics.org
National Center on Physical Activity and Disability
www.ncpad.org
National Consortium on Physical Education and Recreation for Individuals with
Disabilities
www.ncperid.org/

3. Organisational Network

3.1. International Level

The International Federation of Adapted Physical Activity (IFAPA), founded in Quebec,
Canada, in 1973, has conducted conferences during odd-numbered years since 1977.
Conferences, traditionally called International Symposia on Adapted Physical Activity
(ISAPA), have been held in 6 of 7 of the world's regions (listed below). IFAPA's official
research journal is *Adapted Physical Activity Quarterly (APAQ)*, published by Human
Kinetics. An affiliate of ICSSPE, IFAPA promotes partnerships and collaboration.

The IFAPA Board's 20 voting members are organised as follows:

Executive Committee: President, President-Elect, Past-President, Vice President,
Secretary, Treasurer.
Regional Representatives: 2 each from regions with IFAPA-affiliated organisations:
Asia, Europe, North America; and 1 each from other regions: Africa, Oceania, Mid-
dle East, South/Central America.
Resource Specialists: APAQ Editor, Disability Community Liaison and Student Rep-
resentative.

3.2. Regional Level

IFAPA has seven regions: Africa, Asia, Europe, Middle East, North America, Oceania and South/Central America. Three regions have affiliated organisations:

Asian Society of Adapted Physical Education and Exercise (ASAPE)
European Federation of Adapted Physical Activity (EUFAPA)

North American Federation of Adapted Physical Activity (NAFAPA)
Regional organisations are each unique in composition, by-laws, constitution and other areas. Conferences occur during even-numbered years.

3.3. National Level

Several countries have national APA or APE organisations, maintain websites, conduct annual or biennial conferences and publish professional journals, generally in their native language but often with abstracts in English. A growing number of APA and APE textbooks are published nationally, in the preferred language of the country. Illustrative of countries with strong national organisations are Brazil, Finland, Norway, Japan, Korea and USA.

3.4. Specialised Centres (Illustrative)

Australian Sport Centre, Belconnen.
Beitostolen Healthsports Centre, Beitostolen, Norway.
European APA professional preparation, in Leuven, Belgium
      www.kuleuven.ac.be/thenapa
Stoke Mandeville Sports Centre, Aylesbury, England.
Swedish Development Centre for Disability Sport (SUD), University of Gavle.
University of Alberta, Steadward Centre for Personal and Physical Achievement,
      Edmonton, Canada.
Wingate University, Zinman College for Physical Education and Sports, Israel.

3.5. Specialised International Degree Programmes

The European Master's Degree in Adapted Physical Activity (EMDAPA) commenced in 1991 and now involves over 30 European universities. English is the official language. www.kuleuven.ac.be/thenapa/education/index.htm Erasmus Mundus is an extension of EMDAPA in which students study in Leuven and at one other European partner university for a 12-month program. Recently this program has been extended to non-European countries.

The Diplôme Européen Universitaire en Activité Physique Adaptée (DEUAPA), in contrast, rotates its headquarters among several sites and uses French as its official language. The DEUAPA certificate requires four stages of study: program at home university, intensive program, hosted exchange program abroad and thesis

at home university. Over 20 universities cooperate in DEUAPA.
European Doctorate in APA. Universities in Norway, Belgium, and Czech Republic provide doctoral-level APA specialisations.

North American Degree Programs welcome international students into Master's and Doctoral APA programs. There is no central coordinating centre. Illustrative programs are:

McGill University, Montreal, Quebec, Canada
University of Alberta, Edmonton, Canada
Ohio State University, Columbus, OH, USA
Oregon State University, Corvallis, OR, USA
Texas Woman's University, Denton, TX (coeducational), USA
University of Virginia, Charlottesville, VA, USA
Indiana University, Bloomington, IN, USA

The best known program in South America is the State University of Rio Claro, UNESP, in Brazil mauerberg@rc.unesp.br

4. Appendix Materials

4.1. Terminology

See earlier sections for terminology.

4.2. Position Statements

United Nations Convention on the Rights of People with Disabilities 2006: IFAPA Position Statement www.IFAPA.biz
On highly qualified APA personnel and on inclusion, see www.aapar.org

4.3. Varia

See earlier sections for varia.

4.4. Free Statement

Not applicable.

Biomechanics

Tony Parker and Michael McDonald

Contact

Prof. Dr. Tony Parker
Queensland University of Technology
Institute of Health and Biomedical Innovation
60 Musk Avenue
Kelvin Grove Qld 4059
Australia
Phone: +61 7 3138 6173
Fax: +61 7 3138 6039
Email: t.parker@qut.edu.au

1. General information

1.1. Function

Biomechanics involves research and analysis of the mechanisms of living organisms. Research and analysis can be conducted on multiple levels and represents a continuum from the molecular, wherein biomaterials such as collagen and elastin are considered, to the tissue, organ and whole body level. Some simple applications of Newtonian mechanics can supply correct approximations on each level, but precise details demand the use of continuum mechanics.

Aristotle wrote the first book on biomechanics, 'De Motu Animalium', translated as 'On the Movement of Animals'. He saw animals' bodies as mechanical systems, but also pursued questions that might explain the physiological difference between imagining the performance of an action and actually doing it. Some simple examples of biomechanics research include the investigation of the forces that act on limbs, the aerodynamics of animals in flight, the hydrodynamics of objects moving through water and locomotion in general across all forms of life, from individual cells to whole organisms.

Although the human body is an extremely complex biological system composed of trillions of cells, it is subject to the same fundamental laws of mechanics that govern simple metal or plastic structures.

The essence of biomechanics is therefore, a synthesis of biology and mechanics that seeks to understand and explain movement, particularly human movement. Biomechanics is often referred to as the link between structure and function.

1.2. Body of Knowledge

Although biomechanics is relatively young as a recognised field of scientific inquiry, biomechanical considerations are of interest to several different scientific disciplines and professional fields. As such, biomechanists may have academic backgrounds in areas such as zoology, orthopaedics, cardiology, sports medicine and biomedical engineering, with the commonality being an interest in the biomechanical aspects of the structure and function of living organisms.

This diversity of scientific and professional background is reflected in the broad range of topics investigated by biomechanists, within the general theme of analysing the motion of a living organism and the effect of forces on it. The biomechanical approach to movement analysis can be qualitative, with movement observed and described, or quantitative, meaning that some aspect of the movement will be measured.

There are many areas of biomechanics and biomechanical research and these have been categorised into developmental; occupational; rehabilitative; and exercise and sports biomechanics research domains (Housh et al., 2003).

Developmental biomechanics focuses on the evaluation of fundamental movement patterns during performance of gross and fine motor skills. This contributes to understanding of motor skill development typically associated with movement patterns such as walking, jumping, throwing and catching for individuals of different ages. It has resulted in the description of a typical pattern of movement for a specific activity, categorised according to age. The data gathered may be used as a reference by developmental biomechanists, to determine the level of movement ability and allow comparisons to be made across ages. The information may also be used to facilitate any remediation strategies (Hutchinson and Wynn, 2004).

Occupational biomechanics focuses on providing a safe and efficient working environment, both indoors and outdoors. For example, this may include the development of better safety equipment such as helmets, shin guards and footwear to protect workers from any work related hazards that can cause injuries and even death (Thuresson et al., 2005). The development of safe and biomechanically efficient behaviour and appropriate distribution of workload is important in minimising the risk of overuse injuries of both upper and lower limbs. Optimising the match between the worker and his/her tools or equipment is also an area in which the occupational biomechanist may contribute to the ergonomics team in the workplace setting.

The purpose of rehabilitative biomechanics is to observe and analyse the movement patterns of individuals who are either injured, disabled or both and to contribute to the provision of appropriate interventions which will enable the injured or disabled person to regain normal function (Willems et al., 2005). Through rehabilitative biomechanical research, exercise equipment and supplementary aids such as canes, crutches and walkers and substitution devices such as prostheses and wheelchairs can be developed and used for rehabilitation purposes (Yakimovich et al., 2006).

Exercise and sports biomechanists apply their knowledge to the analysis and enhancement of sports performance and the development of strategies to reduce the risk of sports related injury. This extends to the development and use of various devices and exercise equipment to improve fitness components such as strength, endurance, flexibility and speed. Considerable research effort has also been applied to the development and improvement of athletic equipment, clothing and footwear, designed to enable the athlete to cope with the specific demands of particular sports and ultimately to reduce the risk of injury without detriment to performance.

As the field of biomechanics continues to evolve, new areas of engagement arise and opportunities for collaboration with other disciplines continues to increase. The areas of computer modelling and movement simulation techniques, robotics and the developments in sensor technology and 'smart' materials are examples of the many emerging areas within the field.

1.3. Methodology

Qualitative, quantitative and predictive biomechanical analysis methods are three approaches most commonly used in biomechanics.

Qualitative biomechanical analysis techniques involve systematic observation and introspective judgement of the quality of human movement and this approach is often used to inform the most appropriate intervention to improve performance. (Knudson and Morrison, 2002). Observation models such as phase analysis and temporal analysis models, as well as the critical feature analysis model, form the basis for this type of analysis.

The use of commercially available video software enables video recordings to be compared against a standard of performance by means of split screen viewing, making this model an extensive but successful and economical approach.

A major deficiency with qualitative approaches relates mainly to the lack of agreement amongst biomechanists in ascertaining a standard description and in identifying an adequate number of identifiable principles.

Advancements in technological capabilities have resulted in the ability to record, display and evaluate dynamic movements, both kinematically and kinetically, in real time. Methods used for quantitative biomechanical analysis are similar to the qualitative methods in terms of approach, but without their subjectivity as they use a range of data collection instruments to capture, observe and evaluate performance. The essential component in using quantitative analysis method is the selection of key variables.

Depending on the particular application, both qualitative and quantitative analysis techniques rely on observation or recorded data from real movement that provides information on the characteristics of the movement being analysed. In contrast, predictive biomechanical analysis methodology uses simulation models of the human anatomical structure to mathematically calculate and predict an ideal performance, thus allowing hypothetical questions to be investigated systematically. Predictive analysis methods can be applied to most movements, provided there are advanced computer simulation programs to fit the nature of that particular movement.

1.4. Relationship to Practice

There are numerous examples within the scientific literature of research concerned with the application of biomechanics to practical situations across the broad range of professional areas identified earlier. While it is not possible to provide a comprehensive account here, readers are referred to the various literature sources identified later for more detailed information.

As the proportion of elderly people within the population increases, an applied area that presents significant challenge to the biomechanist is the study of mobility impairment in the elderly. Age related decrements in muscle strength, dynamic postural stability and movement speed and accuracy are associated with the relatively high incidence of falls and hip fracture in the elderly. Biomechanists are working with other specialists to investigate the mechanisms associated with falls and the biomechanical risk factors for injury in this population. Applied research into the value of protective equipment and safe environments is also contributing to the development of a range of injury prevention strategies (Gapeyeva et al., 2006) Research into the field of occupational biomechanics has included identification of risk factors for musculoskeletal injury and the effect of mechanical loading on joint structures. New sensor technologies are being used to quantify risk factors and there is increasing recognition of the physical and psycho-social determinants of musculoskeletal injury and low back pain (Splittstoesser et. al., 2007).

Biomechanists' contribution to sports medicine research has also been extensive in the area of mechanisms of injury and in the development and evaluation of protective equipment such as knee and ankle braces.

References:

Gapeyeva, H., et al., (2006). Differences in gait and isokinetic strength and power characteristics of knee extensor muscles in women aged 50 and 70 years. *Gait and Posture*, 24, Supplement 2, S272-S273.

Housh, T.J., et al., (2003). *Introduction to exercise science. (2nd Ed.)*. San Francisco : Benjamin Cummings.

Hutchinson, M.R. and Wynn, S. (2004). Biomechanics and development of the elbow in the young throwing athlete. *Clinics in Sports Medicine*, 23(4) 531-544.

Knudson, D.V. and Morrison, C.S. (2002). *Qualitative analysis of human movement*. Champaign, IL.: Human Kinetics.

Splittstoesser, R.E., et. al., (2007). Spinal loading during manual materials handling in a kneeling posture. *Journal of Electromyography and Kinesiology*, 17 (1) 25-34.

Thuresson, M., et. al., (2005). Mechanical load and EMG activity in the neck induced by different head-worn equipment and neck postures. *International Journal of Industrial Ergonomics*, 35(1) 13-18.

Willems, T., et al., (2005). Relationship between gait biomechanics and inversion sprains: a prospective study of risk factors. Gait and Posture, 21(4) 379-387.

Yakimovich, T., et. al., (2006). Preliminary kinematic evaluation of a new stance-control knee–ankle–foot orthosis. *Clinical Biomechanics*, 21(1) 1081-1089.

2. Information Sources

2.1. Journals

The following journals contain articles related to various aspects of biomechanical investigation.

Clinical Biomechanics (United Kingdom)
Bone (New York, USA)
Computer methods and programs in biomedicine (Amsterdam)
Computers methods and programs in biomedicine (Holland)
Electroencephalography and clinical neurophysiology (Limerick)
Gait and Posture (Oxford)
Injury (Holland)
Journal of Applied Biomechanics (Illinois)
Journal of Back and Musculo-skeletal Rehabilitation (Holland)
Journal of Biomechanical Engineering (New York)
Journal of Biomechanics (New York)
Journal of Electromyography and Kinesiology (Holland)
Journal of Human Movement Studies (London)
Journal of Sport Sciences (London/New York)
Medicine and Science in Exercise and Sport (United States)

2.2. Reference Books, Encyclopaedias

Greenwood, R.J., et al. (Eds.). (2003). *Handbook of neurological rehabilitation.* Hove (England) Psychology Press.
Hall, S.J. (2003). *Basic Biomechanics (4th Ed.).* Boston, McGraw Hill.
Hamill, J. and Knutzen, K.M. (2003). *Biomechanical Basis of Human Movement (2nd Ed.).* Philadelphia, Lippincott Williams and Wilkins.
Hung, G.K. and Pallis, J.M. (Eds.). (2004). *Biomedical engineering principles in sports.* New York, Kluwer Academic/Plenum Publishers.
Levangie, P.K. and Norkin, C.C. (2005). Joint Structure and Function : A Comprehensive analysis (4th Ed.). Philadelphia, F.A. Davis.
Lovell, M.R., et al. (Eds.). (2004). *Traumatic brain injury in sports : an international neuropsycholgical perspective.* Lisse, Exton, PA, Swets and Zeitlinger.

Narvani, A.A., Thomas, P. and Lynn, B. (Eds.). (2006). *Key topics in sports medicine*. London, Routledge.

Oatis, C.A. (2004). *Kinesiology : The Mechanics and Pathomechanics of Human Movement*. Philadelphia, Lippincott Williams and Wilkins.

Panjabi, M.M. and White, A.A. (2001). *Biomechanics in the musculoskeletal system*. New York, Churchill Livingstone.

Trew, M. and Everett, T. (Eds.). (2005). *Human movement : an introductory text*. Edinburgh, Elsevier/Churchill Livingstone.

Whittle, M. (2002). *Gait Analysis : An introduction*. Boston, Butterworth-Heinemann.

Winter, D.A. (2005). *Biomechanics and motor control of human movement*. Hoboken, New Jersey, John Wiley and Sons.

2.3. Book Series

The American College of Sports Medicine publishes Exercise and Sports Sciences Reviews annually, which reviews current research concerning biomechanics (and other topics) in exercise science. Online information related to the reviews can be obtained from the College's website listed in section 2.6 Internet Sources.

2.4. Conference/Workshop Proceedings

Various international and national societies of biomechanics hold either annual or biennial conferences during which workshops are held by special interest groups within the respective society (refer to the International Society of Biomechanics website in section 2.6 Internet Sources). Certain societies now publish the proceedings from these workshops on their home page on the World Wide Web.

2.5. Data Banks

Data banks for communal use by other biomechanists outside the laboratory or institution gathering the data is uncommon, due mainly to non-compatible equipment, variations in testing protocols and experimental set up. Nevertheless, there are a few websites that contain data banks accessible to biomechanists. These include the International Society of Biomechanics site for movement data, pressure data, musculo-skeletal models and 3-D imaging data and the Clinical Gait Analysis website for limited data sets on gait analysis for various patient populations (see section 2.6 Internet Sources).

2.6. Internet Sources

The following Internet sources are available for use in relation to various aspects of biomechanics:

American College of Sports Medicine
    www.acsm.org
American Society of Biomechanics (ASB)
    www.asbweb.org/
Australia New Zealand Society of Biomechanics
    http://www.anzsb.asn.au/
Canadian Society of Biomechanics (CSB/SCB)
    www.health.uottawa.ca/biomech/csb/
Clinical Gait Analysis
    www.univie.ac.at/cga/index.html
European Society of Biomechanics (ESB)
    www.esbiomech.org
International Society of Biomechanics
    http://isbweb.org/
International Society of Biomechanics in Sports
    www.uni-stuttgart.de/External/isbs/

3. International Networks

3.1. International Level

The International Society of Biomechanics (ISB) was founded in 1973 to promote the study of all areas of biomechanics at the international level, although special emphasis is given to the biomechanics of Human Movement. The Society encourages international contacts amongst scientists, promotes the dissemination of knowledge and forms liaisons with national organisations. The Society's membership includes scientists from a variety of disciplines including anatomy, physiology, engineering (mechanical, industrial aerospace, etc.), orthopaedics, rehabilitation

medicine, sport science and medicine, ergonomics and electro-physiological kinesiology. Society activities include the organisation of biennial international conferences, publication of congress proceedings and a biomechanics monograph
series. Newsletters are distributed quarterly and the Society sponsors scientific
meetings related to biomechanics. It is also affiliated with the Journal of Biomechanics, the Journal of Applied Biomechanics, Clinical Biomechanics, the Journal
of Electromyography and Kinesiology and Gait and Posture.

The ISB also supports technical and working groups for the purpose of advancing
knowledge in specialised areas within the field of biomechanics. Currently, active
technical sections include computer simulation, shoulder biomechanics, footwear
biomechanics and 3-D motion analysis.

The World Commission of Science and Sports (WCSS) was established in 1967
and is a working group of the International Council for Sports Science and Physical Education (ICSSPE). The WCSS covers a range of sports, which currently are
Cricket, Football, Golf, Swimming, Shooting Sports, Racket Sports and Winter
Sports. Details of each are found under the relevant web page (see section 2.6
Internet Sources). In general, each sport area holds its own scientific congress
and ensures that the proceedings are published and available to the international
community. The WCSS has a formal link with the Journal of Sports Sciences in
which abstracts from each Congress are published.

The WCSS has held International Symposia since the inaugural meeting in 1970 in
Brussels. The most recent Congresses were the 5th World Congress on Science
and Football in Lisbon and the 3rd World Congress on Science and Racquet Sports
in Paris, held in 2003.

3.2. Regional Level

Europe

The European Society of Biomechanics (ESB) was founded at a meeting of 20
scientists from 11 countries in Brussels in 1976. Biomechanics was defined as
"the study of forces acting on and generated within a body and of the effects of
these forces on the tissues, fluids or materials used for diagnosis, treatment or

research purposes". The primary goal of the ESB is "to encourage, foster, promote and develop research, progress and information concerning the science of Biomechanics".

The first scientific meeting and General Assembly of the membership was held in Brussels in 1978 and regular meetings have occurred since this time. The 15th Biennial conference was held in Munich in 2006 and the next conference is scheduled for Lucerne 6 – 9 July, 2008.

North America

The Canadian Society of Biomechanics (CSB/SCB) was formed in 1973. The purpose of the Society is to foster research and the interchange of information on the biomechanics of human physical activity. The main activity of the CSB/SCB is the organisation of a biannual Scientific Conference held in the years opposite to those of the International Society of Biomechanics. A newsletter is published periodically and is now in web format. CSB/SCB is affiliated with the International Society of Biomechanics (ISB).

The American Society of Biomechanics (ASB) was founded in October, 1977 with the aim of providing a forum for the exchange of information and ideas among researchers in biomechanics. The term biomechanics is defined by the Society as the study of the structure and function of biological systems using the methods of mechanics. The mission of the ASB is to encourage and foster the exchange of information and ideas among biomechanists working in different disciplines and fields of application. These include the biological sciences, exercise and sports science, health sciences, ergonomics and human factors, and engineering and applied science. There are several regional and national associations affiliated with the ASB including the American College of Sports Medicine, the American Society of Mechanical Engineers and the Orthopaedic Research Society.

Australia/New Zealand

The Australia New Zealand Society of Biomechanics (ANZSB) was founded in 1996 as a forum for biomechanists in Australia and New Zealand to communicate and present their research. The Society acknowledges and encourages a diverse range of disciplines among its members. Many sub-areas within biomechanics have evolved within the organisation including, for example, cardiovascular and respiratory biomechanics, rehabilitation biomechanics, sport biomechanics, bone and hard tissue biomechanics, connective tissue biomechanics, orthopaedic biomechanics and cellular and molecular biomechanics. ANZSB aims to provide a forum for all areas of biomechanics to exchange ideas and experiences within the Oceania region. The First Australasian Biomechanics Conference was held in Sydney, Australia, in February 1996, with subsequent conferences held in 1998 (Auckland, New Zealand), 2000 (Gold Coast, Australia), 2002 (Melbourne) and 2004 (Sydney). The last conference was held in February, 2007 in Auckland, New Zealand.

3.3. National Level

There are numerous national associations/societies of biomechanics including those listed below, which are affiliated with the ISB:

American Society of Biomechanics
Australia New Zealand Society of Biomechanics
Brazilian Society of Biomechanics
British Association of Sport and Exercise Sciences
Bulgarian Society of Biomechanics
Canadian Society of Biomechanics
Chinese Society of Sports Biomechanics
Comisia de Biomecanica Inginerie si Informatica (Romania)
Czech Society of Biomechanics
Japanese Society of Biomechanics
Korean Society of Sport Biomechanics
Polish Society of Biomechanics
Russian Society of Biomechanics
Societ de Biom, Canique (France).

3.4. Specialised centres

Not applicable.

3.5. Specialised International Degree Programmes

Not applicable.

4. Appendix Material

4.1. Terminology

The terminology used in biomechanics draws from the disciplines of human anatomy, physiology, pathology, physics, engineering, rehabilitation and general medical practice.

4.2. Position Statement(s)

Position statements on topics of current and vital interest can be found on the website of the American College of Sports Medicine at www.acsm.org

4.3. Varia

Not applicable.

4.4. Free Statement

Not applicable.

Coaching Science

Uri Schaefer and Mark Wertheim

Contact

Dr. Uri Schaefer
Ministry of Science, Culture and Sport
Israel Sport Authority
Hamasger 14 Street
61575 Tel-Aviv
ISRAEL
Email: uris@most.gov.il

Dr. Mark Wertheim
Wingate Institute of Physical Education and Sport
42902 Netanya
ISRAEL
Phone: +972 9 8639544
Fax :+972 9 8639513
Email: markw@wingate.org.il

1. General Information

General Coaching Science/Theory pertains to what is common in coaching all sports, while Sports-Specific Coaching Science/Theory deals with coaching a specific sport. This chapter is confined to General Coaching Science/Theory.

The change from the modern concept of the science of coaching to coaching science shows a trend towards a discipline based on specialised knowledge, unique research methods and structured training spheres.

This recent science focuses on two areas:

1. Coaching as a process for target acquisition in competitive sport within an organised ompetition system; and
2. Coaching as a process of different physical activities and in a range of sport fields where the primary aim is health, weight reduction, healthy body and mind.

Coaching science is the kernel of sport sciences (Hohman, Lames and Letzelter, 2003). Today, more than ever, there is a methodological base with an understanding that coaching theory both as a theory and as a methodology is not just applicable to elite sport.

1.1. Historical Developments

Until the 20th century, coaching was done by instinct, tradition and personal trial and error. A real theory began to develop (mainly in measurable sports, such as track and field, and on the subject of developing physical ability) when coaches called upon experts in exercise physiology for advice. By the 1950s, coaching theory came of age, initially in the Soviet Union and its satellites, with the work of Matveyev et al., on coaching theory, especially on periodisation of training. The East-West competition, with attendant government support, fuelled both coaching practice and the development of coaching theory. In the 1970s, upgrading coaching by means of increased volume had reached its limits and increasing intensity was called upon; but this, too, was nearing its bounds. So, in the 1980s, coaching theory (and practice) turned to periodical shifting of emphasis on different elements of ability, on increased specificity of exercise and on ergogenic aids (initially of metabolic, later of hormonal nature).

Coaching science, which began as a copy of successful methods of coaches, today focuses on the question of "why", via interdisciplinary research and asks the questions of what is the best method for the effectiveness of coaching over the short, medium and long terms in the areas of preventative, sport functioning and rehabilitation.

The search for the scientific base accompanies the theory and structured methodology (the theory of coaching), which cannot exist in disciplinary sciences that do not have a direct base on which to work (Wertheim, 2004). Coaching sciences today are not based on pedagogical principles, as they were in the past, and on the physiological principles of a few years ago, but on the interaction between them with the understanding of the connection between the educational process in coaching and the physiological process.

Nowadays, ergogenic aids are being frowned upon and coaching theory turns to emphasising psychological training, biofeedback, "delayed response" training and, with the shift from government to commercial funding, to developing coaching theory for the leisure-sports athlete.

1.2. Function

Coaching sports is the act of leading/guiding an athlete or team to maximise performance in the chosen sport, especially (from intermediate level upwards) at the most important competition of the period. Coaching science provides guidelines for effectively carrying out this task.

The many aspects of coaching include instruction in the technique of the sport, improving the physical fitness of the athletes, guiding the athletes' nutrition, developing tactics for effective application of the sports' techniques, preparing the athlete or team mentally, measuring training loads and recuperation means, providing technical, tactical and mental assistance during competition, and analysing past competitions to develop future objectives on the basis of this analysis. When preparing for tournaments to be held abroad, the coach may need to address many potential problems, for example, acclimatisation. In addition, coaches also try to identify talented youngsters and to assist them to identify and develop their opportunities in sport. Doing all this rationally and effectively requires knowledge provided by coaching science/theory.

The concept of coaching in modern terms relates to structured planning and systematic approaches that puts into practice types of coaching, contents and forms of coaching and the methods of implementation of coaching (exercises and training), aims of coaching whilst relating to scientific knowledge, practices and follow up.

Coaching is an open process for the beginner, advanced and elite athlete, for the student, the youngster and the elderly.

1.3. Body of Knowledge

Coaching science/theory brings together data from exercise physiology, biomechanics, sports psychology, sports medicine, sports sociology, kinanthropometry, motor learning, sport pedagogy, etc., as well as empirical data from coaching in all sports. From a coaching aspect, these data can be combined to form a multi-disciplinary, applicative discipline.

Coaching science as an interdisciplinary, integrative science based on the systematic collection and collation of significant data from coaching areas, sourced from competition, laboratory based experiments and the fields or play/participation. It is no coincidence that this area is called "systematic results".

The list of sources of knowledge is not conclusive and developments in various areas of human study may open up further sources for coaching to draw upon.

1.4. Methodology

Data "imported" from the various disciplines mentioned above have, of course, been formed under the methodology of the relevant science so they must then be checked, empirically and logically, for their relevance to coaching. A common pitfall occurs when data gained from general populations is used for coaching elite athletes, whose physical characteristics and mental attitude are entirely different.

Data gleaned from coaching practice depend, to a large extent, on the existence of orderly documentation of that practice. In most cases however, the data will not satisfy the usual criteria applied to scientific inquiries: the population would be too small for statistical significance; a control group may not be used as a comparison, so "double blind" studies are not tenable. This is why some experts in the field prefer the term "Coaching Theory" to "Coaching Science".

Interdisciplinary research into sport problems is the way of the future. Research will take into account the complex interactions between the mechanical and physiological systems, cognition and emotion, social groupings, as well as political and economic factors. These areas of research will have direct impact and practical consequences on talent identification, adherence to training and development programs, as well as prevention and treatment of sport injuries.

The scientific character of coaching theory will become more pronounced: the measurement of an athlete's state of training will become more accurate and reliable. The revolution in microelectronics, leading both to more powerful instruments and much lower costs, enables much better objectivity in the process of training. Thus, coaches and scientists will develop a greater degree of precision in understanding exact training loads, which the coach will use to achieve more rational and effective coaching, and the scientist will gain a deeper understanding of the relation between cause and effect in training.

The massive involvement of the electronic media in virtually every sport – both amateur (sport-for-all), competitive and top-level – will probably entail an information specialist as part of the team which will include the athlete, the coach, sports scientists and sport medicine practitioners.

Coaching science, like all applicable sciences, amasses its knowledge through practical, supervised experiences using empirical science after follow up and/or using hypotheses that have been thoroughly assessed.

The focus is multi-disciplinary and simultaneous (application, behaviour, physiological profile, etc) whilst parallel to this is a collection of developing statistics in order to point to a working model, its influence and effectiveness over a period of time.

1.5. Relationship to Practice

Coaching science is the main subject in the process of coach education. In the past, we have had self-educated coaches (usually former athletes), who coached mainly by intuition. Sometimes they would employ aides to inform them of developments in coaching theory. This will no longer be sufficient - in fact, it already isn't. Just as the annual plan must be based on coaching theory, so too, the periodical adjustments of that plan and the day-to-day details for fulfilling the plan require understanding of the rationale – again, of coaching science/theory.

A central characteristic of coaching science and the theory of coaching is the connection between the field of work of the coach and the trainee (through training and relevant knowledge). From here stems the reason that the field is defined as a science, as a theory and as a methodology of directing the process and development of coaching with the understanding of the needs, demands and reality.

Supervision in the coaching process is done using valid tests and practical tools to assess the results and then to adapt them to the process of planned coaching.

Central to this assessment is the interaction between the coach and the trainee, where coaching is the field of activity in which relationships and abilities develop.

Top-level athletes (defined as realistic contenders for medals in Olympic Games and/or world championships) in leading countries, financed by the state and/or commercial sponsors, train under close to ideal conditions. The quest for higher levels of performance leads to innovations in the process of training and thus to deeper understanding of coaching and further development of coaching science/theory. While initially, such innovations may be beyond the scope available to intermediate, much less leisure-time athletes, in time, these innovations "percolate down" to all levels of sport training.

Examples of such innovations are:

- Altitude training to enhance aerobic endurance;
- ECG-style monitoring of heart rate during performance;
- Biofeedback in mental training;

- Measurement of lactic-acid concentration to determine optimal running/swimming speed in training and competition;
- Simulation camps as part of preparation for major competitions;
- The use of computer technology and of videos for analysis and instant playback of training sessions and competition;
- The understanding and application of the process of recovery from training.

There is increasing involvement of commercial firms in competitive sport, via sponsorship of special tournaments and outstanding athletes. At the same time, governments are decreasing their involvement. This is leading to a new competition timetable, which is less favourably oriented towards the athlete peaking at world championships and/or Olympic Games. This presents a challenge to coaching theory – to develop guidelines for sensible annual planning in this changing environment, which will facilitate proper preparation of the athlete for main events while ensuring that the athlete maintains both competitive fitness and long-term health and well-being.

1.6. Future Perspectives

Two developments seem to be shaping the road that coaching theory is going to take in the future: the revolution in microelectronics and the retreat of governments from direct involvement in the pursuit of achievement in sport, with commercial firms taking their place.

As electronic devices become both more sophisticated and more attainable (cheaper), much more use will be made of them to ensure objectivity of the process of training, on-line observation and analysis of the effects of training. Monitoring heart rate during exercise and feeding the observation into electronic data-processing instruments is already with us. It is only a question of time until non-invasive ways will be found to do the same with physiological markers of anaerobic exercise. Detailed measurements of the athlete's body (e.g. limbs and internal organs) would be fed into simulation programs, which would serve to personalise technique for optimal results.

The increasing involvement of commercial firms in competitive sport has been discussed in section 1.5 Relationship to Practice.

Basic questions are:

1. How do you create an optimal relationship between the above mentioned areas?
2. How can you direct the coaching session according to the results of the competition?
3. How does coaching influence or not influence the competition?
4. How do you reduce the gap between ability (potential) and current performance of the athlete/group?
5. How does indepth understanding of the ability to perform influence the aims of the coaching session in a decisive way?
6. How does high-tech activity (e.g. microelectronics) impact/influence the training process and athlete's potential?

In the near future, coaching systems in every field of activity will be closely monitored and documented over a period of time. This process will enable monitoring of the influence of the coaching systems (not just physical) and will check if the accompanying different scientific areas really have a direct and simultaneous impact.

2. Information Sources

2.1. Journals

Coaching science/theory is addressed in many journals dealing with sport science in general, and more specifically in journals devoted to sports coaching. Among the journals containing high numbers of articles are:

Coaching and Sports Science Journal (Rome: Italian Society of Sports Science, 1996-present);

Leistungssport (Elite Sport) (Mainz: Deutscher Sportbund, 1971-present, in German);

Teorija I Praktika Fizicheskoj Kul'tury (Theory and Practice of Physical Culture) (Moscow: Russian State Committee on Physical Culture and Tourism, Russian Academy of Physical Culture, 1925-present, in Russian);

Research Quarterly for Exercise and Sport (Reston, VA: American Association for
    Health, Physical Education, Recreation and Dance, 1929-present);
Biology of Sport (Warsaw: Institute of Sport, 1983-present);
Journal of Sports Sciences (London: E&FN Spon, 1982-present);
Theorie und Praxis der Körperkultur (Theory and Practice of Physical Culture) (Ber-
    lin: Sportverlag, 1951-1990, in German);
Theorie und Praxis Leistungssport (Theory and Practice of Elite Sport) (Berlin:
    Sportverlag, 1962-1990, in German);
Training und Wettkampf (Training and Competition) (Berlin: Sportverlag, 1962-
    1990, in German);

To these journals, one must add the various journals of the sciences contributing
to coaching theory mentioned in section 1.2. Function above.

There are also more practice-oriented journals, such as:

Coaching Focus (Leeds: United Kingdom National Coaching Foundation, 1985-
    present);
Olympic Coach (Colorado Springs: United States Olympic Committee, 1991-
    present);
Sports Coach (Belconnen: Australian Coaching Council, Inc., 1977-present);
Sport Pulse (Limerick, Ireland: National Coaching and Training Centre, 1993-p-
    resent);
Kinesiology (Zagreb, Croatia: University of Zagreb, 1971-present);
China Sport Coaches (Beijing, All-China Sports Federation, 1992-present).

Most specific sports have their own coaching journals, often published by their
    governing bodies, which of course also cover coaching science.

2.2. Reference Books, Encyclopaedias, etc.

The basic dictionary is:

Thiess, G., Schnabel, G. and Baumann, R. (Eds.). (1980). *Training von A bis Z*.
Berlin: Sportverlag.

This has been updated by:

Schnabel, G., Harre, D., Krug, J. and Borde, A. (Eds.). (2003). *Trainingswissenschaft: Leistung - Training - Wettkampf.* München: Berlin: Sportverlag.
Thiess, G. (Ed.). (1987). *Leistungsfaktoren in Training und Wettkampf.* Berlin: Sportverlag.

A general sports-sciences oriented reference is:

Roethig, P. (Ed.) (1987). Dictionary of Sport Science (German-English-French). Schorndorf: Karl Hofmann.

Perhaps the most comprehensive book today is:

Harre, D. et al. (Eds). (1994). Trainingswissenschaft. Berlin: Sportverlag.

This is the 4th edition of the East German Coaching Science book, which was also translated into English (1982).

One of the "classic" books on the subject is:

Matveyev, L.P. (1965). *Problema Periodizatsii Sportivnoy Trenirovki (Problems of the Periodisation of Athletic Training).* Moscow: Fizkultura I Sport.

The "state of the art" is presented in:

Platonov, V.N. (1997). *Obshchaja Teorija Podgotovki Sportsmenov v Olimpichkom Sporte (General Theory of Athletes' Preparation in Olympic Sports).* Kiev: Olimpiyskaja Literatura.

Good recent handbooks in English for coaches on General Coaching Science/ Theory include:

Ben-Melech, Y. (1998). *Training for Top Performance.* Cape Town: Gariep.
Crisfield, P., Cabral, P. and Carpenter, F. (Eds.). (1996). *The Successful Coach: Guidelines for Coaching Practice.* Headingley: National Coaching Foundation.

Pyke, F.S. (1991). *Better Coaching: Advanced Coach's Manual.* Canberra: Australian Coaching Council.

Carmeli, E., Wertheim, M. and Werner, S. (2000). *Geriatric Rehabilitation Model as a Controlled Training Process.* Physiotherapy, 4(1).

Carmeli,E. and Wertheim, M. (2001). Handverletzungen bei Jugendlichen und erwachsenen Sportklettern. *Deutsche Zeitschrift Für Sportmedizin,* 52(10) Germany.

Wertheim, M. (2000). Die Ausbildung des Sportlehrers zum Trainer im Wettkampfsport unter besonderer Berücksichtigung der Integration der Trainingswissenschaft. *Leistungssport* (2). Germany.

2.3. Book Series

American Coaching Effectiveness Program. (Champaign, Ill.: Leisure Press 1984-94, Leisure Kinetics 1989-92).

National Coaching Certification Program. (Various places, Various publishers).

Studienbrief der Trainerakademie Köln (Study Letters of the Coaching Academy in Cologne). Schorndorf: Karl Hofmann 1988-present.

2.4. Proceedings

Tenenbaum, G. and Eiger, D. (Eds.). (1991). *Coach Education: Proceedings of the Maccabiah-Wingate International Congress.* Netanya: E. Gil.

Tenenbaum, G. and Raz-Liebermann, T. (Eds.). (1993). *Proceedings: 2nd Maccabiah- Wingate International Congress on Sport and Coaching Sciences.* Netanya: Wingate Institute.

Coach Education - 1st International Coach Education Summit July 1995. Leeds U.K. National Coaching Foundation.

Lidor, R., Eldar, E. and Harari, I. (Eds.). *Proceedings of the 1995 AIESEP Congress: Bridging the Gaps between Disciplines, Curriculum and Instruction.* Netanya: Wingate Institute, Zinman College of PE and Sports Sciences.

Coach Education towards the 21st Century: 2nd International Coach Education Summit: September 21-24, 1997. Netanya, Israel: Wingate Institute – Nat Holman School for Coaches and Instructors.

Walkuski, J.J., Wright, S.C. and San, S.T.K. (Eds.). (1997). *AESEP Singapore 1997 World Conference on Teaching, Coaching and Fitness Needs in Physical Edu-*

cation and the Sports Sciences: Proceedings. Singapore: School of Physical Education, Nanyang Technological University.

Coaching New Zealand et al. (1996). *Partners in Performance National Conference 11-13 October, 1996 - Proceedings of the combined national conferences of Sports Medicine New Zealand and Sport Science New Zealand.*

Canadian Olympic Association: Coaching Association of Canada 1991: Ottawa: Fitness and Amateur Sport.

11th Annual Conference on Counselling Athletes (1994). *Assisting today's athletes toward peak performance.* Springfield: Springfield College.

1994 the year of the coach (1994). National Coaching Conference proceedings, National Convention Centre, Canberra, Australian Sports Commission, p. 10-12.

Marconnet, P. et al. (Eds.). (1996). *First annual congress, frontiers in sport science, the European perspective, May 28-31: Book of abstracts.* Nice: European College of Sport Science.

Fitzgerald, A. (Ed.). (1990). *Olympic education: breaking ground for the 21st century: proceedings of the USOA XIII Evergreen State College 1989.* Colorado Springs: United States Olympic Committee Education Council.

UK sport: partners in performance. (1993). *The contribution of sport science, sports medicine and coaching to performance and excellence.* Book of abstracts, s.l., Sports Council.

For excellence. A Symposium on Canadian High Performance Sport (1990). February 12, 13, 14, 1989: proceedings, February 1990. Ottawa: *Fitness and Amateur Sport.*

Coaching Association of Canada: Association canadienne des entraineurs, Gloucester, Ontario, 1992, 1 binder.

Developing the young athlete: 3rd Elite Coaches Seminar, Australian Institute of Sport, Canberra, 29 November - 2 December 1990. Canberra: Australia Coaching Council.

Beyond Barcelona: (proceedings of the) 4th Elite Coaches Seminar, Australian Institute of Sport, Canberra 27 to 29 November 1992, Canberra, Australian Coaching Council, 1992.

2.5. Databases

Sport Discus (Sport Information Resource Center, Canada, multilingual).
Atlantes (Instituto Andaluz del Deporte, Spain, in Spanish).

Heracles (Institut National du Sport et de l'Education Physique, France, in French).
Spolit (Bundesinstitut für Sportwissenschaft, Germany, in German).

2.6. Internet Sources

Coaching & Officiating unit of the Australian Sports Commission (ASC)
    www.ausport.gov.au/participating/coaches
Coaches Corner (Gatorade Sports Science Institute, USA)
    www.gssiweb.com
Coaching Association of Canada
    www.coach.ca/
Sports Coach (B. Mackenzie, United Kingdom)
    www.brianmac.demon.co.uk/
Wingate Institute for Physical Education & Sport (Israel)
    www.wingate.org.il
United States Olympic Committee – Olympic Coach
    rose.snyder@usoc.org and peter.davis@usoc.org

3. Organisational Network

3.1. International Level

The International Council for Coach Education (ICCE), represents the various bodies concerned with education of coaches and application of Coach Science/Theory in daily work all over the world. It was founded in 1997. The objectives of the ICCE are to promote Sport Coaching in general, and as a profession in particular, by various means, specifically, by "promoting and utilising research in the field of training and competition", "exchanging knowledge in the field of coaching", "disseminating information about curricula, qualifying standards etc.", "professional publication in the field of coaching education" (ICCE Objectives). In 2002, 21 countries were represented in ICCE, which holds conferences every second year (one year before the summer and winter Olympic Games, in the city in which the Games are to be held).

Other international organisations concerned with sports and physical education, such as AIESEP (Association Internationale des Ecoles Supérieures d'Education Physique), FIEP (Fédération Internationale d'Education Physique), ICHPER-SD (International Council for Health, Physical Education, Recreation, Sport and Dance) and the International Olympic Committee (through the aid it provides by way of the Olympic Solidarity program) all use and promote Coaching Science/Theory as a secondary part of their activities.

The Governing Bodies (International Federations) of the various sports do the same with regard to sports-specific Coaching Science/Theory.

3.2. Regional Level

The European Union has taken the initiative to design a 5-tier scheme of coaching levels. Many other countries, including Canada have adopted a similar scheme for their National Coaching Certification Program. This scheme provides a framework for the use of Coaching Science/Theory, from the rudimentary stage of the Level 1 Coach up to the academic-degree level of the Level 5 Coach, who utilises the most sophisticated and advanced facets of Coaching Science.

3.3. National Level

Several countries teach, advance and promote Coaching Science/Theory in their national sports centres, in which both coach education and top-level sport training is centralised. Examples are:

- Australia, through the Australian Coaching Council;
- Canada, through the Coaching Association of Canada;
- China, in its Coaching Department of the All-China Sports Federation;
- France, in its INSEP (Institut National du Sport et Education Physique);
- Germany, in its Cologne association of the Trainerakademie and Deutsche Sporthochschule and in - Leipzig (Leipzig University and the Institut für Angewandte Trainingswissenschaft);
- Great Britain, through the National Coaching Foundation;
- Hungary, in the Budapest University of Physical Education;
- Israel, in the Nat Holman School for Coaches and Instructors at the Wingate

Institute for Physical Education and Sport;
- Switzerland, in its Magglingen Sports Centre in Biel; and
- United States, through its coaching leadership of the Olympic Training Center at Colorado Springs.

In addition, countries are developing Coaching Science/Theory through their Sport Governing Bodies, University Physical Education and Sport Colleges, Sport Institutes etc.

3.4. Specialised Centres

See details under section 3.3 National Level.

3.5. Specialised International Degree Programmes

Several universities in the former Eastern Bloc e.g. Kiev, Budapest and Leipzig, educate coaches in a 4-year academic program which includes a thorough grounding in Coaching Science/Theory.

The Level 5 coaching program of the European Union is similar and is carried out in accredited universities. It takes 4 full years and includes 2400 study hours, half of which are general sports science studies and the other half sports-specific studies. In addition, the students must have 2 years of practical coaching experience.

4. Appendix Material

4.1. Terminology

Most of the terminology of Coaching Science/Theory has been developed in the countries of the former Eastern Bloc. Today, there is an overall consensus on the terminology, although some inconsistencies and illogicalities remain. As examples, use of the terms "mesocycle" and "macrocycle" differ between different authors; "training means" is sometimes defined narrowly but sometimes very broadly; "training methods" sometimes refers to the spectrum "continuous-interval-repetition" only, while on other occasions it will include the "pyramid" method, "whole" and "part" technical learning methods, etc. In English, "strength" is used in places where "force" would be appropriate, while in German, "Leistung" is used for "power", "achievement" and "performance". The difference between "power", "speed-strength", "strength-speed" and "explosive strength" is often unclear. But these inconsistencies and illogicalities are minor flaws and usually the context makes the meaning abundantly clear.

4.2. Position Statement

"Coaching Science is a branch of Sports Science, which deals with the theoretical foundations and methodological forming of athletic training, subject to the aim of achieving physical perfection, developing athletic performance as well as success in athletic competition." (Schnabel et al. 1986).

4.3. Varia

Not applicable.

4.4. Free Statement

Sports Coaching Science in general, and the education of coaches in particular, are currently helping in the creation of a new profession: Sports coaching, which is coming of age as a fully-fledged profession. As the public today has more leisure time than previously, there is a need to use this time effectively and sensibly. There is a demand of parents, teachers, physicians and the public at large, that care be taken of children's health and well-being, and that they may be enabled to grow and develop according to their gifts and preferences. This necessitates, much more than in the past, that sports coaching be based on moral, scientific and professional foundations and that those engaged in coaching be educated professionally, even academically, in the opinion of some people, just as members of other professions are prepared for theirs.

With rapid developments in information technology, both of software and hardware, coaching education will increasingly turn to various forms of "e-learning" in order to, on one hand, reduce the expenses incurred in coach education and, on the other hand, to provide wider access of students of coaching to the world's leading authorities on the subject. These new methods of teaching will also serve to disseminate more rapidly and more widely the most recent advances in coaching know-how.

4.5. Update

New Tendencies in Training Science for Children

1. Scientific focus on and ways of research in children's training.
. Different structures for motor coordination
   Ability for coordination under pressure of time , place and complex
   Ability for coordination under pressure of time, place, complex and with the ability of control and adaptation as a process in training.
2. Training, planning in combination with computer sciences, long term planning.
3. Research on competition: From analysing the performance to simulating the successful strategy.
4. The role of intellectual processes increasing as a result of the new tendencies.
5. New ways in the education of coaches and special instructors for children.

Kinanthropometry

Lindsay Carter

Contact

Dr. Lindsay Carter
School of Exercise and Nutritional Sciences
San Diego State University
San Diego, CA, USA 92182-7251
Email: lindsay.carter@sdsu.edu

1. General Information

1.1. Historical Developments

Kinanthropometry is the study of human body measurement for use in anthropological classification and comparison. Among the techniques used in modern kinanthropometry, anthropometry has the longest history.

Artists and sculptors used dimensions of the human body in absolute and proportional ways, for example, Leonardo da Vinci (1452-1519) showed this in his many works, especially the "Vitruvian Man" and Sigmund Elsholtz (1623-1688) was probably the first to use "anthropometry" in its contemporary meaning. In 1628, Gerard Thibault, wrote an extensive book, "L'Académie de l'Espée", on body dimensions and fencing success. Adolphe Quetelet (1796-1874), described the Quetelet Index (now commonly called the Body Mass Index) and was known for his application of statistical methods to human body dimensions.

Anthropologists and archaeologists also have a long tradition of the use of anthropometry, particularly relating to skeletal measurement. Late in the 19th century, several meetings were held to agree upon standards of measurement for anthropometry. The 13th International Congress of Prehistory, Anthropology and Archaeology, held in Monaco, 1906, is recognised for the first agreements. A subsequent meeting in Geneva, 1912, at the 14th International Congress of Anthropology and Archaeology, supplemented the agreements at Monaco. A copy of this Report is reproduced in Alex Hrdlička, Practical Anthropometry, Philadelphia: The Wistar Institute (1939). The book also contains descriptions of instruments and techniques of measurement.

In 1914, Rudolph Martin formalised the methods in his book "Lehrbuch der Anthropologie", and in subsequent revisions up until the late 1950s (with co-author, K. Saller). The methods of this German school dominated anthropometry in general during the first half of the 20th century. This influence spread into the United Kingdom and was soon seen incorporated into Sport Science and Sport Medicine research in North America.

In September 1978, kinanthropometry became an officially-recognised discipline in its own right at a meeting of the Research Committee (RC) of the International

Council of Sport Science and Physical Education (ICSSPE) in Brazil, when a proposal was presented to form the International Working Group on Kinanthropometry (IWGK) under the auspices of ICSSPE. The term kinanthropometry, rather than simply anthropometry, was chosen to emphasise that this particular group of scientists was interested in the application of anthropometry to movement just as much as it was interested in the measurements themselves.

Today, kinanthropometry continues to grow in popularity and application with senior practitioners found in all five continents and with its own international society, the International Society for the Advancement of Kinanthropometry (ISAK), to promote and foster its goals.

1.2. Function

The aim of kinanthropometry is to improve understanding of the gross functioning of the human body by measurement of its size, shape, proportions and composition and relating these to health, exercise and performance. A central interest is that of physical performance, in particular, though not limited to, sport performance. By examining the relationship between body measurements and aspects of performance, kinanthropometry helps in optimising training to improve performance, and also helps to reduce injuries. It is useful for children, to aid in the early recognition of athletic potential, and to examine the impact of early training on their growth and maturation. It serves an important function in assessing the relationship between exercise, nutrition and health, from the requirements of normal growth to the effects of ageing on the body, to the evolution and characteristics of the expression of different disease processes in the body. Gross functioning may also refer to applications other than sport: kinanthropometry is ideally suited to ergonomics, the optimisation of the fit between worker and workplace. A further important function of kinanthropometry is to improve, validate and standardise techniques for the measurement of the human body.

1.3. Body of Knowledge

The root words in kinanthropometry refer to movement, humans and measurement. In less simplistic terms, it is the study of human size, shape, proportion, composition, maturation and gross function (Ross, 1978). The discipline has a long history, since height and weight, the two simplest and most commonly used measures in kinanthropometry, have been measured for many centuries. Increasing sophistication led to the modern fields of anthropometry and biometry, and much has been written on these topics. The classic reference is "Lehrbuch der Anthropologie" (Martin and Saller, 1957), but Rudolph Martin's earlier work (1914) and the description of the measurements carried out in the International Biological program (Weiner and Lourie, 1969) are important additions. ISAK has modified the detailed descriptions by Ross and Marfell-Jones (1991) and Norton et al. (1996), to produce a new manual, International Standards for the Anthropometric Assessment, (ISAK, 2001 and 2006), in an effort to bring uniformity to the techniques in anthropometry. Since 1996, ISAK has operated an international Anthropometry Accreditation Scheme with four levels of expertise. To date, about 2 500 people from more than 20 countries have been certified.

Our quantitative knowledge of human physique and composition includes the classic work of Sheldon (1954). Somatotyping was made more rigorous in a series of publications by Carter and Heath, starting in 1966 and culminating in their definitive volume in 1990. No discussion of body shape and proportion would be complete without reference to the brilliant work of D'Arcy Thompson (On Growth and Form, 1917) and the application of allometry to growth by Huxley (1932). A comprehensive summary of our knowledge relating anthropometry to human growth is given by Edith Boyd (1980) in her treatise "Origins of the Study of Human Growth".

Techniques for the assessment of body composition are now of use in many diverse fields. Researchers have compiled a large body of knowledge regarding the relationship between body composition and both health and sport performance. Thousands of papers have been published on factors affecting body fat alone. However, direct information obtained from cadaver studies is still quite limited (Clarys, et al. 1984), but is now supplemented by CT and MRI scans for a better understanding of the anatomical compartments. An important component of kinanthropometry concerns the measurement of a wide range of physical performance variables such as muscular strength, power, fitness and flexibility.

1.4. Methodology

Research in kinanthropometry is essentially quantitative in nature, using standard methodologies and analytical procedures. Typically studies are descriptive (cross-sectional) or experimental (involving an intervention). Considerable advances have been achieved by cross-sectional studies that compare, for example, athletes in different sports, people in different states of health, the relationship between lifestyle variables and physique and composition variables. Interventions may test hypotheses related to the effects of such variables as exercise, in its various forms, on performance or variables related to body composition, body proportions, growth, etc.

1.5. Relationship to Practice

The scientific literature abounds with reports of research describing applications of kinanthropometry and it is not possible to give a comprehensive account here. However, some important applications will be given.

Kinanthropometry has been used to detail normal growth patterns in children and then to examine factors affecting growth, in particular exercise and nutrition. In this context, kinanthropometric assessment of sexual maturity has demonstrated the effects of athletic training on puberty as well as on the reproductive cycle of athletic women.

Kinanthropometry has been used to examine performance related variables in world class athletes (e.g. Carter, 1984; Carter and Ackland, 1994; Rienzi et al., 1998). It has been used in a wide range of cultural settings to investigate factors affecting nutritional status (e.g. Himes, 1991), and has been applied extensively in western countries in studies investigating health aspects of atypical fatness, ranging from extreme obesity to the emaciation of anorexia nervosa, and to the effects of exercise on the body's fat distribution. The role of different types of physical activity in the health of the skeleton is an area of much contemporary interest, and new, sophisticated techniques for assessing bone, such as dual-energy X-ray absorptiometry and magnetic resonance imaging, are now included in the spectrum of kinanthropometric methods. These have been applied to populations including athletes, postmenopausal women and children, to provide support for the hypothesis that regular weight-bearing exercise is important for skeletal integrity.

Kinanthropometry has many applications in medicine. It has been applied to genetic studies, to examine the physique and body composition correlates of certain chromosomal configurations. Similarly, it has been applied in describing physical correlates of different disease states and for helping evaluate therapeutic strategies. It has important applications in the area of public health, from its general role in providing normative data for guiding individuals who wish to make lifestyle changes to improve their health, to the detection of those at risk for such common problems as cardiovascular disease.

With the continuing emergence of evidence that regular physical activity improves health status and extends life expectancy, it appears that kinanthropometric research and its applications will continue to expand as they have done for the past several decades.

1.6. Future Perspectives

Given its application to both health and sport, we will see an increased use of kinanthropometry in the future as scientists, clinicians and coaches demand up-to-date information on norms and ranges for their populations of interest. These practitioners need ready access to data and simple, easily-applied techniques which will provide them with meaningful information and help them perform their jobs better. For their work to be of value, kinanthropometrists will need to continue the development of such techniques to ensure that they are practicable and affordable as well as meaningful. ISAK's role in this future will be to provide such data access as well as continuing anthropometric information, education and training so that its members and graduates will be able to provide measurement and interpretation services to address demand. ISAK also sees a major role in providing an international forum for kinanthropometrists to share their skills and findings. As technology daily increases the ability to communicate globally, the opportunity to obviate duplication of effort in isolated pockets around the world and thus improve knowledge progression increases in parallel. With ISAK acting as a conduit for anthropometric information transfer and dissemination, that progress can be significantly enhanced.

References

Boyd, E. (1980). *Origins of the Study of Human Growth*. Portland: University of Oregon Health Sciences Center.

Clarys, J.P., Martin, A.D. and Drinkwater, D.T. (1984). *Gross tissue weights in the human body by cadaver*. Human Biology, 54(3), 459-73.

Carter, J.E.L. (Ed.). (1984). *Physical Structure of Olympic Athletes-Part 2*. Basel: Karger Verlag.

Carter, J.E.L. and Ackland, T.R. (Eds.). (1994). *Kinanthropometry in Aquatic Sports*. Champaign, IL.: Human Kinetics.

Carter, J.E.L. and Heath, B.H. (1990). *Somatotyping - Development and Applications*. Cambridge: Cambridge University Press.

Himes, J.H. (1991). *Anthropometric Assessment of Nutritional Status*. New York: Wiley-Liss.

Huxley, J.S. (1932). *Problems in Relative Growth*. London: Methuen.

Marfell-Jones, M., Olds, T., Stewart, A. and Carter, L. (2006). *International Standards for Anthropometric Assessment* (2006). Potchefstroom, SA.

Martin, R. (1914). *Lehrbuch der Anthropologie*. Jena: Fischer.

Martin, R. and Saller, K. (1957). Lehrbuch der Anthropologie. Stuttgart: Fischer.

Norton, K., Whittingham, N., Carter, L., Kerr, D., Gore, C. and Marfell-Jones, M. (1996). Measurement Techniques in Anthropometry. In: Norton, K. and Olds, T. (Eds.). *Anthropometrica*. Sydney: UNSW Press.

Rienzi, E., Mazza, J.C., Carter, J.E.L. and Reilly, T. (Eds.). (1998). *Futbolista Sudamericano de Elite: Morfologia, Analisis del Juego y Performance*. Buenos Aires: Biosystem Servicio Educativo.

Ross, W.D. (1978). Kinanthropometry: an emerging scientific specialisation. In: Landry, F. and Orban, W.A.R. (Eds.). (1978). *Biomechanics of Sports and Kinanthropometry*, Vol. 6, Miami: Symposium Specialists.

Ross, W.D. and Marfell-Jones, M.J. (1991). Kinanthropometry. In: MacDougall, J.D.; Wenger, H.A. and Green, H.J. (Eds.). *Physiological Testing of the High-Performance Athlete, pp*. 233-308, Champaign, IL: Human Kinetics.

Sheldon, W.H. (1954). *Atlas of Men*. New York: Harper.

Thompson, D.A.W. (1917). *On Growth and Form*. Cambridge: Cambridge University Press.

Weiner, J.S. and Lourie, J.A. (1969). Human Biology: *A guide to practical field methods*. Philadelphia: Davis.

2. Information Sources

2.1. Journals

Journal of Sports Sciences
African Journal for Physical , Health Education, Recreation & Dance
American Journal of Human Biology
Annals of Human Biology
American Journal of Physical Anthropology
British Journal of Sports Medicine
Medicine & Science in Sports and Exercise
Revista Brasileira de Ciencia e Movimento, Brazil
Revista de Medicina Ciencias y Deportes, Argentina

This is only a selection. There are many others, such as journals on nutrition, that carry articles on kinanthropometry.

2.2. Reference Books, Encyclopaedias, etc.

Dirix, A., Knuttgen, H.G. and Tittel, K. (Eds.). (1988) *The Olympic Book of Sports Medicine, Vol. 1*, Chap. 6.1, pp. 233-265. Oxford: Blackwell Scientific.

MacDougall, J.D., Wenger, H.A. and Green, H.J. (Eds.). (1991). *Physiological Testing of the High-Performance Athlete*, Chap. 6, pp. 233-308. Champaign, IL: Human Kinetics.

Martin, R. and Saller, K. (1957). *Lehrbuch der Anthropologie*. Stuttgart: Fischer.

Handbooks:

Ackland, T.R., Elliott, B.C. and Bloomfield, J. (Eds.). (2006). *Applied Anatomy and Biomechanics in Sport, 2nd Edition*. Melbourne: Blackwell.

Ackland, T.R., Elliott, B.C. and Bloomfield, J. (Eds.). *Building a Champion Athlete: Applied Anatomy and Biomechanics in Sport.* (In Press) Champaign, Il: Human Kinetics.

Gore, C. et al. (1998). *Accreditation in anthropometry - a curriculum guide*. Belconnen: Australian Sports Commission.

Esparza Ros, F. (Ed.). (1993). *Manual de Cineantropometria*. Madrid: Monographias FEMEDE.

Eston, R. and Reilly, T. (Eds.). (2001). *Kinanthropometry and Exercise Physiology Laboratory Manual: Tests, Procedures and Data (second edition), Vol. 1: Anthropometry.* London: Routledge

Marfell-Jones, M., Olds, T., Stewart, A. and Carter, L. *International Standards for Anthropometric Assessment (2006).* (2nd Ed.) Potchefstroom, SAF. Also published in Spanish as: Estándares Internacionales para la Valoración Antropométrica (2005).

Norton, K. and Olds, T. (Eds.). (1996). *Anthropometrica.* Sydney: UNSW Press.

Ross, W.D. (1996). Anthropometry in Assessing Physique Status and Monitoring Change. In: Bar-Or, O. (Ed.). *The Child and Adolescent Athlete.* London: Blackwell Science.

2.3. Book Series

Medicine and Sport Sciences. Karger Verlag Basel, New York. (Series Editors: Borms, J.; Hebbelinck, M. and Hills, A.P.):

Vol. 16 – Carter, J.E.L. (Ed.). (1982). Physical Structure of Olympic Athletes. Part I: The Montreal Olympic Games Anthropological Project. Basel: Karger.

Vol. 18 – Carter, J.E.L. (Ed.). (1984). Physical Structure of Olympic Athletes. Part II: Kinanthropometry of Olympic Athletes. Basel: Karger.

Vol. 20. - Kemper, H.C.G. (Ed.). (1985). Growth, Health and Fitness of Teenagers. Basel: Karger.

Vol. 44 – Jürimäe, T. and Hills, A.P. (Eds.). (2001). Body Composition Assessment in Children and Adolescents. Basel: Karger.

2.4. Congress/Workshop Proceedings

The kinanthropometry community has gathered as a group (either as the International Working Group on Kinanthropometry, IWGK or ISAK), or as part of a major congress in 1976 (Montréal), 1978 (Leuven), 1982 (Brisbane), 1984 (Eugene, Oregon), 1986 (Glasgow), 1988 (Cheonan), 1990 (Brussels), 1992 (Málaga), 1994 (Victoria, B.C.), 1996 (Dallas), 1998 (Adelaide), 2000 (Brisbane), 2002 (Manchester), 2004 (Thessaloniki) and 2006 (Melbourne).

The following congress proceedings are in the order they were published:

Landry, F. and Orban, W.A.R. (Eds.). (1978). *Biomechanics of Sports and Kinanthropometry, Vol. 6*, Miami: Symposium Specialists.

Ostyn, M., Beunen, G. and Simons, J. (Eds.). (1980). *Kinanthropometry II*. Baltimore: University Park Press.

Day, J.A.P. (Ed.). (1986). *Perspectives in Kinanthropometry*. Champaign, IL: Human Kinetics.

Reilly, T., Watkins, J. and Borms, J. (Eds.). (1986). *Kinanthropometry III*. London: E.E. & F.N. Spon.

Duquet, W. and Day, J.A.P. (Eds.). (1993). *Kinanthropometry IV*. London: E. & F.N. Spon.

Bell, F.I. and Van Gyn, G.H. (Eds.). (1994). *Access to Active Living, Kinanthropometry section*. Victoria, B.C.: University of Victoria.

Norton, K., Olds, T. and Dollman, J. (Eds). (2000). *Kinanthropometry VI*. Underdale, SA: International Society for the Advancement of Kinanthropometry (ISAK).

De Ridder, H. and Olds, T. (Eds.). (2003). *Kinanthropometry VII*. Potchefstroom, SAF: ISAK.

Reilly, T. and Marfell-Jones, M. (Eds.). (2003). *Kinanthropometry VIII*. London: Routledge.

Marfell-Jones, M., Stewart, A. and Olds, T. (Eds.). (2006). *Kinanthropometry IX*. London: Routledge.

Marfell-Jones, M. and Olds, T. (Eds.). *Kinanthropometry X*. (In press)

2.5. Data Banks and Computer Applications

Olds, T. and Norton, K. (1999). *LifeSize*. (CD-ROM). Champaign, IL.: Human Kinetics.

Goulding, M. (2002). Somatotype v.1.1. Mitchell Park, S Aust: Sweattechnologies.

Ross, W., Carr, R. and Carter, J.E.L. (1999). *Anthropometry Illustrated*. (CD-ROM). Surrey, B.C., Canada: Turnpike Electronic Publications.

Ross, W.D., Carr, R.V., Guelke, J.M. and Carter, J.E.L. (2003). *Anthropometry Fundamentals* (CD-ROM). Surrey, B.C., Canada: Turnpike Electronic Publications.

Carr, R.V. and Ross, W.D. (2005) *Anthropometry Technique*, (Menu Driven DVD). Surrey, B.C., Canada: Rosscraft/Turnpike Electronic Publications.

2.6. Internet Sources

ISAK's Website
  www.isakonline.com
Anthropometric Equipment
  www.rosscraft.ca

3. Organisational Network

3.1. International Level

The science of kinanthropometry is represented at the international level by the International Society for the Advancement of Kinanthropometry (ISAK). The Society was founded in Glasgow in 1986, having grown out of the International Working Group in Kinanthropometry (IWGK), a branch of the then ICSSPE Research Committee. Officers on the Board include a President, Vice-President, Secretary General, Past-President and 5 Board Members. Since 1996, ISAK has promoted an accreditation scheme in anthropometry with four levels of certification (Levels 1-4), with more than 2500 persons certificated by 2007.

There are approximately 260 active members of the Society, from 33 countries. The Society meets biennially and publishes a newsletter (Kinanthreport) three times per year to inform members about developments and issues within the science of kinanthropometry. In addition, the Journal of Sports Sciences is published in association with ISAK. See www.isakonline.com for updates.

3.2. Regional Level

Europe

Europe has several centres of activity involved in anthropometric projects. These include the centres in Belgium, Estonia and Portugal. Within the UK, there is activity focused on paediatric exercise at the University of Exeter, and ageing at Liverpool John Moores University. Anthropometry, using ISAK profiling, is performed at several universities including Heriot Watt (Edinburgh), Portsmouth, Birmingham and the Robert Gordon University (Aberdeen). ISAK courses have been delivered at Levels 1, 2 and 3 in Aberdeen every year since 2002, and Levels 1 and 2 are embedded into the sports science undergraduate program at the Robert Gordon University. Level 1 courses have been delivered at several universities and anthropometric profiling has become standardised procedure for elite soccer players in several premier teams.

Africa

The Kinanthropometry Interest Group of Africa (KIGA) is a resource for projects. Major research projects were conducted at the 1995 (Zimbabwe) and 2005 (Nigeria) All African Games. In South Africa, there have been projects on schoolchildren, athletes and heath concerns.

Latin America

There is a strong interest in kinanthropometry in Latin America. In general, the leaders are affiliated with sports medicine groups, research institutes, government entities or universities. Kinanthropometry has been featured in their national congresses and many workshops and courses have been organised. The groups that are most active in kinanthropometry are in Mexico, Brazil, Argentina, Chile, Uruguay, Puerto Rico and Venezuela. In addition, courses have been held in Panama and the Dominican Republic.

South Korea, Malaysia, Japan

Level 2 and 3 certification courses have been delivered on at least an annual basis for the past five years in Korea. The most recent course there, for example, saw 19 Level 3 completions and 5 Level 2. Level 1 and 2 courses have been delivered in Malaysia for the past six years and in Japan for the past four.

3.3. National Level

Australia

Kinanthropometry is a well-developed specialisation in Australia. In 1993, the Laboratory Standards Assistance Scheme conducted an inaugural workshop for representatives from all universities and sports institutes. As a result, three levels of national accreditation were established based on a standard protocol (ISAK) and standards for technical error of measurement. A textbook was developed and supported by software (LifeSize). In addition, the University of New South Wales implemented the Australian Anthropometric Data Base as a component of LifeSize to collect nation-wide data in a standardised format. Kinanthropometry centres are found at the Australian Institute of Sport and its regional laboratories, sports dietician groups, as well as at many of the major universities and sports centres around the country. The work at the University of South Australia under the direction of Tim Olds is outstanding.

New Zealand

Many ISAK certification courses have been run during the past decade under the direction of Mike Marfell-Jones. The Institute of Sport and Recreation Research New Zealand (ISRRNZ) at Auckland University of Technology Anthropometry Laboratory (Director: Patria Hume) have ongoing extensive testing programs for many sports and special populations.

Malaysia

The first Level 1 ISAK course was run in 2005 in Kuala Lumpur and was organised by Tan Yoke Hwa, President of the Malaysia Dieticians Association. In attendance were 30 dieticians and nutritionists from Malaysian hospitals and the National Sports Council. The course instructors were Drs Deborah Kerr, Gary Slater and Patricia Wong.

Singapore

Since 2001, several Level 1 and 2 courses have been run in Singapore at the Singapore Sports Council. The course instructors were Drs Tim Ackland, Deborah Kerr and Shelley Kay.

India

The National Working Group on Kinanthropometry is based at the National Institute of Sport, Patiala, India. Anthropometry is a part of the program for athletes at the Sports Authority of India, J.N. Stadium, Delhi.

Iran

Through the National Olympic and Paralympic Academy of Iran Department of Anthropometry, kinanthropometry and ISAK Level 1 courses have been introduced to teachers, coaches and sports medicine specialists over the past three years in many courses (Director is Dr. Shahraham Faradjzadeh Mevaloo).

United Kingdom

Several sport governing bodies and regional sports institutes are supporting ISAK-qualified measurements and approximately four Level 1 courses are offered annually by institutions by the available level 3 instructors.

3.4. Specialised Centres

A list of specialised centres, categorised by region, is presented below. Some contact persons are named; and there are several people at other centres.

Europe

Vrije Universiteit Brussel, Faculty of Physical Education and Physiotherapy, Pleinlaan 2, B-1050 Brussels, Belgium.

Centre for Sport and Exercise Sciences, Liverpool John Moores University, Mountford Building, Byrom Street, Liverpool, England, L33AF, UK.

National Sports Academy, Sofia, Bulgaria.

Faculdade de Motricidade Humana, Estrada da Costa 1499-002 Cruz Quebrada – Dafundo, Lisboa, Portugal.

School of Health Sciences, Faculty of Health and Social Care, The Robert Gordon University, Garthdee Road, Aberdeen AB10 7QG, UK. (Contact Dr Arthur Stewart.)

North America

School of Human Kinetics, University of British Columbia, 6081 University Blvd., Vancouver, BC V6T 1Z1, Canada.
College of Physical Education, University of Saskatchewan, Saskatoon, SK, S7N 0W0, Canada.
Rosscraft, 14732A 16-A Ave, Surrey, B.C. V4A 5M7, Canada.
School of Exercise and Nutritional Science, San Diego State University, San Diego, CA 92182-7251, USA.
U.S. Olympic Training Center - Colorado Springs, CO. USA.

Asia and Australasia

University of Western Australia, Biomechanics and Applied Anatomy, Crawley, Western Australia 6009, Australia.
Australian Institute of Sport, Canberra, Australia.
University of South Australia, School of PE, Exercise and Sport Studies, Underdale, Australia.
School of Public Health, Curtin University of Technology GPO Box U1987 Perth WA 6845, Australia.
Universal College of Learning, Private Bag 11022, Palmerston North, New Zealand.
School of Physical Education, University of Otago, Box 56, Dunedin, New Zealand.
Institute of Sport and Recreation Research, Auckland University of Technology, Auckland, New Zealand.
Korean National University of Physical Education, Department of Physical Education, 88-16 Olympic Park, Songpa-Ku, Seoul, South Korea 138-763.
National Olympic and Paralympic Academy of Iran Anthropometry Department, Tehran 16346, Iran.

Latin America

ILSI (International Life Sciences Institute) Institute, Argentina, Santa Fe 1145, 4th floor, Buenos Aires, Argentina. info@ilsi.org.ar
CELAFISCS, Av. Goias, 1400, São Caetano do Sul, Sao Paulo, 09520, Brazil.
Universidad Autónoma de Chihuahua, Facultad de Educación Física y Ciencias del Deporte, Chihuahua, CHIH, Mexico.

Libertator Pedagogical Experimental University, Barquisimeto, Lara State, Venezuela.
University of Puerto Rico, Physical Education and Recreation Department, Rio Piedras, Puerto Rico.

South Africa

School of Biokinetics, Recreation and Sport Science, North-West University (Potchefstroom Campus), Potchefstroom, 2520. (Contact Prof. Hans de Ridder.)

3.5. Specialised International Degree Programmes

Not applicable.

4. Appendix Material

4.1. Terminology

See:
Ross, W.D. and Marfell-Jones, M.J. (1991). Kinanthropometry. In: MacDougall, J.D., Wenger, H.A. and Green, H.J. (Eds.). *Physiological Testing of the High-Performance Athlete*, Chapt. 6, pp. 233-308. Champaign, IL: Human Kinetics.
Norton, K. and Olds, T. (Eds.). (1996). *Anthropometrica*. Sydney: UNSW Press.
Ross, W., Carr, R. and Carter, J.E.L. (1999). *Anthropometry Illustrated* (CD-ROM). Surrey, Canada: Turnpike Electronic Publications.
Marfell-Jones, M., Olds, T., Stewart, A. and Carter, L. (2006). *International Standards for Anthropometric Assessment*. (2nd Ed.) Potchefstroom, SAF: ISAK.

4.2. Position Statement(s)

Not applicable.

4.3. Varia & 4.4. Free Statement

Not applicable.

Motor Behaviour: Motor Development, Motor Control and Motor Learning

Darlene A. Kluka

Contact

Prof. Dr. Darlene A. Kluka
Barry University
School of Human Performance and Leisure Sciences
Department of Sport and Exercise Sciences
11300 NE 2nd Ave.
Miami Shores, Florida 33161-6695
USA
Phone: +1 305 899 3549
Email: DKluka@mail.barry.edu

1. General Information

Motor behaviour involves human volitional movement (action) that includes motor control, motor development and motor learning. Motor development is the study of the sequential, continuous age-related processes involving changes in movement behaviour (Haywood and Getchell, 2005). Motor control involves the study of movements and postures and the mechanisms that underlie it (Rose, 1997). Motor learning refers to a multifaceted set of internal processes that affect relatively permanent changes in human performance through practice, provided the change cannot be attributed to a human's maturation, a temporary state, or instinct (Kluka, 1999).

1.1. Historical Development

The theoretical foundations of motor development have evolved from three different perspectives: maturational, information-processing and ecological. Biological development, through maturation and central nervous system development, has been emphasised by scholars categorised as maturationists (Gesell, 1928, 1954; McGraw, 1943). Information-processing advocates view an individual's capacity to assimilate sensory information from environment as a primary contributor to motor development (Schmidt, 1989; Clark and Whitall, 1989). Ecological theorists emphasise that it is the interaction of the human, the environment and the task that are critical to motor development (Kugler, Kelso and Turvey, 1980, 1982).

For the past several decades, research initiatives conducted in motor control have resulted in the development of several theoretical models, including reflex theories, hierarchical theories and dynamic systems theory. Some of the earliest recorded investigations involved those of Sherrington (1906) related to stimulus-response coupling for action. Psychologists in the 1920s and 1930s (Skinner, 1938; Thorndike, 1927), viewed movement pattern acquisition as a link in an action chain that was triggered by an external stimulus and visually observed. Hierarchical theories focus upon all aspects of movement planning and execution and include the central nervous system hierarchy. Becoming popular in the 1960s and 1970s (Keele, 1968; Greene, 1972; Schmidt, 1976; Shapiro, 1978), motor programs, consisting of motor commands from the brain's highest level through the musculature, were postulated as controlling human action. More recently, a very different approach to motor control has evolved. By the 1960s, the relationship of the performer and the environment in which action occurred became important to understanding motor control (Bernstein, 1967; Gibson, 1966). Since the 1980s, there has been much interest in dynamic systems theory that provides an alternative to previous motor control theories. The theory suggests that human movement results from the body's self-organisation of the performer's environment and the task demands (Sheridan, 1984; Turvey and Carello, 1981, 1988).

Motor learning theories have been derived to explain how motor skill acquisition is achieved. The first theories involved the development of memory representations to guide human action. Two of the most popular have been the closed-loop theory (Adams, 1971) and the schema theory (Schmidt, 1975). A relatively new approach to the understanding of motor learning involves the ecological theory of perception and action (Gibson, 1966, 1979; Turvey, 1974, 1977). The dynamic relationship between the performer and the environment in the learning process becomes paramount to the discussion. Other models constructed to explain the process of motor learning include Fitts' Three Stages (1964) and Gentile's Two-Stage model (1972).

1.2. Function

There are presently three major bodies of knowledge which involve research and practice in motor behaviour. The first involves structural and functional constraints and transitions relative to physical growth, maturation and aging in human action (motor development). The second is built upon the neurosciences, and describes the neural structures, processes, functions and effects that undergo changes in performance through motor control. The third involves basic tenets upon which motor skill acquisition can be built, utilising relationships between sensory systems and objects, surfaces and events in the environment (motor learning).

1.3. Body of Knowledge

Motor behaviour includes the specialised areas of motor development, motor control and motor learning, each of which contributes to our understanding of the mental structures and processes that produce skilled human action. The field traditionally encompasses research concerned with how humans develop, learn and control complex motor skills. An increasing number of researchers are investigating the effects of anxiety, motivation, relaxation and other sport psychology topics on fundamental and developing neural and cognitive processes. The area includes the sequential, continuous age-related processes involved in human action (motor development), relatively permanent action changes not directly attributable to aging (motor learning) and the nervous system's control of muscles that produce skilled and coordinated action (motor control).

1.4. Methodology

The methods used for each of the specialised motor behaviour areas are distinct. Each of the areas includes research conducted in laboratory settings, using novel tasks that elicit simple responses or must limit the number of variables being investigated. This type of research has served as the basis for more ecologically-based investigations. From this type of research, investigators have been able to determine performance-based characteristics of novice and elite performers in a variety of real-world environments. Additional types of ecologically-based research designs have been formulated, including the integration of motor development, motor control and motor learning to understand children with special challenges such as Down Syndrome, Attention Deficit Hyperactivity Disorder (ADHD) and cerebral palsy.

Motor control and motor learning research designs include methods that directly measure human neural processes through the use of brain scanning devices such as EEG, PET, MRI and fMRI. They also include devices that record sensory movement, including eye tracking, auditory stimulus processing and tactile sensitivity.

Motor behaviour studies also include dynamic systems approaches by utilising mathematical, engineering, neural network, biomechanical and thermodynamic models to describe, analyse and interpret motor behaviour. Motor development studies may also include longitudinal investigations focusing on various parameters throughout portions of the lifespan.

1.5. Relationship to Practice

The field of motor behaviour provides detailed information for those in teaching, coaching, athletic training, sports medicine and rehabilitation, human factors and ergonomics, physiotherapy and other areas that seek information about the relationship between the brain and body and the effects of maturation on motor skill acquisition and performance.

1.6. Future Perspectives

During the last half of the 20th Century, space research was referred to as the "final frontier". We know more about sending humans into space to distant planets and space stations orbiting the earth than we know about the human brain and how it functions. Brain research and, specifically, brain research relating to motor behaviour, may well become "THE frontier" for the first half of the 21st Century that will then develop related fields involving artificial intelligence and robots. As new technologies are discovered, more investigations will be conducted about human perception, attention, memory, decision making and their roles in the continuous age-related processes involved in human action, relatively permanent action changes not directly attributed to aging and the nervous system's control of muscles that produce skilled and coordinated movement. In addition to these areas, other areas of research interest will include feedforward and feedback perspectives and anticipation, prediction and timing. Performance measurement of vision (perception and decision making) during human action will be of particular interest, along with the development of instrumentation that is real-time, accurate and valid. Future approaches to motor behaviour study will involve the inclusion of multi-disciplinary research teams focusing upon the dynamics of space, time and the individual.

2. Information Sources

2.1. Journals

Adapted Physical Activity Quarterly
Applied Cognitive Psychology
Behaviour and Brain Sciences
Developmental Medicine and Child Neurology
Experimental Brain Research
Human Movement Science

Human Perception and Performance
Human Performance
International Journal of Sports Vision
International Journal of Volleyball Research
Journal of Experimental Psychology
Journal of Exercise and Sport Psychology
Journal of Motor Behaviour
Journal of Neuroscience
Journal of Sport Sciences
Motor Control
NeuroReport
Neuron
Perceptual and Motor Skills
Research Quarterly for Exercise and Sport
Vision Research

2.2. Reference Books, Encyclopaedias, etc.

Bard, C., Fleury, M. and Hay, L. (Eds.). (1990). *Development of eye-hand coordination across the lifespan*. Columbia, SC: University of South Carolina Press.

Bloedel, J.R., Ebner, T.J. and Wise, S.P. (1996). *The acquisition of motor behavior in vertebrates*. Cambridge, MA: MIT Press.

Cockerill, I.M. and MacGillivary, W.W. (Eds.). (1999). *Vision and sport*. Cheltenham, UK: Stanley Thomas.

Coker, C.A. (2004). *Motor learning and control for practitioners*. St. Louis, MO: McGraw-Hill Publishing.

Dordo, P., Bell, C. and Harnad, S. *Motor learning and synaptic plasticity in the cerebellum*. Cambridge, UK: Cambridge University Press.

Davids, K., Button, C. and Bennett, S. (2008). *Dynamics of skill acquisition*. Champaign, IL: Human Kinetics.

Davids, K., Williams, J. and Williams, M. (1998). *Visual perception and action in sport*. London, UK: Routledge.

Davis, W.E. and Broadhead, G.D. (2006). *Ecological task analysis perspectives on movement*. Champaign, IL: Human Kinetics.

Haywood, K.M. and Getchell, N. (2005). *Life span motor development*, 4th ed. Champaign, IL: Human Kinetics.

Honeybourne, J., Mangan, J.A. and Galligan, F. (2006). *Acquiring skill in sport: An introduction*. London, UK: Routledge.

Kluka, D. (1999). *Motor Behavior: From learning to performance*. Stamford, CT: Wadsworth Publishing Co.

Latash, M.A. (1998). *Neurophysiological basis of movement*, 2nd ed. Champaign, IL: Human Kinetics.

Loran, D.F.C. and Mac Ewen, C.J. (1995). *Sports vision*. Oxford, UK: Butterworth-Heninman, Ltd.

Magill, R.A. (2006). *Motor learning: Concepts and applications*, 8th ed. Madison, WI: McGraw-Hill.

Malina, R.M. and Bouchard, C. (2004). *Growth, maturation, and physical activity*, 2nd ed. Champaign, IL: Human Kinetics.

Milner, A.D. and Goodale, M.A. (1995). *The visual brain in action*. Oxford, UK: Oxford University Press.

Piek, J. (2006). *Infant motor development*. Champaign, IL: Human Kinetics.

Piek, J. (1998). *Motor behavior and human skill*. Champaign, IL: Human Kinetics.

Riehle, A. and Vaadia, E. (2004). *Motor cortex in voluntary movements*. London, UK: Routledge.

Rose, D.J. (1997). *A multilevel approach to the study of motor control and learning*. Needham Heights, MA: Allyn and Bacon.

Schmidt, R.A. and Lee, T.D. (2005). *Motor control and learning: A behavioral emphasis*, 4th ed. Champaign, IL: Human Kinetics.

Schmidt, R.A. and Wrisberg, C.A. (2008). *Motor learning and performance*, 4th ed. Champaign, IL: Human Kinetics.

Schmidt, R.A. and Wrisberg, C. (2008). *Motor learning and performance presentation package*. Champaign, IL: Human Kinetics.

Shumway-Cook, A. and Wollacott, N. (2002). *Motor control: Theory and practical applications*. Philadelphia, PA: Lippincott, Williams and Wilkins.

Starkes, J. and Ericsson, A. (2003). *Expert performance in sports*. Champaign, IL: Human Kinetics.

Stein, P., Grillner, S., Selverston, A. and Stuart, D. (Eds.). (1997). *Neurons, networks, and motor behavior*. Cambridge, MA: MIT Press.

Wiley, A. and Astill, S. (2007). *Instant notes in motor control, learning, and development*. London, UK: Routledge.

Wing, A.M., Haggard, P. and Flanagan, J.R. (Eds.). (1996). *Hand and brain: The neurophysiology and psychology of hand movements*. San Diego, CA: Academic.

Wulf, G. (2006). Attention and motor skill learning. Champaign, IL: Human Kinetics.

Zelaznik, H. (1996). *Advances in motor learning and control.* Champaign, IL: Human Kinetics.

2.3. Book Series

Attention and performance. Bradford Book: MIT Press

Tutorials in motor behavior. Amsterdam: Elsevier Science

Latash, M.A. (1998). *Progress in motor control: Volume 1.* Champaign, IL: Human Kinetics.

Latash, M.A. (2002). *Progress in motor control: Volume 2.* Champaign, IL: Human Kinetics.

Latash, M.A. and Levin, M. (2004). *Progress in motor control: Volume 3.* Champaign, IL: Human Kinetics.

2.4. Conference/Workshop Proceedings

Current research in motor control. (2000 - 2007) Polish Scientific Physical Education Association: University School of Physical Education in Katowice.

Proceedings of International Joint Conference on Artificial Intelligence (1996 – present).

Proceedings of North American Society for Psychology of Sport and Physical Activity (2000 – present).

2.5. Data Banks

Main data banks used to locate works in motor behaviour are SPORT DISCUS and INDEX MEDICUS and are accessible via the internet, for a price.

2.6. Internet Sources

Motor Behaviour:

University of Auckland, New Zealand, motor behaviour laboratory
www.ses.auckland.ac.nz/

University of Calgary, Canada, Neuromotor Psychology laboratory
  www.kin.ucalgary.ca/2002/profiles/neuromotor.asp
Louisiana State University, Baton Rouge, USA, motor behaviour laboratory
  http://mb.lsu.edu
Purdue University, Lafayette, Indiana, USA, motor behaviour program
  www.cla.purdue.edu/hk/motorbehavior/motor%20behavior/Home.html
Radford University, Radford, Virginia, USA, motor behaviour laboratory
  www.radford.edu/~mobelab/motor_behavior_laboratory_coordi.htm
University of Memphis, USA, motor behaviour laboratory
  https://umdrive.memphis.edu/g-HSS/HSS_WEB/MBL/index.html
University of Michigan, Ann Arbor, USA, motor behaviour centre
  www.kines.umich.edu/research/cmbpd.html
University of Tennessee, Knoxville, USA, motor behaviour laboratory
  http://web.utk.edu/~sals/resources/motor_behavior_laboratory.html
University of Utah, Provo, motor behaviour research laboratory
  www.health.utah.edu/clinics/motor.html
Motor Development:
Purdue University, USA, infant motor development laboratory
  www.cla.purdue.edu/hk/discovery/infant.motor.htm
University of Texas at San Antonio, USA, motor development clinic
  http://kah.utsa.edu/research_facilities.htm#mdc
University of Michigan, USA, motor development laboratory
  www.kines.umich.edu/research/chmr/motdev

Motor Control:

Arizona State University, USA, motor control laboratory
  www.asu.edu/clas/espe/MClab/motorcontrolwebpage.html
McGill University, Canada, motor control laboratory
  www.psych.mcgill.ca/labs/mcl/Lab-Home.html
Newcastle Motor Control Laboratory, UK
  www.staff.ncl.ac.uk/stuart.baker/
Penn State Motor Control Laboratory, USA
  www.kinesiology.psu.edu/research/laboratories/mcl/index.html

University of Maryland, College Park, USA, cognitive neuroscience laboratory
www.hhp.umd.edu/KNES/research/cmb.html

Motor Learning:

Iowa State University, USA, motor control and learning research laboratory
www.kin.hs.iastate.edu/research/labs/Ann's%20lab%20(MCL%20web%20
page%2010%202005).htm

3. Organisational Network

3.1. International Level

International Society for Ecological Psychology (ISEP).
International Society of Sport Psychology (ISSP).
International Association of Applied Psychology (IAAP).
Society for Neuroscience (SNS).
Society for the Neural Control of Movement and Cognitive Neuroscience Society.

3.2. Regional Level

Federation Eduopeene de Psychologie des Sports et des Activites Corporelles /
European Federation of Sport Psychology (FEPSAC).
North American Society for the Psychology of Sport and Physical Activity (NASP-
SPA).
South American Society of Sport Psychology, Physical Activity and Recreation
(SASSPPAR).

3.3. National Level

American Optometric Association Sports Vision Section.
Australian Association for Exercise and Sport Science (AAESS).
Canadian Society of Psychomotor Learning and Sport Psychology (CSPLSP).
French Society of Sport Psychology (FSSP).
National Association for Sport and Physical Education (NASPE) of AAHPERD – Mo-
tor Development/Learning Academy.

3.4. Specialised Centres

Cognitive Engineering Laboratory, University of Toronto, Canada.

Neuromotor and Psychology Laboratory, University of Calgary, Canada.

Neural Control Laboratory, University of Waterloo, Canada.

Marseille Mouvement et Perception Laboratory, Marseille, France.

Planck Institute for Psychological Research, Cognition and Action, Munich, Germany.

Motor Control Laboratory, University School of Physical Education, Katowice, Poland.

Center for the Ecological Study of Perceiving and Acting, University of Connecticut, USA.

Center for Complex Systems, Florida Atlantic University, USA.

Cognitive and Neural Systems, Boston University, USA.

Cognitive and Linguistic Sciences, Brown University, USA.

Haskins Laboratories, University of Connecticut, USA.

Human Performance Center, United States Air Force Academy, Colorado Springs, Colorado, USA.

Human Performance Center, United States Military Academy, West Point, New York, USA.

Motor Learning Laboratory, University of Virginia, Charlottesville, USA.

Motor Control Laboratory, Pennsylvania State University, USA.

Motor Behaviour Laboratory, Radford University, USA.

University of Maryland, College Park, Cognitive Neuroscience Laboratory, USA.

University of Michigan, Ann Arbor Motor Behaviour Center, USA.

3.5. Specialised International Degree Programmes

Presently, there appears to be no international degree programs in the field. There are numerous PhD programs located throughout parts of North America (Canada and the United States) and there are other programs located throughout Europe, Australia and New Zealand.

4. Appendix Material

4.1. Terminology

Anticipation/prediction timing; apparent motion; attentional focus/styles; childhood/adolescence/adulthood/late adulthood; contextual interference; feedback/feedforward; levels of performance skill; long term athlete development; memory/forgetting; modeling; neural networks; motivation; neuromuscular control; practice structure/organisation; quality of life in lifespan development; sensory information; schema; skill acquisition; speed/accuracy; stages of learning; talent development; transfer.

4.2. Position Statement(s)

None available.

4.3. Varia

Not applicable.

4.4. Free Statement

None available.

Philosophy of Sport

Michael McNamee

Contact

Prof. Dr. Michael McNamee
University of Wales Swansea
Department of Philosophy, Law and History in Healthcare
School of Health Science
SA2 8PP Singleton Park, Swansea
UNITED KINGDOM
Phone: +44 1242 544028
Email: m.j.mcnamee@swansea.ac.uk

1. General Information

1.1. Historical Development

The International Association for the Philosophy of Sport (IAPS) is the only international, scholarly agency explicitly and exclusively devoted to the subject. The Association was established in Boston, USA, on 28 December, 1972, as the Philosophic Society for the Study of Sport (PSSS). It has staged annual meetings across the world since 1973.

1.2. Function

The philosophy of sport is concerned with the conceptual analysis and interrogation of key ideas and issues of sports and related practices. At its most general level, it is concerned with articulating the nature and purposes of sport. The philosophy of sport not only gathers insights from the various fields of philosophy as they open up our appreciation of sports practices and institutions, but also generates substantive and comprehensive views of sport itself. The philosophy of sport is never fixed: its methods require of practitioners an inherently self-critical conception of intellectual activity, one that is continuously challenging its own preconceptions and guiding principles both as to the nature and purposes of philosophy and of sports.

1.3. Body of Knowledge

Being a form of philosophical discourse, the philosophy of sport embodies the formal and contextual character of philosophy broadly defined. Unlike the natural or biomedical sciences (of sport), philosophers are more apt to generate research that is overtly reflective of its non-theory neutrality. Just as with the humanities and social sciences of sport, intellectual progress can be made in philosophy without presupposing an idea of linear development - or at least largely shared view of cumulative, commensurable knowledge - that is assumed within the natural or biomedical sciences of sport.

The philosophy of sport then, is characterised by conceptual investigations into the nature of sport and related concepts, areas and professions. It draws upon and develops many of the diverse branches of the parent discipline, philosophy, and reflects a broad church of theoretical positions and styles. It has most specifically interrogated substantive issues in the following sub-fields of philosophy as exemplified within sport and related human activities involving the use of the body in human practices and institutions:

- aesthetics (e.g. can aesthetic sports have objective judging?);
- epistemology (e.g. what does knowing a technique entail?);
- ethics (e.g. what, if anything, is wrong with gene doping?);
- logic (e.g. are constitutive and regulative rules distinct?);
- metaphysics (e.g. are humans naturally game playing animals?);
- philosophy of education (e.g. do dominant models of skill-learning respect phenomenological insights?);
- philosophy of law (e.g. can children give consent to performance enhancing drugs?);
- philosophy of mind (e.g. is mental training distinguishable from mere imagination?);
- philosophy of rules (e.g. are regulative rules not moral in character?);
- philosophy of science (e.g. is it true that only natural sciences of sport deliver the truth?), and
- social and political philosophy (e.g. are competitive sports a hostage to a capitalist world-view?).

Within these diverse fields, there has been a tendency for one philosophical tradition to dominate: analytical philosophy. This is not to deny that continental philosophy has not developed a sport philosophical literature. Indeed, the labels themselves are somewhat misleading – and both, being traditions of Western philosophy, take no significant account of Eastern philosophy, which has spawned a significant volume of sport philosophical literature, notably in Japan.

Given that philosophical research is always and everywhere internally related to the expression of ideas, the idiom of that expression somewhat shapes the boundaries of what can be said. In contrast to the idea that the biomedical sciences of sport represent a universal language housed in technical rationality ("the" scientific method), philosophers working in the continental tradition have largely developed research within the fields of existentialism, hermeneutics and phenomenology. Although the label is itself driven by geographical considerations (the work emanated from communities of scholars in France, Germany and Continental Europe), one finds philosophers of sport right across the globe drawing upon those traditions. Similarly, analytical philosophy, though the dominant Anglo-American tradition of Western Philosophy is misleading in the sense that some of its founding fathers were indeed from Continental Europe. The drawing of distinctions to represent our experience of the world, however, is common to all schools or traditions of philosophical and sport philosophical endeavour. Given the dominance of the analytic tradition – and the English-speaking counterparts of it – a few more specific words are required in order to make sense of recent developments in the philosophy of sport.

Analytical philosophy emerged as an essentially conceptual enquiry whose aim was foundational. It is often captured in Locke's famous remark about philosophical work being akin to an underlabourer working in the garden of knowledge. As a second-order activity, its central aim was to provide secure foundations for other disciplines by articulating their conceptual geography. Its pre-eminence was captured by the insistence that conceptual work precedes all proper empirical enquiry. Its exponents were equipped with the analytical tools of dissecting concepts for constituent criteria, drawing conceptual distinctions by their logical grammar and seeking fine-grained differences in their employment. The discipline of philosophy reduced in some quarters to the detailing of ordinary linguistic usages and necessary and sufficient conditions in order to detect the proper meaning of concepts others had to operate with and between. Despite this "new" direction, there remained a strong sense of continuity here with the ancient past. Philosophers such as Plato and Aristotle were also concerned with marking distinctions, bringing clarity where before there was puzzlement or, worse, commonsensical acquiescence.

Many philosophers argue now that we are in a period of post-analytical philosophy. What this means is not entirely clear. We are living through a period of exciting intellectual development in the subject, which is very much reflected in the Philosophy of Sport. While careful attention to conceptual analysis will always be an essential component of the philosophers' toolkit, research-driven analyses of the key concepts of sports, games and play, have to a clear extent declined. Of much greater prevalence in the contemporary literature has been the development of substantive axiological issues ranging from social and political philosophy of sport to the rapidly growing field of ethics of sport. Philosophers have been clear about the need to throw off the cloak of apparent neutrality of analytical philosophy in favour of arguing for substantive positions in terms of the "commodification" of sports, their "commercialistion" and their "corruption". The development of substantive normative positions has proceeded in addition, rather than in opposition, to the careful articulation of precisely what those concepts logically entail. If these debates have also raged in the social scientific literatures, then it is clear that academics in this portion of the philosophy of sport have made their own important contributions, premised on a clear understanding of the potentially diverse conceptualisations of sport. Similarly, in ethics, philosophers of sport have attempted to argue for the aptness of different moral philosophical theories to capture sports' nature and the nature of sporting actions therein. In these fields, philosophers have generated new ideas about the contested nature of sports ethics itself – whether as contract, or duty/obligation, or utility, or virtue. In doing so, they have often connected with the empirical research of other bodies of knowledge that would have been unimaginable to the "ordinary language philosophers' who saw themselves neutrally dissecting the linguistic usage of others through much of the last fifty years.

1.4. Methodology

Although early analytical philosophers saw themselves elucidating the concepts others used in their sports talk and research, there is a clear sense in which we can say the empirical researchers of the natural and social sciences and the humanities have themselves become much more sophisticated in their conceptual approaches to sports related research. One of the traditional roles of the philosophers of sport, that is, to clear the conceptual ground for others to carry out their research, has diminished, though it is never likely to disappear altogether. In politics, as in ethics and other branches of study, there will always be disputes about what constitutes "democratic processes" or "good character", for these debates cannot be eliminated from the field itself. Yet the convergence of the conceptual and empirical cuts both ways. Philosophers of sport themselves are paying much greater attention to the processes and outcomes of empirical research. Nevertheless, their focus remains exclusively conceptual in character. Every philosopher worthy of the name still seeks to get things right – even if there is no clear and undisputed sense of what the truth of matters might be. Its task is, through dialogue, to aim at the truth by close attention to valid argumentation entailing the clear explication of ideas that aim towards truth. In this sense, philosophy does not try to be pure, nor do philosophers of sport attempt to view sports as if they were in a position of complete neutrality, as is presupposed in positivistic research. The old philosophical ideal of philosopher as an ideal spectator embodies a view of sports worlds from nowhere in particular within those worlds. Such a view has largely disappeared in contemporary philosophy of sport. In a clear sense, then, philosophy is returning to its ancient promise to bring wisdom to bear on important matters that concern us (in sports) and not merely to the detailed technical analysis of key concepts.

1.5. Relationship to Practice

The diversity of practices that fall within the compass of different schools and traditions of philosophy means that there is not a universal method to characterise the philosophy of sport. It is impossible, therefore, to state unequivocally what relations hold between philosophising and practice. While there will always be a portion of philosophical scholarship in sport that is more abstract (whether in the analytical, Continental or Eastern traditions), there is a growth of more applied work in the fields of axiology. Increasingly, philosophers are making contributions to national and international sports policy development, along with pressure groups where the need for the knowledge and skills of argumentation philosophers characteristically bring to bear on challenging normative issues is clear. Examples of such applied work include research into diverse conceptions of equity in operation with respect to categories such as gender and race; arbitrating between proper and improper means of performance enhancement and genetic engineering; linking the shared terrains between philosophy of sport and other social practices such as medicine or other key identity constituting aspects such as disability; and illuminating the fascistic tendencies of elite sports or the xenophobia of modern sporting nationalism. Many of these issues would have been unthinkable to philosophers fifty years ago but are increasingly becoming part of the standard work of philosophers of sport.

1.6. Future Perspectives

Given the breadth of scholarship and research across the Association, it is not possible to specify definitive directions that will apply to all future research. Recent special issues within the field's journals interrogate issues in the ethics of sports medicine; disability sports; Olympism as a philosophy of sport and life; and a consideration of the scholarship of Bernard Suits whose ground breaking book "The Grasshopper" in 1967 significantly helped to establish the field.

2. Information Sources

2.1. Journals

Philosophers of sport have tended to publish their research in a wide variety of outlets from scientific to professional journals. Many philosophers have published their work in national and international multi-disciplinary journals on sport. Equally, it is very common for philosophers to publish in national and international social scientific sports journals. IAPS' own publication is the Journal of the Philosophy of Sport (humankinetics.com/JPS/journalAbout.cfm), which has been published annually since 1974, and bi-annually since 2001. It is currently edited by Prof. John Russell (jrussell@langara.bc.ca). In addition, the British Philosophy of Sport Association has produced its own Journal, *Sport, Ethics and Philosophy* since 2007, which is published tri-annually. The middle issue of each year is devoted to an area of particular interest or a dedicated monograph. (www.tandf.co.uk/journals/authors/rsepauth.asp). Its editor is Dr Mike McNamee (m.j.mcnamee@swansea.ac.uk). Both journals consider the full range of philosophic issues pertinent to sport, irrespective of the school of thought from which it emerges. They are tightly refereed and internationally indexed.

2.2. Reference Books, Encyclopaedias, etc.

The philosophical literature concerning sport is extensive. Historically important and contemporary books in the field notably include the following:

Abe, S. (Ed.). (1977). *The Philosophy of Physical Education*, (sixth edition). Tokyo: Shoyo Shoin.

Arnold, P.J. (1988). *Education, Movement and the Curriculum.* Lewes: Falmer.

Arnold, P.J. (1977). *Meaning in Movement,* Sport and Physical Education. London: Heinemann.

Best, D. (1974). *Expression in Movement and the Art.* London: Lepus.

Best, D. (1978). *Philosophy and Human Movement.* London: Allen and Unwin.

Brohm, J.-M. (1978). Sport - *A Prison of Measured Time.* London: Ink Links.

Caillois, R. (1961). *Man, Play, and Games.* New York: Free Press.

Cantelon, H. & Gruneau, R.S. (Eds.)(1982). *Sport, Culture and the Modern State.* Toronto: University of Toronto Press.

Davis, E.C. & Miller, D.M. (1961). *The Philosophic Process in Physical Education* (Second edition). Iowa: W C Brown.

Fraleigh, W.P. (1984). *Right Actions in Sport: Ethics for Contestants.* Champaign, Ill: Human Kinetics.

Galasso, P.J. (Ed.). (1988). *Philosophy of Sport and Physical Activity Issues and Concepts.* Toronto: Canadian Scholarsbb Press.

Gerber, E. W. and Morgan, W.J. (Eds.). (1979). *Sport and the Body: A Philosophical Symposium* (2nd ed.). Philadelphia: Lea and Febiger.

Gruneau, R.S. (1999). *Class, Sports, and Social Development,* (second edition). Champaign, Ill: Human Kinetics.

Grupe, O. and Dietmar, M. (1988). *Lexikon der Ethik im Sport.* Verlag Karl Hofmann: Schorndorf.

Haag, H. (1991). *Sportphilosophie.* Frankfurt: Diesterweg-Sauerländer.

Haag, H. (Ed.). (1996). *Handbuch Sportphilosophie.* Schondorf: Verlag K. Hofmann.

Hargreaves, J. (Ed.). (1991). *Sport, Culture and Ideology.* Cambridge: Polity.

Harper, W.A., Miller, D.M., Park, R.J & Davis, E.C. (1977). T*he Philosophic Process in Physical Education,* (3rd ed.). Philadelphia: Lea and Febiger.

Hoberman, J.M. (1984). *Sport and Political Ideology.* Austin: University of Texas Press.

Huizinga, J. (1970). *Homo Ludens: A Study of the Play Element in Culture.* London: Paladin.

Hyland, D.A. (1994). *Philosophy of Sport.* Maryland: University Press of America.

Hyland, D.A. (1984). *The Question of Play.* Washington: University Press of America.

Keating, J.W. (1978). *Competition and Playful Activities.* Washington: University Press of America.

Kleinman, S. (Ed.). (1986). *Mind and Body: East Meets West.* Champaign: Human Kinetics.

Kretchmar, R.S. (1994). *Practical Philosophy of Sport.* Champaign: Human Kinetics.

Landry, F. and Orban, W.A.R. (Eds.). (1978). *Philosophy, Theology and History of Sport and of Physical Activity.* Quebec: Symposia Specialists.

Lenk, H. (1969). *Social Philosophy of Athletics.* Illinois: Stipes Publishing.

Lenk, H. (Ed.). (1983). *Topical Problems of Sport.* Schorndorf: Verlag Karl Hofmann.

Loland, S. (2001). *Fair Play: a moral norm system.* London: Routledge.

Loland, S., Skirstad, B. and Waddington, I. (Eds.). (2005). *Ethics, Pain and Sport,* London: Routledge.

Lumpkin, A., Stoll, S.K. and Beller, J.M. (1999). *Sport Ethics: Applications for Fair Play.* (second edition). Boston: McGraw Hill.

McFee, G. (2003) *Sport, Rules and Values,* London: Routledge.

McIntosh, P.C. (1978). *Fair Play: Ethics in Sport and Education.* London: Heinemann.

McNamee, M.J. and Parry, S.J. (Eds). (1998). *Ethics and Sport.* London: Routledge.

McNamee, M.J. (2005). *Philosophy and the sciences of exercise, health and sport,* London: Routledge.

McNamee, M.J., Olivier, S and Wainwright, P. (2006). *Research ethics in exercise, health and sport sciences,* London: Routledge.

McNamee, M.J. (2007). *Philosophy, risk and adventure sports,* London: Routledge.

McNamee, M.J. (2008). *Sports, Virtues and Vices,* London: Routledge.

Metheny, E. (1965). *Connotations of Movement in Sport and Dance.* Iowa: W C Brown.

Metheny, E. (1968). *Movement and Meaning.* New York: McGraw Hill.

Miah, A. (2004). *Genetically Engineered Athletes,* London: Routledge.

Mihalich, J.C. (1982). *Sports and Athletics: Philosophy in Action.* Totowa: Littlefield Adams.

Morgan, W.J. (1994). *Leftist Theories of Sport: A Critique and Reconstruction. Urbana:* University of Illinois Press.

Morgan, W.J. (Ed.). (1979). *Sport and the Humanities: A Collection of Original Essays.* Knoxville: University of Tennessee Press.

Morgan, W.J. and Meier, K.V. (Eds.). (1988). *Philosophic Inquiry in Sport.* Champaign: Human Kinetics.

Morgan, W.J. (Ed.). (2007). *Ethics in sports,* Champaign; IL: Human Kinetics (2nd edition)

Morgan, W.J. (2006). *Why sports morally matter,* London: Routledge.

Osterhoudt, R.G. (1978). *An Introduction to the Philosophy of Physical Education and Sport.* Champaign, Ill: Stipes.

Osterhoudt, R.G. (1991). *The Philosophy of Sport: An Overview.* Champaign, Ill: Stipes.

Postow, B.C. (Ed.). (1983). *Women, Philosophy, and Sport.* New York: Scarecrow Press.

Rigauer, B. (1982). *Sport and Work.* New York: University of Columbia Press.

Sheridan, H., Howe, L. and Thompson, K. (2007). Sporting reflections, Aachen: Meyer and Meyer.

Simon, R.L. (1991). *Fair Play: Sports, Values, and Society.* Colorado: Westview Press.

Simon, R.L. (1996). *Sports and Social Values.* Colorado: Westview Press.

Slusher, H.S. (1967). *Man, Sport and Existence: A Critical Analysis.* Philadelphia: Lea and Febiger.

Suits, B. (1978). *The Grasshopper; Games, Life and Utopia.* Toronto: University of Toronto Press.

Tambooer, J. and Steenbergen, J. (2000). *Sport Filosofie.* Leende: Davon.

Tamburrini, C. (2000). *The "Hand of God": Essays in the philosophy of Sports.* Gothenburg: University of Gothenburg Press.

Tamburrini, C and Tannsjo, T (Eds.). (2005). *The Genetic design of winners,* London: Routledge.

Tannsjo, T. and Tamburrini, C. (2000). *Values in Sport.* London: Routledge

Thomas, C.E. (1983). *Sport in a Philosophic Context.* Philadelphia: Lea and Febiger.

Vanderwerken, D.L. and Wertz, S.K. (Eds.) YEAR. *Sport Inside Out: Readings in Literature and Philosophy.*

Vander Zwaag, H.J. (1985). *Toward a Philosophy of Sport. Fort Worth:* University of Texas Press.

Volkwein Caplan, K. (2004). *Sport, Culture and Physical Activity,* Aachen: Meyer and Meyer.

Walsh, A. and Guillianotti, R. (2006). *Commercialism, Commodification and the*

Corruption of Sport, London: Routledge.

Webster, R.W. (1965). *Philosophy of Physical Education.* Iowa: W C Brown.

Weiss, P. (1969). *Sport: A Philosophic Inquiry. Carbondale*: Southern Illinois University Press.

Wertz, S.K. (1994). *Talking a Good Game: Inquiries into the Principles of Sport,* Texas: Southern Methodist University Press.

Zeigler, E.F. (1984). *Ethics and Morality in Sport and Physical Education* (second edition). Englewood Cliffs: Prentice Hall.

Zeigler, E.F. (1977). *Physical Education and Sport Philosophy.* Englewood Cliffs: Prentice Hall.

Zeigler, E.F. (1968). *Problems in the History and Philosophy of Physical Education and Sport.* New Jersey: Prentice Hall.

2.3. Book Series

A series on philosophical and social scientific ethics of sport is edited by McNamee, M. J. and Parry, S. J. under the title "Ethics and Sport" and is published by Routledge. A new book series titled Sport, Culture and Society has been published by Meyer and Meyer Sport, Aachen, Germany since 2000. This series, developed by and edited by K. Volkwein (USA), K. Gilbert (Australia), and O. Schantz (France), is interdisciplinary and cross-cultural in nature, including a sport philosophical approach, and focuses on current and controversial topics in sport in the world.

2.4. Congress/Workshop Proceedings

Proceedings of some Annual Meetings of the International Association for the Philosophy of Sport (IAPS), formerly Philosophic Society for the Study of Sport, have been published in various forms on an irregular schedule since 1973. IAPS has also published a newsletter, IAPS News, tri-annually since 1987.

2.5. Data Banks

IAPS has published several versions of a comprehensive bibliography concerning the philosophy of sport in its Journal of the Philosophy of Sport. This bibliography is periodically updated.

2.6. Internet Sources

Recommended Websites for philosophy:

The IAPS web site, containing membership, journal and conference information
    http://iaps.glos.ac.uk/
University of Idaho, Center for Ethics
    www.educ.uidaho.edu/center_for_ethics/
The Philosopher's Magazine
    www.philosophers.co.uk
Papers from the Twentieth World Congress of Philosophy, in Boston, Massachu-
setts from August 10-15, 1998
    www.bu.edu/wcp/MainSpor.htm
EpistemeLinks.com – Philosophy Resources
    www.epistemelinks.com/
Guidebook for Publishing/ Philosophy: Journals
    www.smith.edu/~jmoulton/jend.htm
British Philosophy of Sport Association
    www.philosophyofsport.org.uk/

3. Organisational Network

3.1. International Level

IAPS has world-wide membership. It has both held meetings and/or has had rep-
resentation regularly at the World Congress of Philosophy. It is a member of the
International Council of Sport Science and Physical Education (ICSSPE) and the
Fédération Internationale des Sociétés de Philosophie.

3.2. Regional Level

Regional/continental affiliates of the IAPS are now under development in East Asia,
Oceania/Southeast Asia, Central/Southern/Eastern Europe, United Kingdom/Ire-
land and Scandinavia.

3.3. National Level

Japan has a long standing, formally developed, national organisation devoted to the philosophy of sport. The British Philosophy of Sport Association was instituted in 2002 (www.philosophyofsport.org.uk/) which, because of the active involvement of scholars across Continental Europe, spawned the European Association for the Philosophy of Sport, whose constitution is subject to ratification at its meeting in 2008. In North America, regular meetings have been held in Canada and the USA. In the United States, the American Alliance for Health, Physical Education, Recreation and Dance (AAHPERD) has a chapter in their organisation devoted to Sport Philosophy - the Academy of Sport Philosophy under NASPE (National Association for Sport and Physical Education). The Academy of Sport Philosophy is represented with lectures, workshops and symposia during the annual AAHPERD meetings and publishes in the AAHPERD sponsored journals *Research Quarterly and JOPERD (Journal of Physical Education, Recreation and Dance)*. In Germany, there is a Sektion Sportphilosophie of the Deutsche Vereinigung für Sportwissenschaft" (DVS). Other countries have similar organisational patterns at the National Level.

3.4. Specialised Centres

The major centres for advanced study and research (in alphabetical order) in the philosophy of sport are:

- Department of Exercise and Sport Science, Pennsylvania State University, University Park, Pennsylvania, USA.
- Department of Philosophy, University of Leeds, England.
- Department of Philosophy, Trinity College, Hartford, Connecticut, USA.
- Department of Philosophy, Texas Christian University, Fort Worth, Texas, USA.
- Department of Philosophy, History and Law, School of Health Science, Swansea University, Wales, UK.
- Department of Physical Education and Recreation, Victoria University of Technology, Melbourne, Victoria, Australia.
- Department of Physical Education and Sport, State University of New York College at Brockport, Brockport, New York, USA.
- Deutsche Sporthochschule Köln, Cologne, Germany.
- Division of Health, Physical Education, Recreation, and Dance, Center for Eth-

ics, University of Idaho, Moscow, Idaho, USA.
- Faculty of Human Movement Sciences, Vrije Universiteit, Amsterdam, the Netherlands.
- Faculty of Kinesiology, University of Western Ontario, London, Ontario, Canada.
- nstitut für Sportwissenschaft, Freie Universität Berlin, Berlin, Germany.
- Institute of Health and Sport Sciences, University of Tsukuba, Tsukuba, Japan.
- Leisure and Sport Research Unit/School of Sport and Leisure, University of Gloucestershire, England.
- Nippon College of Education, Tokyo, Japan.
- Norwegian University of Sport Sciences, Oslo, Norway.
- School of Sport, University of Wales Institute Cardiff, United Kingdom.
- Stockholm Bioethics Centre, University of Stockholm, Sweden.

3.5. Specialised International Degree Programmes

A collaborative program called a Master's of Arts in Physical Education has been established between Bienville University (USA) and Christian-Albrechts-University of Kiel (Germany). The goal of this program is to provide students the unique opportunity to be educated from two socio-cultural perspectives in the same field of study. This program includes education in the philosophy of sport.

Additionally, topical seminars in sport philosophy at Doctoral level are offered infrequently for scholars from around the world through the Norwegian University of Sport and Physical Education, Oslo, Norway.

4. Appendix Material

4.1. Terminology

See section 2.6 Internet Sources for more information.

4.2. Position Statement(s)

The purpose of the International Association for the Philosophy of Sport is: to stimulate, encourage and promote study, research and writing in the philosophy of sporting (and related) activity; to demonstrate the relevance of philosophic thought concerning sport to matters of professional concerns; to organise and conduct meetings concerning the philosophy of sport; to support and to cooperate with local, national and international organisations of similar purpose; to affiliate with national and international organisations of similar purpose; and to engender national, regional and continental affiliates devoted to the philosophic study of sport (from the Constitution of IAPS).

4.3. Varia

Not applicable.

4.4. Free Statement

Not applicable.

Sociology of Sport

Joseph Maguire

Contact

Prof. Dr. Joseph Maguire
School of Sport & Exercise Sciences
Loughborough University
LE11 3TU Loughborough
United Kingdom
Phone: +44 1509 223328
Fax: +44 1509 223971
Email: J.A.Maguire@lboro.ac.uk

1. General Information

The sociology of sport, while grounded in sociology, also encompasses research in history, political science, social geography, anthropology, social psychology and economics. Also, the new off-shoots of sociology such as, cultural studies, postmodernism, media studies and gender studies, are well represented in the field. Sociology of sport is both a theoretically driven, and an empirically grounded sub-discipline. It overlaps with, and is informed by, work on the body, culture and society more broadly.

1.1. Function

The aims of the sociology of sport are:
- to critically examine the role, function and meaning of sport in the lives of people and the societies they form;
- to describe and explain the emergence and diffusion of sport over time and across different societies;
- to identify the processes of socialisation into, through and out of modern sport;
- to investigate the values and norms of dominant, emergent and residual cultures and subcultures in sport;
- to explore how the exercise of power and the stratified nature of societies place limits and possibilities on people's involvement and success in sport as performers, officials, spectators, workers or consumers;
- to examine the way in which sport responds to social changes in the larger society;
- to contribute both to the knowledge base of sociology more generally and also to the formation of policy that seeks to ensure that global sport processes are less wasteful of lives and resources.

The sociology of sport also seeks to critically examine common sense views about the role, function and meaning that sport has in different societies. By challenging 'natural' and taken-for-granted views about sport, sociologists seek to provide a more social and scientifically adequate account that can inform both the decisions and actions of people and the policy of governments, non-government organisations (NGOs) and sport organisations.

Although, as in sociology more generally, there are several different perspectives from which to examine the relationship between sport, cultures and societies, sociologists of sport do have certain assumptions in common. For example, sociologists, whether they examine the 'micro' or 'macro' aspects of sport, seek to embed their research in the wider cultural and structural context.

In the context of sport sciences, sociologists of sport seek to generate knowledge that will contribute to 'human development' as opposed to 'performance efficiency'. That is, they seek to critically examine the costs, benefits, limits and possibilities of modern sport for all those involved, rather than focus on the performance efficiency of elite athletes. Those sociologists working with sociology departments examine sport much in the same way they would examine religion, law or medicine - to highlight aspects of the general human condition.

Sociology of sport, then, seeks not only to contribute to its parent discipline, but also to changing the sports world. With respect to the latter, research seeks to debunk popular myths about sport, critically appraise the actions of those more powerful groups involved in sport and critique and inform social policy about sport.

1.2. Body of Knowledge

Although the first texts on sociology of sport appeared in the 1920s, this sub-discipline did not develop until the early/mid 1960s in Europe and North America. A small number of scholars from both physical education and sociology formed the International Committee for the Sociology of Sport (ICSS) in 1965. From this point on, symposia, conferences and congresses were held annually and theoretical and empirical work was presented. Researchers from different sociological backgrounds began to develop sociological definitions of sport, conduct pioneering work in different aspects of sport and develop undergraduate, Master's and doctoral courses and programs. Research areas include: sport and socialisation; sport and social stratification; sport subcultures; the political economy of sport; sport and deviance; sport and the media; sport, the body and the emotions; sport violence; sport politics and national identity; and sport and globalisation.

On this basis, the sub-discipline has now developed a sophisticated understanding of how people become involved in sport; what barriers they face; and how gender, class, ethnicity and sexual relations work in sport. In addition, scholars have developed considerable knowledge about how sport is mediated, contoured by a complex political economy and bound up in global identity politics.

Over the last thirty years, theoretical and empirically based case studies have been developed on various sports in different societies. The sub-discipline has various edited works, handbooks and textbooks from North America and Europe. Sociology of sport has also been established in Asia and more recently scholars from South America and Africa are using a sociological perspective to help make sense of the social problems that beset sport and to understand how sport illuminates wider sociological issues.

1.3. Methodology

The range of research methodologies used in the sociology of sport are the same as those used in sociology and in other social sciences, and are frequently characterised as qualitative or quantitative methodologies. Preferred methodologies have changed over time and vary by place. Perhaps more importantly, methodologies are often quite distinctly related to the theoretical perspective employed by the researcher. For example, those perspectives that tend to treat data as social facts (e.g., functionalist and Durkheimian approaches) tend to employ quantitative methodologies, e.g. survey research, content analysis and statistical analysis. Those perspectives that see social data in more relative terms (e.g., symbolic, interactionist and postmodern approaches) tend to employ more qualitative methodologies, e.g. discourse analysis, ethnography and interviews. Many of the critical, Eliasian/figurational and cultural studies perspectives employ a range of methodologies, and select those methods as appropriate to the data being collected, including historical methodologies. In their simplest form, multiple methodologies are employed in order to confirm the reliability of data (e.g., Denzin's triangulation technique) and are more often employed because of the complexity of social data and the recognition that a single method, such as a questionnaire, provides limited insight, in itself, to complex social behaviour.

1.4. Relationship to Practice

Sociology of sport, as noted, seeks to contribute to our understanding of sport and also to inform policy that will make the sports experience less wasteful of lives and resources. Sociologists of sport have sought to achieve this latter aim in several ways:

- by offering expert advice to government agencies, public enquiries and commission reports on areas such as drugs, violence and health education;
- by acting as an advocate for athletes' rights and responsibilities;
- by providing research for groups who seek to challenge inequalities of gender, class, ethnicity, age and disability, particularly with respect to access, resources and status;
- by promoting human development as opposed to performance efficiency models within physical education and sport science; and
- by encouraging better use of human and environmental resources and thus ensuring that there is a sporting future for generations to come.

2. Information Sources

2.1. Journals

The number of journals continues to grow in this area. The following journals are either devoted to sociology of sport or contain a high proportion of papers written from a sociological perspective:

International Review for the Sociology of Sport (IRSS), quarterly;
Sociology of Sport Journal (SSJ), quarterly;
Journal of Sport and Social Issues (JSSI), quarterly;
Sport in Society, quarterly;
Leisure Studies;
Japanese Journal of Sociology of Sport
Soccer and Society
Football Studies

2.2. Reference Books, Encyclopaedias, etc.

Coakley, J. and Dunning, E (Eds.). (2000). *Handbook Of Sport Studies. London*: Sage.

Levinson, D. and Christensen, K. (Eds.). (1996). Encyclopedia Of World Sport: *From Ancient Times To The Present.* Santa Barbara: Abc-Clio. 796.03.

Oglesby, Carole (Ed.). (1998). *Encyclopedia Of Women And Sport In America.* Phoenix: Oryx.

Boyle, R. and Haynes, R. (2000). *Power Play: Sport, the media and popular culture.* London: Longman.

Brohm, J.-M. and Marc Perelman (2006). *Le Football une peste Emotionnelle.* Paris: Folioactuel Inedit.

Cashmore, E. (2005). *Making Sense of Sports*, 4th edition. London: Routledge. 306.483.

Coakley, J. (2007). *Sport in Society*. 9th edition. Boston, Ma: McGraw-Hill.

Digel, H. (1995). *Sport in a Changing Society. Sport Science Studies* Volume 7. Schorndorf, Germany: Verlag Karl Hofmann.

Elias, N. and Dunning, E. (Eds.). (1986). *Quest for Excitement.* Oxford: Basil Blackwell.

Gruneau, R. (1999). *Class, Sports, and Social Development.* Champaign, Il: Human Kinetics.

Guttmann, A. (1978). *From Ritual to Record: The nature of modern sports.* Columbia University Press.

Jarvie, G. and Maguire, J. (1994). *Sport and Leisure in Social Thought*, London: Routledge.

Loy, J.W., Kenyon, G.S. and McPherson, B.D. (Eds.). (1981). *Sport, Culture and Society: A reader on the sociology of sport.* Philadelphia: Lea and Febiger.

Maguire, J. (2005). Power and Global Sport: Zones of Prestige, Emulation and Resistance. Routledge: London.

Maguire, J. and Young, K. (Eds.). (2002). *Research in the Sociology of Sport* Elsevier Press.

Maguire, J. and Nakayama M. (Eds.). (2006). Japan, Sport and Society: Tradition and Change in a Globalizing World London: Routledge.

Ohl, F. (Ed.). (2006). Sociologie du sport. *Perspectives Internationales et mondialisation*, Paris: PUF.

Putnam, D. (1999). *Controversies of the Sports World*. London: Greenwood
    Press.
Rigauer, B. (1969). *Sport and Work*. New York: Columbia University Press.
Rowe, D. (2004). *Sport, Culture and the Media*, 2nd edition, Maidenhead, UK:
    Open University Press.
Rowe, D. (Ed.). (2004). *Critical Readings: Sport, culture and the media*. Maiden-
    head, UK: Open University Press.
Scambler, G. (2005). *Sport and Society: History, power and culture*. Maidenhead:
    Open University Press.

2.3. Book Series

Jennifer Hargreaves and Ian McDonald (Eds). Routledge, Critical Studies in Sport.

2.4. Congress/Workshop Proceedings

First World Congress for the Sociology of Sport, Seoul, Korea, 2001.
Second World Congress for the Sociology of Sport, Koln, Germany, 2003.
Third World Congress for the Sociology of Sport, Buenos Aires, Argentina, 2005.
Fourth World Congress for the Sociology of Sport, Copenhagen, Denmark, 2007.
Fifth World Congress for the Sociology of Sport, Kyoto, Japan, 2008.

2.5. Data Banks

SportDiscus
    www.sirc.ca/products/sportdiscus.cfm

2.6. Internet Sources

Listservs, web sites and bibliographical search centres are developing rapidly. The
official internet site for ISSA is www.issa.otago.ac.nz/links.php

For more information, please contact the following e-mail addresses:
isa@sis.ucm.es
sporthist@pdomain.uwindsor.ca
nassserv@listserv.bc.edu

Laboratoire de recherche Activités Physiques et Sportives et Sciences Sociales
  http://u2.u-strasbg.fr/laboaps/index.htm
International Sport and Culture Association
  www.isca-web.org
International Society for the History of Physical Education and Sport (ISHPES)
  http://www.ishpes.org/
Fédération Internationale d'Education Physique (FIEP)
  www.fiep.net/index.asp?l=en&i=13
International Society for Comparative Physical Education and Sport (ISCPES)
  www.iscpes.org/

3. Organisational Network

3.1. International Level

The sociology of sport is internationally represented by the International Sociology
of Sport Association (ISSA, formerly ICSS, founded in 1965), which also publishes
the *International Review for the Sociology of Sport*. This body is a research committee of the International Sociological Association (ISA) and also an official member of ICSSPE's Associations Board.

At present, there are 200 members from different parts of the globe. ISSA holds
annual conferences, including congresses in conjunction with the World Congress
of Sociology and its own World Congresses held in Korea, Germany, Argentina and
in 2007, in Copenhagen, Denmark.

As the international 'umbrella' organisation, ISSA consults with national and regional groups. Some national groups are federated with the national sociological
association of that country, or with a sport science /physical education organisation. Either through ISA or ICSSPE, such groups have a direct link to ISSA. There
are also regional groups in areas such as Asia and North America. NASSS, the
North American Society for the Sociology of Sport, which publishes the *Sociology
of Sport Journal*, is the most well known regional group. European researchers
are also linked to the European College of Sport Science and the recently formed
European Association for Sociology of Sport.

ISSA Executive Board, 2008-2011

President
Professor Steven Jackson
School of Physical Education
University of Otago
NEW ZEALAND
Email: steve.jackson@otago.ac.nz

General Secretary
Dr Elizabeth Pike
School of Sport, Exercise and Health Sciences
University of Chichester
UNITED KINGDOM
Email: e.pike@chi.ac.uk

Vice President (Conferences)
Associate Professor Chris Hallinan
School of Human Movement, Recreation and Performance
Victoria University; Melbourne
AUSTRALIA
Email: Christopher.Hallinan@vu.edu.au

Vice President (ISA)
Professor Fabien Ohl
Faculté des SSP – ISSEP
Université de Lausanne
SWITZERLAND
Email: fabien.ohl@citycable.ch

Vice President (Promotions and Awards)
Dr Kimberly Schimmel
School of Exercise, Leisure and Sport
Kent State University, Ohio
UNITED STATES OF AMERICA
Email: kschimme@kent.edu

Past President and Vice President (ICSSPE)
Professor Gertrud Pfister
Institute of Exercise and Sport Sciences
University of Copenhagen
DENMARK
Email: gpfister@ifi.ku.dk

3.2. Regional Level

See section 3.1. International Level.

3.3. National Level

Each country has its own sociological association. Sociology of sport groups exist as either part of these parent discipline associations and/or in conjunction with sport science organisations.

3.4. Specialised Centres

The University of Waterloo in Canada was the first prominent centre to be associated with the sociology of sport. There are now some long established research centres in North America (e.g., North Eastern University in Boston, University of Illinois); Europe (e.g., the Norwegian University of Sport, Oslo, and the University of Jyväskylä, Finland); and Asia (e.g. Tsukuba University, Japan, Seoul National University, Korea). For the most part, research tends to be conducted by smaller groups of scholars, sometimes working as individuals. Centres of excellence include Loughborough University, England, the University of Toronto, Canada, the Norwegian University of Sport and the University of Otago, New Zealand.

3.5. Specialised International Degree Programmes

Not applicable.

4. Appendix Material

4.1. Terminology

See any standard dictionary of sociology.
See also:

Cashmore, E. (2000). *Sports Culture.* An A-Z Guide. London: Routledge.
D. Levinson and K. Christensen (Eds.). (1990). *Encyclopaedia of World Sport* (3
    volumes.). Oxford: ABC-CIIO.

4.2. Position Statement(s)

See ISSA documentation on role and function of ISSA. ISSA has also endorsed the
Brighton Declaration on Women and Sport.

4.3. Varia

Not applicable.

4.4. Free Statement

The three main tasks of ISSA are service, advocacy and research. ISSA seeks to
provide a service to members, to represent the interests of sociologists and to
develop the field in areas where, at present, the state of sociological knowledge
about sport is in its infancy. ISSA is also about advocacy. ISSA attempts to intervene in the Sport Worlds of today – using what influence it has to make such
worlds less wasteful of lives and resources. ISSA is also about research – it seeks
to create a knowledge base on which to build future Sport Worlds that can be
similar to today or which can be made anew. Such worlds can enhance the positive aspects of contemporary Sport Worlds, or they can reinforce, or make worse,
what we already experience as negative features. Sociologists of sport therefore
have a part to play in approaches to development through sport and in wider
United Nations initiatives on sport, culture and society.

Sport and Exercise Physiology

Ian Stewart

Contact

Dr. Ian Stewart
Institute of Health and Biomedical Innovation,
Queensland
University of Technology.
Email: i.stewart@qut.edu.au

1.General Information

1.1. Function

Physiology is a discipline of the biological sciences dealing with the function of living organisms and their parts. The study of physiology depends on, and is permeated with, other biological science disciplines such as anatomy, biochemistry, molecular biology and biophysics. This interdependence is based on the fact that the human body follows the natural laws of structure and function, which fall within the domain of these disciplines.

Exercise physiology is a division of physiology that focuses on the functioning of the body during exercise. Physiological responses to exercise depend on the intensity, duration, frequency and modality of the exercise, as well as the interacting environmental circumstances, diet, health and physiological status of the individual.

1.2. Body of Knowledge

The origins of exercise physiology can be dated to the Greek physicians, Herodicus, Hippocrates and Galen, with their work on diet, health, hygiene and physical training. Indeed Galen wrote detailed descriptions of the appropriate intensity of exercise 200 years before the birth of Christ and his work influenced the early anatomists, physicians and physiologists. Throughout the next 2000 years, the work of these generalist scientists and their interest in sport and exercise gave birth to the specific discipline of exercise physiology. There are numerous scientists and physicians who have been influential in the development of the discipline, indeed too many to name here, therefore the reader is referred to McArdle, Katch and Katch's publication *Exercise Physiology* or Brooks, Fahey and Baldwin's *Exercise Physiology* (see section 2.2 Reference Books, Encyclopaedias) for a detailed historical overview.

The body of knowledge derives from its founding disciplines and as such, the initial research articles were published in physiological publications: the American Journal of Physiology (1898-); Physiological Reviews (1921-); and the German publication Internationale Zeitschrift fur angewandte Physiologie enschliesslich Arbeitsphyiologie (1929 – 1973), now titled the European Journal of Applied Physiology. Journals pertaining specifically to sport and exercise physiology were not published until after World War II, for example The Journal of Applied Physiology was first published in 1948 and Medicine and Science in Sports and Exercise in 1969. These publications and those listed in section 2.1 Journals, document an exponential growth in exercise physiology's body of knowledge over the last fifty years.

1.3. Methodology

The research is predominantly quantitative in nature. With interventional investigations of an acute or chronic design and observational investigations utilising cross-sectional or longitudinal design classifying the majority of research.

Investigations have progressed from the macro to the micro level, with whole body, organ system experiments being complemented by cellular responses. Thirty years ago, the focus was at an organ level, mainly due to the ability to instrument and monitor humans during acute and chronic exercise. This research was enhanced by more invasive procedures including the use of muscle-biopsies and radioisotopes, as well as non-invasive imaging technologies, in both humans and heavily instrumented rodents, enabling research at a cellular level. The subsequent use of techniques borrowed from molecular biology including polymerase chain reaction (PCR), x-ray crystallography, mass spectrometry and nuclear magnetic resonance has enabled investigations of the structure, dynamics and interactions of biological molecules at the atomic level. Specifically, PCR has presented the opportunity to genetically profile large groups of athletes in an attempt to identify commonality within their DNA.

As with all disciplines, the methodologies employed by exercise physiologists have developed as technology has advanced, and undoubtedly will continue to do so. As for what the future holds, the reader is referred to Kenneth Baldwin's article, "Research in the exercise sciences: Where do we go from here?" (Journal of Applied Physiology 88: 332-336, 2000).

1.4. Relationship to Practice

Exercise physiology has applications for all individuals from elite athlete to sedentary obese worker, child to octogenarian, and acutely injured to chronic disease populations.

The changing demographics of western civilisation have produced two major application areas: the ageing population and the obesity epidemic. The loss of physiological function associated with ageing has been shown to be slowed, if not reversed, by exercise and physical activity. An increase in sedentary lifestyles has contributed to the obesity epidemic. Obesity associated illnesses; type II diabetes and vascular disease are also placing huge demands on public health systems. Appropriate prescription and monitoring of exercise is critical if these two major influences facing the world are to be controlled.

Age and obesity also combine as an occupational issue. While technology has mechanised numerous manual handling processes, many still remain. Age and obesity have decreased the functional capacity of the worker, placing the worker, and depending on the occupation, colleagues and the public, at an increased risk of injury. Identifying the physiological cost of work tasks within an occupation, screening potential employees and matching workers with appropriate functional capacity to tasks, is an ever increasing role for the occupational physiologist.

Rehabilitation from acute musculoskeletal injuries or from chronic diseases has long been the domain of the allied health professions, including exercise physiology. Cardiac rehabilitation is the most widely recognised clinical application area for exercise physiology, however respiratory, musculoskeletal and other vascular diseases have all been shown to benefit from exercise.

Sports physiology is concerned with developing profiles of individual athletes and teams on a sports specific basis through monitoring and evaluation. The monitoring can then be employed to identify specific strengths and weakness, prescribe appropriate training levels and periodise training programs, assess health status and monitor overtraining, and ultimately maximise the sports potential of each individual athlete. Another branch of sports physiology is involved in detecting athletic performance that has been artificially and illegally enhanced by, for example, anabolic agents, stimulants or blood doping.

2. Information Sources

2.1. Journals

The following journals contain articles related to various aspects of exercise and sports physiology investigation.

Acta Physiological Scandinavica
Applied Physiology, Nutrition, and Metabolism previously Canadian Journal of Applied Physiology
British Journal of Sports Medicine
European Journal of Applied Physiology
European Journal of Sport Science
Exercise and Sport Science Reviews
International Journal of Sports Physiology and Performance
nternational Journal of Sport Nutrition and Exercise Metabolism
Journal of Applied Physiology
Journal of Athletic Training
Journal of Exercise Physiology online
Journal of Exercise Science and Fitness
Journal of Human Movement Studies
Journal of Science and Medicine in Sport
Journal of Sport Sciences
Journal of Strength and Conditioning Research
Medicine and Science in Sports and Exercise
Scandinavian Journal of Medicine and Science in Sports
Strength and Conditioning Journal

2.2. Reference Books, Encyclopaedias, etc.

Numerous reference books exist on the topic of sport and exercise physiology. Listed below are a selection of those that have editions published since the year 2000.

Astrand, P. and Rodahl, K. (2003). *Textbook of work physiology: physiological basis of exercise.* 4 ed. Champaign, IL, Human Kinetics.

Brooks, G.A., Fahey, T.D. and Baldwin, K.M. (2005) *Exercise physiology: human bioenergetics and its applications.* Boston,:McGraw-Hill.,

Brown, S.P., Miller, W.C. and Eason, J.M. (2006). *Exercise physiology: basis of human movement in health and disease.* Philadelphia, Lippincott Williams and Wilkins.

Gore, C.J. (2000). *Physiological tests of elite athletes,* Human Kinetics.

Hale, T. (2003). *Exercise Physiology: a thematic approach.* West Sussex, England: John Wiley.

Hargreaves, M. and Hawley, J. (2003). Hawley. *Physiological bases of sports performance,* McGraw-Hill.

McArdle, W.D., Katch, F.I. and Katch, V.L. (2007). *Exercise physiology: energy, nutrition, and human performance.* 6 ed. Philadelphia, Lippincott Williams and Wilkins.

Powers, S.K. and Howley, E.T. (2007). Howley. *Exercise physiology: theory and application to fitness and performance.* 6 ed. Boston, Mass. McGraw-Hill.

Robergs, R.A. and Keteyian, S.J. (2003). Keteyian. *Fundamentals of exercise physiology: fitness, performance, and health.* Boston, McGraw-Hill.

Wasserman, K., Hansen, J.E., Sue, D.Y., Stringer, W.W. and Whippo, B.J. (2005). Whipp. *Principles of exercise testing and interpretation: including pathophysiology and clinical applications.* 4 ed, Lippincott Williams and Wilkins.

Wilmore, J.H. and Costill, D.L. (2004). Costill. *Physiology of sport and exercise.* 3 ed. Champaign, IL, Human Kinetics.

2.3. Book Series

The International Olympic Committee's Medical Commission, in collaboration with the International Federation of Sports Medicine, publishes the Handbook of Sports Medicine and Science. This is a series of specialist reference volumes designed specifically for the use of professionals working directly with competitive athletes.

2.4. Conference/Workshop Proceedings

Various national organisations publish the proceedings of their annual conferences through supplement volumes of their own journals. Speciality workshops held by interest groups within the respective societies often publish their proceedings on their home page on the World Wide Web.

2.5. Data Banks

No specific data banks exist with regard to exercise physiology. Original research articles can be obtained though appropriate databases ie Medline, ScienceDirect and SPORTDiscus.

2.6. Internet Sources

The following Internet sources are available for use in relation to various aspects of exercise physiology:

American College of Sports Medicine (ACSM)
    www.acsm.org
American Society of Exercise Physiologists (ASEP)
    www.asep.org
Australian Institute of Sport (AIS)
    www.ais.org.au
British Association for Sport and Exercise Science (BASES)
    www.bases.org.uk
Canadian Society of Exercise Physiology (CSEP/SCPE)
    www.csep.ca
European College of Sports Sciences (ECSS)
    www.ecss.de
Gatorade Sports Science Institute (GSSI)
    www.gssiweb.com
International Council of Sports Science and Physical Education (ICSSPE)
    www.icsspe.org
International Federation of Sports Medicine (FIMS)
    www.fims.org
International Society of Exercise and Immunology
    www.isei.dk/
International Union of Physiological Sciences (IUPS)
    www.iups.org
National Strength and Conditioning Association (NSCA)
    www.nsca-lift.org

The Physiological Society
    www.physoc.org
Society of Chinese Scholars on Exercise Physiology and Fitness (SCSEPF)
    www.scsepf.org

3. Organisational Network

3.1. International Level

At the international level, exercise physiology is not organised by a single body but has influences within the Fédération Internationale de Médecine du Sport (FIMS)/ International Federation of Sports Medicine; the International College of Sport Science and Physical Education (ICSSPE); and the International Union of Physiological Sciences (IUPS).

3.2. Regional Level

Europe

The European College of Sports Science was founded in 1995 with the purpose of promotion of sport science at the European level. For that it is dedicated to the generation and dissemination of scientific knowledge concerning the motivation, attitudes, values, responses, adaptations, performance and health aspects of persons engaged in sport, exercise and movement. The College holds annual congresses and publishes both annual bulletins plus a peer reviewed journal, the European Journal of Sport Science.

Britain

The British Association for Sport and Exercise Science (BASES, formerly BASS) was founded in September, 1984 following the dissolution of the Biomechanics Study Group (SBSG), the British Society of Sports Psychology (BSSP) and the Society of Sports Sciences (SSS). BASES' mission is to promote excellence in sport and exercise sciences through evidence-based practice. The Association disseminates information through workshops and its annual conference.

3.3. National Level

The Canadian Society of Exercise Physiology (CSEP/SCPE) was founded at the Pan American Games, Winnipeg, Manitoba in 1967, although it was originally known as the Canadian Association of Sport Sciences. The mission of the Society is to promote the generation, synthesis, transfer and application of knowledge and research related to exercise physiology (encompassing physical activity, fitness, health, nutrition, epidemiology and human performance). CSEP/SCPE holds an annual Scientific Conference and publishes both fitness and scholarly publications including position statements and the peer reviewed journal Applied Physiology, Nutrition and Metabolism.

The American Society of Exercise Physiology (ASEP) is the professional organisation representing and promoting the profession of exercise physiology within America. It is committed to the professional development of exercise physiology, its advancement and the credibility of exercise physiologists. ASEP's objective to broaden the professionalism perspectives and expose students and others to a much wider range of professional thinking and resources, is carried out through the efforts of its online newsletters and journal, the Journal of Exercise Physiology.

The American College of Sports Medicine (ACSM), founded in 1954, promotes and integrates scientific research, education and practical applications of sports medicine and exercise science to maintain and enhance physical performance, fitness, health and quality of life. ACSM holds numerous Scientific Conferences including its Annual Meeting as well as speciality conferences. ACSM publishes position stands, plus peer reviewed journals Medicine and Science in Sports and Exercise, Exercise and Sport Sciences Reviews, ACSM's Health and Fitness Journal and Current Sports Medicine Reports.

The Society of Chinese Scholars on Exercise Physiology and Fitness (SCSEPF) is committed exclusively to the advancement and improvement of exercise physiology and fitness. SCSEPF provides a forum through its annual conference and peer reviewed publication, Journal of Exercise Science and Fitness, for the exchange of information to stimulate discussion and collaboration among exercise physiologists and fitness professionals.

Within Australia, three organisations support sport and exercise physiology: the Australian Association for Exercise and Sport Science (AAESS); the Australian Physiological Society (AuPS); and Sports Medicine Australia (SMA). AAESS is a professional organisation which is committed to establishing, promoting and defending the career paths of tertiary trained exercise and sports science practitioners. AAESS holds biennial scientific conferences to promote the synthesis of knowledge from scientific research to contemporary practice. The objectives of AuPS are to promote the advancement of the science of Physiology and encourage all aspects of research and teaching in this discipline. AuPS disseminates knowledge through quarterly newsletters and annual conferences. SMA is an advisory board for all medical and allied health issues for active people, with safe participation in sport and healthy physical activity at all stages of life being its primary concern. SMA provides continuing education to practitioners, position papers, policies and guidelines to ensure safe participation in exercise. SMA disseminates information through its annual scientific conference and its fitness and scholarly publications. There are numerous other national associations/societies of physiology that support the subdiscipline of exercise physiology, including those affiliated with the International Union of Physiological Sciences.

3.4. Specialised Centres

Not applicable.

3.5. Specialised International Degree Programmes

Not applicable.

4. Appendix Material

4.1. Terminology

The terminology used in sport and exercise physiology draws predominantly from the disciplines of physiology and medicine.

4.2. Position Statement(s)

Position statements on topics of current and vital interest can be found on the web sites of organisational networks listed in section 3 Organisational Network, for example ACSM, CSEP and SMA.

4.3. Varia

Not applicable.

4.4. Free Statement

Not applicable.

Sport and Exercise Psychology

Gershon Tenenbaum, Tony Morris and Dieter Hackfort

Contact

Prof. Dr. Dieter Hackfort
President ISSP
Dean
ASPIRE Academy for Sports Excellence
P.O. Box 22287
Doha, Qatar
Phone: +974 413 6222
Fax: +974 413 6221
Email: dieter.hackfort@gmx.de

1. General Information

Sport and Exercise Psychology is a fundamental sport science discipline. The establishment of this sport science discipline is a consequence of the development and differentiation of disciplines and subgroups of specialists in the scientific community. The localisation of Sport and Exercise Psychology can be described by a triangulation:

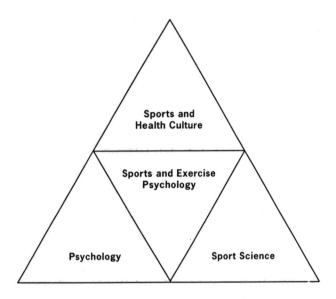

Figure 1: Triangulation of Sport and Exercise Psychology.

1.1. Historical Development

Sport psychology is a relatively young discipline that was developed in the 20th century. Briefly, after the foundation of the first laboratory for experimental psychology in Leipzig (Germany) by Wilhelm Wundt in 1879, a first sport psychology laboratory was established in 1920 by Robert Werner Schulte in Berlin (Germany) and in 1925 by Coleman Griffith in Illinois (USA). Prior to that, in 1913, Pierre de Coubertin, the founder of the modern Olympic Games, organized the First International Congress on the Psychology and Physiology of Sport (the only one). Following a dormant period, sport psychology was developed substantially in the 1960s, in Europe and the US, by university professorships and the foundation of the International Society of Sport Psychology (ISSP) in 1965. The discipline developed in the frame of Sport Science as a function of the increased interest in sport and exercise within the modern society, particularly in elite sport. Continental societies in Europe, North America, Asia, South America, and recently in Africa have been established. These international societies organize congresses in a four yearly cycle, e.g., the ISSP World Congress in Sport Psychology. Numerous national societies have been established around the globe. These societies organize annual conferences and regional meetings. In some international and continental congresses the number of participants approximates 1,000. Furthermore, a respectable number of international and national scientific journals, books, congress proceedings and other scholarly materials (e.g., the ISSP Newsletter, which is now available online) are published. Sport and Exercise Psychology has developed into a prominent research domain, a scientific discipline taught in academic institutes world-wide and a widespread field of application. The practice of sport psychology has seen substantial growth in many sport organisations, and athletes increasingly use psychological knowledge in training and to prepare for competition. Sport and Exercise Psychology is a field of inquiry and application that covers issues, such as mental training, motor learning, imagery, injury rehabilitation, treatment of various mental, emotional, and behavioural disorders, disabilities, gender issues, cross-cultural perspectives, methodologies, and others.

1.2. Function

Sport and Exercise Psychology is a scientific and professional stream of knowledge that focuses on various dimensions of sport and exercise behaviour. The general orientation of the discipline is to describe, explain, predict, and to develop interventional methods to modify intentional organized and purposive behaviour (actions) in sports based on empirical research, qualitative analyses, and ethical standards. More specifically, the intra- and inter-individual, the psycho-physiological, psycho-social, and psycho-ecological foundations, preconditions, configurations, processes, and consequences of actions in sports and exercise are investigated and clarified with the applied mission to optimize sport-related actions (Hackfort, 2006). Various authors and experts focus on selected aspects or emphasize a special purpose or approach, e.g., Weinberg and Gould (1999) defined sport and exercise psychology as the scientific study of people and their behaviours in sport and exercise activities. Rejeski and Brawley (1998) defined sport psychology as the educational, scientific, and professional contributions of psychology to the promotion, maintenance, and enhancement of sport-related behaviour. Rejeski and Thompson (1993) distinguished exercise from sport psychology as being "the educational, scientific and professional contributions of psychology to promoting, explaining, maintaining and enhancing parameters of physical fitness" (p. 5). Anshel (1994) further argued that exercise psychology has become important exploratory ground for sport psychology, thus, their interface yields fertile information. A Position Statement of the European Federation of Sport Psychology (FEPSAC) issued in 1996 claimed that sport psychology "is concerned with psychological foundations, processes, and consequences of psychological regulation of sport-related activities of one or several persons acting as the subject(s) of the activity. The focus may be on behaviour or on different psychological dimensions of human behaviour (i.e., affective, cognitive, motivational, or sensorimotor dimensions)." (p. 221). Accordingly, physical activity can be competitive, recreational, and rehabilitative and the participants can be involved in a variety of activities and exercises. The perspective of Sport and Exercise Psychology is twofold, on the analysis of how actions in sports are regulated/controlled by psychic processes on the one hand and on the other hand on how actions in sports and exercise regulate/control psychic processes (cognitive, motivational, volitional, and affective/emotional processes; see Hackfort & Birkner, 2005) with the intention to enhance sound understanding (theoretical concepts), advance methodological approaches (research methods) and to enlarge effective interventional tools (applied methods).

Applied sport and exercise psychologists are engaged in performance enhancement, counselling, injury rehabilitation, and promotion of physical activity for health maintenance. Research-oriented sport and exercise psychologists develop and test models and theories and undertake scientific investigations to understand sport- and exercise-related actions. Sport psychologists, whether researchers or practitioners, contribute to personal growth in conditions where exercise and sport are performed. A framework proposes that contemporary sport psychologists are engaged in the pursuit of three varied but interlocked activities: theory and research, education, and application. More specifically,

Theory and Research: This stream is concerned with establishing field-driven models and theories that represent the domains of sport and exercise. It is also concerned with mainstream psychological issues and methodology.

Education: This stream is concerned with dissemination of knowledge that can be applied to practitioners, students, and populations with special needs. It also incorporates position stands on various behaviours and ethical issues.

Application: This stream is concerned mainly with applications to a variety of professions such as: coaching, education, clinical practice, athletic training and performance, and injury prevention and rehabilitation.

These three activities are displayed in Table 1 in more detail.

Table 1: Functions of Sport and Exercise Psychology.

Sport and Exercise Psychology

Educational	Theory and Research	Applications
• Psychological principles in teaching physical education • Teaching principles in special populations. • Psychological principles in coaching youth, adult, and elite athletes. • Motivational principles to adhere to motor programs. • Position stands on relevant issues such as: drug abuse, aggression and violence, ethics in sport and psychological benefits of physical activity.	• Cognitive mechanisms. • Motivation and attribution. • Psychophysiology. • Exercise and health aspects. • Personality and individual differences. • Group dynamics and leadership. • Communication. • Emotions. • Arousal, stress and anxiety. • Motor learning, development, control. • Gender issues. • Burnout and over-training. Drop out.	• Performance enhancement. • Exercise and rehabilitation. • Team cohesion. • Clinical (i.e., treatment). • Educational (i.e., teaching, learning). • Leisure and recreation. • Social support (i.e., youth and elite sports, exercise programs)

1.3. Body of Knowledge

Sport and exercise psychology is a unique scientific discipline, but it derives its theoretical and applied perspectives from several links with other "bodies" of knowledge. Those are primarily:

a) Sport Science and Physical Education, including motor learning/development/ control, biomechanics and exercise physiology.
b) Psychology, especially the social and cognitive streams, as well as various sub-disciplines of applied psychology.
c) Health Sciences, especially medical, social and behavioural knowledge.
d) Methodology, in particular measurement, assessment, and evaluation.

In searching for trends in Sport and Exercise Psychology theory and research, Biddle (1997) surveyed all the articles published in the *International Journal of Sport Psychology (IJSP)* and the *Journal of Sport and Exercise Psychology (JSEP; Journal of Sport Psychology JSP* from 1979 to 1987) from 1985 to 1994. The most researched topics were motivation, anxiety, imagery, self-efficacy/confidence, exercise and mental health, and group dynamics. In addition, most of the publications consisted of experimental (30.3%) or survey (38.0%) methodologies, whereas only 13.1% were literature reviews, 8.6% were psychometrics, 5.7% qualitative, 2.7% archival/historical, 1.0% case studies, 0.4% content analysis and 0.2% meta-analyses. Morris (1999) confirmed this pattern in an analysis of IJSP and JSEP from 1979 to 1998. Comparisons of five-year periods indicated that the proportion of correlational (questionnaire/survey) research has increased recently. Biddle noted that most of the participants in the studies were school (14.4%) and college (33.8%) students, whereas only 3.8% were elite athletes. Vealey (1994), on the other hand, showed that in the applied sport psychology journals 52% of the participants were elite or national calibre athletes.

There has been no equivalent evaluation of published research in Exercise Psychology up to now. Interest in Exercise Psychology arises from two sources. Many researchers and practitioners in Sport Psychology were trained in physical education or human movement, so the interests of researchers and practitioners trained through this route have always been broader than competitive sport. The human movement-trained sport and exercise psychologists typically consider the term

"Sport and Exercise Psychology" to be a reflection of the historical development of the study of psychological factors related to all aspects of physical activity. The other, more recent, source of development of Exercise Psychology is Health Psychology. This sub-discipline of Psychology centres on examination of the role of physical activity in the promotion of physical and mental well-being, as well as factors that motivate people to initiate and sustain systematic physical activity. Thus, it seems that Sport and Exercise Psychology is a domain that consists of several inter-related bodies of knowledge, which share motor and physical activity as a common interest. Some studies are theory driven, whereas others are field driven. Most research is applied-oriented (in both sport and exercise) and relatively little is purely theoretical or methodological.

1.4. Methodology

A wide range of research designs (e.g., experimental and non-experimental, laboratory and field studies, longitudinal and cross-sectional, single case and group studies) and methods (e.g., qualitative and quantitative, interview and observational methods, questionnaires/scales and tests) have been employed in Sport and Exercise Psychology research and practice. Morris (1999) examined the methods used in research published in the *IJSP* and the *JSP/JSEP* from the inception of the latter in 1979 to 1998, a 20-year period. By dividing the period into blocks of five years. Thus, he was able to comment on trends. Consistent with Biddle's (1997) results, Morris found that laboratory/field experiments and questionnaire studies were the two most common research methods, followed by theoretical/ methodological papers. In the period from 1993 to 1998, more than half of the published work in these journals was questionnaire studies or psychometric studies of new questionnaires. In their applied work, sport psychologists use questionnaires extensively. Other techniques, such as observation and direct interviews with athletes, coaches, educators, administrators and people with special needs are also common. As the extent of counselling has increased in applied work (e.g., Etzel, Ferrante, & Pinkney, 1996; Murphy, 1995; Van Raalte & Brewer, 1996), use of these introspective techniques has also become more prevalent in research. Recently, the integration of quantitative and qualitative types of data has been recommended, in research and practice, to better account for sport and exercise behaviours, as well as for validation requirements. Ostrow (1996) collected 314 psychometrically-validated tests that have been developed specifically for use in the sport and exercise psychology domain. Duda (1998) has thoroughly reflected the different approaches to measurement in sport and exercise psychology. Computer-assisted testing is not yet popular in Sport and Exercise research and practice. It can be predicted that based on the development of the new technologies, including virtual reality and simulation, that computer-based test- and training programs will receive more attention in Sport and Exercise Psychology.

1.5. Relationship to Practice

In addition to the above outlined (see section 1.1 and Table 1) relationship Sport and Exercise Psychology aims to

a) investigate the preconditions and circumstances for an effective usage or application of relevant knowledge and methods (technology research)
b) examine and check the effectiveness and efficiency of the interventional techniques (evaluation research)
c) ensure that the application of psychological techniques, counselling, and treatment methods are exclusively provided by appropriately educated, trained and competent experts who adhere to the ethical principles of the discipline (Supervision). A brochure about supervision in Sport Psychology was introduced online by the ISSP (ISSP Position Stand) in 2007 and a further "ISSP Statement on Ethical Principals" is now published online via the ISSP homepage (www.issponline.org).

1.6. Future Perspectives

The future developments of Sport and Exercise Psychology should be considered from educational, professional, organisational, and scientific perspectives. Sport and Exercise Psychology can be perceived today as two distinct, though complementary, disciplines. Each developed its unique theoretical concepts and lines of research. Though some common interests and conceptual foundations still exist, the content areas and interests of scientists and practitioners tend to differ. Sport Psychology is concerned with theoretical and practical topics related to sport involvement and an emphasis on psychological factors and processes relevant in high-performance sports, whereas Exercise Psychology concentrates more on health-related issues, and motivation and drop out from exercise regimens, as well

as increasing levels of physical activity among all people. There is a tendency toward increasing differentiation on the one hand and toward multi-perspective and
interdisciplinary approaches on the other hand. Due to these trends an integrative
theoretical concept is needed for a holistic understanding and the action-theory
perspective (see Hackfort, Munzert, & Seiler, 2000; Nitsch, 2000; Schack & Tenenbaum, 2004) is a promising approach to elaborate such a framework.

As sport continues to grow and become a multi-million dollar business, more sport
psychologists join sport organisations, professional teams and clubs to counsel
athletes. Many private and public organisations advertise jobs not previously offered and athletes in professional sports more often hire mental coaches and
sport psychologists for performance enhancement purposes. As a consequence,
many academic institutes world-wide develop sport psychology programmes, and
national societies, such as the AASP (Association for Applied Sport Psychology)
in the USA or the ASP (Association of Sport Psychology) in Germany , provide
certificates to acknowledge securing professional services. This trend will further
develop and certification, accreditation, and quality management procedures have
to be developed to ensure high standard professional services provided by accepted and recognized experts.

Sport Psychology came to a mature stage in which field-driven and field-specific
theories and models are emerging. Applied sport psychology is also widening the
scope of issues addressed form a psychological perspective. Performance enhancement techniques still dominate the applied sport literature, but many other
educational and clinical aspects draw attention and research. The range of topics
expands along with in-depth investigations and scientific quality. The dissemination
of that knowledge is increasing with respect to interested and relevant people,
e.g., coaches, referees, and officials in sports. Furthermore, based on the new
information and communication technologies the preconditions for dissemination
of knowledge, continuous and further education in Sport and Exercise Psychology are improving (see also the ISSP Position Stand on "Sport Psychology and
the Internet") and more and more interested people have access to information
resources.

Sport and Exercise Psychology has already been integrated within national and
international psychological and sport science societies, and the number of mem-

bers joining grows steadily. Many scientific forms, professional meetings, and interdisciplinary groups develop. This trend will continue and expand in the future as the preconditions for information exchange via the internet is improved and the opportunities to communicate are enlarged. This way the body of experience in the discipline will be enriched by a lot of people and for a broader community in Sport and Exercise Psychology.

References

Anshel, M.H. (1994). *Sport psychology: From Theory to Practice.* Scottsdale, AZ: Gorsuch Scarisbrick.

Biddle, S. (1997). Current Trends in Sport and Exercise Psychology Re-search. *The Sport Psychologist,* 10, 63-68.

Hackfort, D. (2006). A conceptual framework and fundamental issues for investigating the development of peak performance in sports. In D. Hackfort & G. Tenenbaum (Eds.), *Essential processes for attaining peak performance* (pp. 10-25). Morgantwon, WV: FIT.

Hackfort, D., & Birkner, H.-A. (2005). An action-oriented perspective on exercise psychology. In D. Hackfort, J.L. Duda & R. Lidor (Eds.), Handbook of Research in Applied Sport and Exercise Psychology: International Perspectives (pp. 351-373). Morgantown, WV: Fitness Information Technology.

Hackfort, D., Munzert, J., & Seiler, R. (Eds.). (2000). *Handeln im Sport als handlungstheoretisches Modell.* Heidelberg: Asanger.

Hardy, L., & Fazey, J. (1998). The inverted-U hypothesis: *A Catastrophe for Sport Psychology.* Leeds, UK: British Association of Sport and Exercise Sciences.

Heyman, S. (1987). Counselling and Psychotherapy with Athletes: Special Considerations. In J.R. May & M.J. Asken (Eds.) *Sport Psychology: The Psychological Health of the Athlete* (pp. 135-136). New York: PMA Publishing.

Lidor, R., Morris, T., Bardaxoglou, N. & Becker, B. (2001). *World Sport Psychology Sourcebook* (3rd ed.). Morgantown, West Virginia: Fitness Information Technology.

Morris, T. (1997). *Psychological Skills Training: An Overview.* Leeds, UK: British Association of Sport and Exercise Sciences.

Morris, T. (1999). The message of methods: Developing Research Methodology in Sport Psychology. In G. Si (Ed.) *Proceedings of the 3rd International Congress of the Asian South Pacific Association of Sport Psychology,* Wuhan, China: ASPASP.

Nitsch, J.R. (2000). Handlungstheoretische Grundlagen der Sportpsychologie. In
H. Gabler, J.R.

Nitsch, & R. Singer (Hrsg.), *Einfuehrung in die Sportpsychologie. Teil 1:
Grundthemen* (Bd. 2, 3. Aufl., S. 43-164). Schorndorf: Hofmann.

Rejeski, W.J., & Brawley, L.R. (1998). Defining the Boundaries of Sport Psychology. *The Sport Psychologist*, 2, 231-242.

Rejeski, W.J., & Thompson, A. (1993). Historical and Conceptual Roots of Exercise Psychology. In P. Seraganian (Ed.), *Exercise Psychology: The Influence
of Physical Exercise on Psychological Processes*, (pp. 3-38). New York: John
Wiley & Sons.

Sachs, M.L., Burke, K.L., & Butcher, L.A. (1995). *Directory of Graduate Programs
in Applied Sport Psychology* (4th ed.). Morgantown, WV: Fitness Information
Technology.

Sachs, M.L., Burke, K.L., & Gomer, S. (1998). *Directory of Graduate Programs
in Applied Sport Psychology* (5th ed.). Morgantown, WV: Fitness Information
Technology.

Salmela, J.H. (1992). *The World Sport Psychology Sourcebook* (2nd ed.). Champaign, IL: Human Kinetics.

Straub, W.F., & Williams, J.M. (1984). *Cognitive Sport Psychology*. Lansing, NJ:
Sports Science Associates.

Schack, T., & Tenenbaum, G. (2004). The construction of action: New perspectives n movement sciences. Part 1 and 2. International Journal of Sport and
Exercise Psychology, 2 (3 and 4, Special Issues).

Vealey, R.S. (1994). Knowledge Development and Implementation in Sport Psychology: A Review of the Sport Psychologist, 1987-1992. *The Sport Psychologist*, 8, 331-348.

Weinberg, R.S., & Gould, D. (1999). *Foundations of Sport and Exercise Psychology* (2nd ed.). Champaign, IL: Human Kinetics.

2. Information Sources

2.1. Journals

The main English language journals in the field of Sport and Exercise Psychology are
the following:

International Journal of Sport and Exercise Psychology (official journal of the ISSP)
International Reviews in Sport and Exercise Psychology
Journal of Applied Sport Psychology
Journal of Clinical Sport Psychology
Journal of Performance Enhancement
Journal of Sport Behavior
Journal of Sport and Exercise Psychology
Psychology of Sport and Exercise
The Sport Psychologist

Related journals, in which Sport and Exercise Psychology contributions are regularly published, include:

Australian Journal of Science and Medicine in Sport
Canadian Journal of Sport Sciences
Journal of Human Movement Studies
Journal of Motor Behavior
Journal of Sport Sciences
New Zealand Journal of Health, PE, and Recreation
Perceptual and Motor Skills
Physical Activity and Health
Research Quarterly for Exercise and Sport

Depending on its theme, Sport and Exercise Psychology material is also published in journals focusing on Sports Medicine, Exercise Physiology, Strength and Conditioning, and Coaching. Journals in languages other than English that publish sport and exercise psychology articles include (some titles are translated):

Atac Universitatis Carolinae Gymnica (Czech Republic)
Apunts de Medecina de l'Esport (Spain)
Apunts de Educación Física (Spain)
Boletim da Sociedade Portuguesa de Psicologia Desportiva (Portugal)
Bulletin Scientifique et Technologique du Sport (Tunisia)
Bulletin Suisse des Psychologues (Switzerland)
Geneeskunde en Sport (Netherlands)

Idrottsforskaren (Sweden)
Japanese Journal of Sport Psychology (Japan)
Journal of Human Movement Sciences (Netherlands)
Journal of Sport Psychology (Greece)
Journal of Sport Sciences: Theory and Practice (Greece)
Korean Journal of Physical Education (Korea)
Kultura Fizyczna (Poland)
Leistungssport (Germany)
Medicina dello Sport (Italy)
Movement (Israel)
Movimento (Italy)
Scandinavian Journal of Sport Sciences (Scandinavia)
Sciences et Motricité (France)
Sciences et Technologie des Activités Physiques et Sportives (France)
Sciences et Technologie (Algeria)
SDS – Rivista di Cultura Sportiva (Italy)
Sportmedizin (Austria)
Sportwissenschaft (Germany)
Sport Wyczynowy (Poland)
Teorie a Praxe Telense Vychovy (Czech Republic)
Teoriia: Praktika Fizicheskoi Kultury (Russia)
Testnevelesi Foiskola Kozlemenyek (Hungary)
Thai Journal of Sports Science (Thailand)
Vaprosi na Fiziceskate Kultura, Psihologia (Bulgaria)
Wychowanie Fizyczne: Sport (Poland)
Zeitschrift für Angewandte Sozialpsychologie (Austria)
Zeitschrift für Sportpsychologie (Germany)

2.2. Reference Books, Encyclopaedias, etc.

Texts in sport psychology:

Andersen, M.B. (Ed.), (2000). *Doing Sport Psychology.* Champaign, IL: Human Kinetics.
Andersen, M.B. (Ed.), (2005). *Sport Psychology in Practice.* Champaign, IL: Human Kinetics.

Anshel, M.H. (1994). *Sport Psychology: From Theory to Practice.* (2nd ed.). Scottsdale, AZ: Gorsuch Scarisbrick.

Begel, D., and Burton, R.W. (Eds.), (2000). *Sport Psychiatry.* New York: Wiley.

Bull, S. (Ed.), (1999). *Adherence Issues in Sport and Exercise Psychology.* Chichester, UK: Wiley.

Bull, S.J., Albinson, J.G., and Shambrook, C.J. (1996). *The Mental Game Plan.* (2nd ed.). Morgantown, WV: Fitness Information Technology.

Burton, D., and Raedeke, T.D. (2008). *Sport Psychology for Coaches.* Champaign, IL: Human Kinetics.

Carron, A.V., Hausenblas, H.A., & Eys, M.A. (2005). *Group Dynamics in Sport.* (3rd ed.). Morgantown, WV: Fitness Information Technology.

Cox, R.H. (1998). *Sport Psychology: Concepts and Applications.* (4th ed.). Dubuque, IA: Wm C. Brown.

Duda, J.L. (Ed.), (1998). *Advances in Sport and Exercise Psychology Measurement.* Morgantown, WV: Fitness Information Technology.

Etzel, E.F., Ferrante, A.P., and Pinkney, J.W. (Eds.). (1996). *Counselling College Student-athletes: Issues and Interventions.* (2nd ed.). Morgantown, WV: Fitness Information Technology.

Feltz, D.L., Short, S.E., and Sullivan, P.J. (2008). *Self-Efficacy in Sport.* Champaign, IL: Human Kinetics.

Gardner, F., and Moore, Z. (2006). *Clinical Sport Psychology.* Champaign, IL: Human Kinetics.

Gill, D.L. (2000). *Psychological Dynamics of Sport* (2nd ed.). Champaign, IL: Human Kinetics.

Gill, D.L. (2008). *Psychological Dynamics of Sport* (3rd ed.). Champaign, IL: Human Kinetics.

Goldberg, A.S. (1998). *Sports Slump Busting.* Champaign, IL: Human Kinetics.

Hackfort, D., Duda, J.L., and Lidor, R. (Eds.). (2005). *Handbook of Research in Applied Sport and Exercise Psychology: International Perspectives.* Morgantown, WV: Fitness Information Technology.

Hackfort, D. and Tenenbaum, G. (Eds.), *Essential processes for attaining peak performance* . Morgantown, WV: FIT.

Hagger, M.S., and Chatzisarantis, N.L.D. (Eds.), (2007). *Intrinsic Motivation and Self-Determination in Exercise and Sport.* Champaign, IL: Human Kinetics.

Hardy, L., Jones, G., and Gould, D. (1996). *Understanding Psychological Preparation for Sport: Theory and Practice of Elite Performers.* Chichester, UK:

John Wiley and Sons.

Hill, K.L. (2001). *Frameworks for Sport Psychologists: Enhancing Sport Performance*. Champaign, IL: Human Kinetics.

Horn, T.S. (Ed.). (1992). *Advances in Sport Psychology (1st ed.)*. Champaign, IL: Human Kinetics.

Horn, T.S. (Ed.). (2002). *Advances in Sport Psychology (2nd ed.)*. Champaign, IL: Human Kinetics.

Horn, T.S. (Ed.). (2008). *Advances in Sport Psychology (3rd ed.)*. Champaign, IL: Human Kinetics.

Schilling, G. (Ed.). (1992). *Sport Science Review: Sport Psychology*. Champaign, IL: Human Kinetics.

Jackson, S. and Czikszentmihalyi, M. (1999). *Flow in Sports*. Champaign, IL: Human Kinetics.

Jarvis, M. (1999). *Sport Psychology*. London, UK: Jarvis.

Jowett, S., and Lavallee, D. (Eds.), (2007). *Social Psychology in Sport*. Champaign, IL: Human Kinetics.

Krause, D.R. (2001). *Mastering Your Inner Game*. Champaign, IL: Human Kinetics

Kremer, J., and Moran, A.P. (2008). *Pure Sport:Practical Sport Psychology*. Hove, UK: Routledge.

Leith, L. M. (2006). *The Psychology of Coaching Team Sports: A Self-Help Guide*. Toronto: Sport Books Publisher.

LeUnes, A.D. and Nation, J.R. (1995). *Sport Psychology: An Introduction*. (2nd ed.). Chicago: Nelson-Hall.

Lidor, R. and Bar-Eli, M. (Eds.), (1999). Sport Psychology: *Linking Theory and Practice*. Morgantown, WV: Fitness Information Technology.

Lidor, R., and Henschen, K.P. (Eds.), (2005). *The Psychology of Team Sports*. Morgantown, WV: Fitness Information Technology.

Lidor, R., Morris, T., Bardaxoglou, N., and Becker, B. (2001). *World Sport Psychology Sourcebook* (3rd ed.). Morgantown, West Virginia: Fitness Information Technology.

Liggett, D.R. (2000). Champaign, IL: Human Kinetics.

Liukkonen, J, Vanden Auwele, Y, Vereijken, B., Alfermann, D., and Theodorakis, Y. (2007). Psychology for Physical Educators. Champaign, IL: Human Kinetics.

Moran, A.P. (1996). *The Psychology of Concentration in Sport Performers: A Cognitive Analysis*. UK: Psychology Press.

Morris, T., Spittle, M, and Watt, A.P. (2005). *Imagery in Sport*. Champaign, IL:

Human Kinetics.

Morris, T., and Summers, J. (Eds.), (2004). *Sport Psychology: Theory Applications and Issues (2nd ed.)*. Brisbane, Australia: Jacaranda Wiley.

Morris, T., Terry, P., and Gordon, S. (Eds.), (2007). *Sport and Exercise Psychology: International Perspectives*. Morgantown, WV: Fitness Information Technology.

Murphy, S.M. (Ed.). (1995). *Sport Psychology Interventions*. Champaign, IL: Human Kinetics.

Murphy, S. (Ed.) (2005). *The Sport Psych Handbook*. Champaign, IL: Human Kinetics.

Orlick, T. (2000). *In Pursuit of Excellence* (3rd ed.). Champaign, IL: Human Kinetics.

Ostrow, A.C. (Ed.), (1996). *Directory of Psychological Tests in the Sport and Exercise Sciences*. Morgantown, WV: Fitness Information Technology.

Pargman, D. (Ed.), (1999). *Psychological Bases of Sport Injury* (2nd ed.). Morgantown, WV: Fitness Information Technology.

Pargman, D. (Ed.), (2007). *Psychological Bases of Sport Injury* (3rd ed.). Morgantown, WV: Fitness Information Technology.

Roberts, G.C. (Ed.) (2001). *Advances in Motivation in Sport and Exercise*. Champaign, IL: Human Kinetics.

Porter, K. (2003). *The Mental Athlete*. Champaign, IL: Human Kinetics.

Rotella, B., Boyce, B.A., Allyson, B., and Savis, J.C. (1998). *Case Studies in Sport Psychology*. Sudbury, MA: Jones and Bartlett.

Seraganian, P. (Ed.). (1993). *Exercise Psychology: The Influence of Physical Exercise on Psychological Processes*. New York: John Wiley & Sons.

Sheikh, A.A. and Korn, E.R. (1994). *Imagery in Sports and Physical Performance*. Amityville, NY: Baywood Publishing.

Silva, J.M. and Weinberg, R.S. (1984). *Psychological Foundations of Sport*. Champaign, IL: Human Kinetics.

Singer, R.N., Murphey, M. and Tennant, L.K. (Eds.), (1993). *Handbook of Research on Sport Psychology*. New York: Macmillan.

Singer, R.N., Hausenblas, H., and Janelle, C.M. (Eds.), (1993). *Handbook of Research on Sport Psychology* (2nd ed.). New York: Wiley.

Skinner, J.S., Corbin, C.B., Landers, D.M., Martin, P.E. and Wells, C.L. (Eds.). (1989). *Future Directions in Exercise and Sport Science Research*. Champaign, IL: Human Kinetics.

Smith, D, and Bar-Eli, M. (Eds.), (2007). *Essential Readings in Sport and Exercise Psychology*. Champaign, IL: Human Kinetics.

Taylor, J., and Wilson, G. (Eds) (2005). *Applying Sport Psychology: Four Perspectives*.

Champaign, IL: Human Kinetics.

Tenenbaum, G. (Ed.), (2001). *The Practice of Sport Psychology*. Morgantown, WV: Fitness Information Technology.

Tenenbaum, G., and Eklund, R. (Eds.), (2007). *Handbook of Sport Psychology* (3rd ed.). New York, NY: Wiley.

Van Raalte, J.L., and Brewer, B.W. (Eds.) (2002). *Exploring Sport and Exercise Psychology* (2nd ed.). Washington, DC: American Psychological Association.

Weinberg, R.S., and Gould, D. (2007). *Foundations of Sport and Exercise Psychology*. (4th ed.). Champaign, IL: Human Kinetics.

Weiss, M.R. (2004). Developmental *Sport and Exercise Psychology: A Lifespan Perspective*. Morgantown, WV: Fitness Information Technology.

Williams, J.M. (Ed.). (2005). *Applied Sport Psychology: Personal Growth to Peak Performance* (5th ed.). New York: McGraw Hill

Texts in exercise psychology:

Biddle, S., Fox, K., and Boutcher, S. (Eds.), (2000). *Physical Activity and Psychological Well-being*. London, UK: Routledge.

Biddle, S.J.H. and Mutrie, N. (1991). *The Psychology of Physical Activity and Exercise: A Health-Related Perspective (1st ed.)*. London: Springer-Verlag.

Biddle, S.J.H. and Mutrie, N. (2007). *The Psychology of Physical Activity and Exercise: A Health-Related Perspective (2nd ed.)*. London: Springer-Verlag.

Biddle, S., and Mutrie, N. (2001). *The Psychology of Physical Activity: An Evidence-Based Approach*. London: Routledge.

Buckworth, J., and Dishman, R.K. (2002). *Exercise Psychology*. Champaign, IL: Human Kinetics.

Berger, B., Pargman, D., and Weinberg, R.S. (2006). *Foundations of Exercise Psychology (2nd ed.)*. Morgantown, WV: Fitness Information Technology.

Dishman, R.K. (Ed.), (1988). *Exercise Adherence: Its Impact on Public Health*. Champaign, IL: Human Kinetics.

Dishman, R.K. (Ed.), (1994). *Advances in Exercise Adherence*. Champaign, IL: Human Kinetics.

Fox, K.R. (Ed.), (1997). *The Physical Self: From Motivation to Well-Being*. Champaign, IL: Human Kinetics.

Leith, L.M. (1994). *Foundations of Exercise and Mental Health*. Morgantown, WV: Fitness Information Technology.

Leith, L.M. (1998). *Exercising your Way to Better Mental Health.* Morgantown, WV: Fitness Information Technology.

Morgan, W.P. (Ed.), (1997). *Physical Activity and Mental Health.* Washington, DC: Taylor and Francis.

Willis, J.D., & Campbell, L.F. (1992). *Exercise Psychology.* Champaign, IL: Human Kinetics.

2.3. Book Series

In the late 1980's, the British Association of Sport and Exercise Sciences developed a series of monographs covering major themes in the sport and exercise sciences. A range of titles have been published, including one in sport psychology by Hardy and Fazey (1987) on catastrophe theory related to anxiety in sport. Another early title was on psychological skills training (Morris & Bull, 1990).

This was revised by Morris (1997). Fitness Information Technology publishes a series of edited volumes on issues of fundamental relevance entitled International Perspectives on Sport and Exercise Psychology edited by Hackfort and Tenenbaum and a series entitled "Sport Psychology Library". This series is consumer-oriented, aimed at a wide audience of participants and professionals. The first titles were released during 1999. The current titles in this series are:

Baker, J., and Sedgwick, W. (2005).). *Sport Psychology Library: Triathlon.* Morgantown, WV: Fitness Information Technology.

Burke, K.L., and Brown, D. (2003). *Sport Psychology Library: Basketball – The Winning Edge is Mental.* Morgantown, WV: Fitness Information Technology.

Lasser, E.S., Borden, F., and Edwards, J. (2006). *Sport Psychology Library: Bowling - The Handbook of Bowling Psychology.* Morgantown, WV: Fitness Information Technology.

Cogan, K.D., and Vidmar, P. (1999). *Sport psychology library: Gymnastics.* Morgantown, WV: Fitness Information Technology.

Van Raalte, J.L., and Silver-Bernstein, C. (1999). *Sport psychology library: Tennis.* Morgantown, WV: Fitness Information Technology.

2.4. Congress/Workshop Proceedings

The proceedings of the ISSP (International Society of Sport Psychology), FEP-SAC (Fédération European de Psychologie du Sport et Activité Corporelle), and ASPASP (Asian and South Pacific Association of Sport Psychology) congresses, since their establishment, can be found in University libraries around the world. These congresses are held every four years. In recent years, abstracts from the annual North American congress of the Association for Applied Sport Psychology (AAASP) have been published annually in a special issue of the *Journal of Applied Sport Psychology* and the abstracts form the North American Society for the Psychology of Sport and Physical Activity (NASPSPA) annual conference have been published annually in a special issue of the *Journal of Sport and Exercise Psychology*. Although proceedings from the international conferences comprised abstracts for many years, a tradition developed in the 1990s for short papers to be submitted. Congresses of ISSP in 1993, 1997, 2001, and 2005, FEPSAC in 1995, 1999, 2003, and 2007,and ASPASP in 1999, 2003, and 2007 include such papers.

2.5. Data Banks

The main data bank used to locate works in sport and exercise psychology is SPORT DISCUS by SIRC (Sport Information Resource Centre) based in Canada. Publications in sport and exercise psychology can also be surveyed in data bases such as: MEDLINE, Current Contents, PSYCLIT, Austrom: AUSPORT, ContentsFirst, ArticleFirst, FastDoc, Periodical Abstracts, Social Sciences Abstracts.

2.6. Internet Sources

The ISSP website www.issponline.org. contains information on the recent publications of the ISSP, its newsletters, position stands, the managing council members, a short introduction on the function of sport psychology and further information. Additional recommended sites where information on sport psychology can be found are www.aasponline.org (website of the American-based Association of Applied Sport Psychology), www.naspspa.org (website of the North American Society for the Psychology of Sport and Physical Activity), www.apa47.org (website of the American Psychological Association Division 47, which focuses on exercise and sport psychology, and www.fepsac.com (website of the European Federation of Sport Psychology).

3. Organisational Network

Few formal and systematic examinations have been conducted of the status of Sport and Exercise Psychology around the world. Around 1980, ISSP sponsored the collection of information about organizations, practices and people active in sport and exercise psychology world-wide. This led to the publication of the *World Sourcebook of Sport Psychology* (Salmela, 1981). Salmela updated the *Sourcebook* and a second edition was published in 1992. ISSP commissioned a third edition, which was published in 2001. The recent edition of the *Sourcebook* contains the best global information on Sport and Exercise Psychology. Salmela (1992) reported that, in 1981, there were 1,320 individuals from 39 countries around the world who were active in the field of Sport Psychology. By the time information was collected for the second edition of the *Sourcebook*, the number of active sport psychologists had doubled and the number of countries with some active involvement had increased to 61. Thus, a 48.4% increase in the number of sport and exercise psychologists world-wide was noted between 1981 and 1992. The increase in the number of countries having some level of Sport Psychology activity during these years was 56.4%. Today this increase continues, although the rate is slowing, as the ceiling in terms of the number of countries not active is approached. The number of individuals involved in activities related to Sport and Exercise Psychology in active countries often expands in a pulsating manner, influenced by such factors as national initiatives in sport and the development of new university programmes or accreditation frameworks. Nonetheless, Lidor et al. reported that the number of individuals involved in Sport Psychology, in the 48 countries that provided reports for the 2001 edition of the *Sourcebook*, was more than double the number cited by Salmela in the second edition in 1992. Lidor et al. observed that the substantial numbers of individuals active in this field today supports the claim that Sport and Exercise Psychology is now a global profession.

According to Salmela (1992), North America has the largest number of individual sport psychologists (43.1%) followed by Europe (21.4%), whereas Latin America, Asia, Australia, and Africa together constitute 35.5%. Updated statistics show some big changes to these figures, resulting from a substantial development of the discipline in countries in South America and Asia. The South American total in 1992 was five sport psychologists, but only one country provided information. Seven nations replied in preparation of the 2001 edition of the *Sourcebook*, providing an estimate of 362 individuals. Caution must be applied to the 1992 figure here, which probably involves some under-reporting. The Asian figures might be more accurate. From six countries in 1992, a total of 511 individuals was reported. By 2001, the number of countries involved had increased to 11 and the total number of people interested in Sport Psychology was estimated to be 1,360. Direct comparisons are not possible between 1992 and 2001 for the USA because the single estimate in 1992 of 750 sport psychologists has been replaced by a total of 2,629 members in three major societies, the North American Society for the Psychology of Sport and Physical Activity (NASPSPA), which had 562 members, the Association for the Advancement of Applied Sport Psychology (AAASP, which became AASP, the Association of Applied Sport Psychology in 2006), with 1,029 members, and the American Psychological Association (APA) Division 47, Exercise and Sport Psychology, which reported 1,038 members. The total (2,629), derived by adding the figures provided by these three groups, certainly includes individuals who are involved with two or even all three groups. It is still reasonable to estimate that the number of people actively involved in Sport and Exercise Psychology in the USA has more than doubled in the last 10 years. In Europe there was a total of 1,142 individuals from 19 countries in the 1992 edition of the *Sourcebook*. This increased to 2,010 from 22 countries by the 2001 edition. Two countries, Russia and Italy, showed large decreases, but these were more than compensated for elsewhere, especially in the United Kingdom, where the total increased from 40 to 525. This indicates that geometric-type growth is not a prerogative of the so-called developing countries. Further statistics from the 1992 edition of the *Sourcebook* indicated that full-time sport psychologists could be found in 83% and 75% of the countries in Western Europe and Scandinavia respectively. In Africa and the Middle-East, these figures decrease to 64%, in Asia to 50%, and in Latin America to 38%. The authors of the 2001 edition did not consider such estimates because of the dubious reliability of information on countries not listed in the book. It should be noted that the majority of the sport psychologists around the world combine their sport psychology work with other jobs, mainly as academics, administrators, or in other specialisms of psychology (Salmela, 1992).

3.1. International Level

In 1965, the International Society of Sport Psychology (ISSP) was established in Rome, Italy on the initiative of Dr. Ferrucio Antonelli. This accelerated the development of continental societies, such as FEPSAC in Europe, the ASP (Association for Sport Psychology) in Germany, and NASPSPA, and CSPLSP (Canadian Society of Psychomotor Learning and Sport Psychology) in North America. The advent of applied Sport and Exercise Psychology was the main impetus for the development of AAASP in the USA in 1985. SOBRAPE (South American Society of Sport Psychology, Physical Activity and Recreation) in South America and AS-PASP (Asian and South Pacific Association of Sport Psychology) in Asia and the South Pacific (including Australia and New Zealand), are the other two active, regional groups, with encouragement from ISSP. ASPASP was proposed in 1988. Appropriately, it was formally established at the 1989 ISSP Congress in Singapore. In 2004, the first International Congress of Sport Psychology in Africa was organized in Marrakech (Morocco). The foundation of an African Society was initiated and supported by the ISSP at that event. It should be noted that the International Association of Applied Psychology (IAAP) has recently established a section dedicated to sport psychology, independent of the other traditional sections. The trend of sport and exercise becoming a recognised part of major, mainstream psychology organizations is typical in many local and national sport and exercise psychology societies and organizations around the world. For example, in 1991, sport and exercise psychologists in Australia established their own College of Sport Psychologists within the Australian Psychological Society. More recently, a Division of Exercise and Sport Psychology was created in the British Psychological Society. A number of European Sport Psychology groups have also linked with the parent discipline.

3.2. Regional Level

Relying on Lidor et al. (2001), the distribution of countries by geographical regions in 2000 was as shown in Table 2.

Table 2: National Sport Psychology Societies and Individual Membership.

Geographical Region	Number of Societies	Number of Individuals
Oceania	2	131
Asia	11	1,360
Middle East & Africa	3	30
South America	7	362
North America	4	2,750
Scandinavia	4	379
Western Europe	11	1,828
Eastern Europe	8	182
Total	49	6,901

This reflects an increase from the year 1989 to 2000 of 8 in the number of countries that reported having national societies and a massive development of 4,908 in the number of individuals involved in sport psychology, as reported by those national groups. Given the overestimation in the USA, due to multiple membership of societies, it would seem that the numbers in Western Europe and North America are comparable and more substantial than other regions. This is not surprising, because, historically, the development of the discipline originated in these two regions (see above 1.1). Nonetheless, the number of people actively involved in sport psychology in the Asian South Pacific region (Asia plus Oceania) appears to be increasing rapidly, since the continental body, ASPASP, was formed.

3.3. National Level

Readers are referred to the third edition of the Sport Psychology Sourcebook, pp. 29-232 (Lidor et al., 2001), for specific details about the history and organisational structures of sport psychology in 48 nations around the globe. Since 1992, the number of sport psychology societies has continued to increase.

3.4. Specialised Centres

Sport and exercise psychology is a discipline that is taught in academic institutions around the world. In most countries, sport and exercise psychology is taught in departments of Sport Science, Kinesiology and/or Human Movement Studies. Australia is an exception, where professional training in sport and exercise psychology, which is recognised by the legal and scholarly organisations in the profession, is offered only in psychology departments and only for graduate students who have an undergraduate degree in psychology (comprising a four-year course). At the same time, sport and exercise psychology subjects are included in many Human Movement, Physical Education, Sports Science and Sports Studies undergraduate programs. Across the world, the relationship between psychology and sport psychology in educational institutions and professional organisations varies greatly. In some countries, university psychology departments teach sport psychology, in some they ignore it and in yet others they actively shun the sub-discipline. Similarly, many national psychology societies still exclude sport psychology, although the inclusion of sport psychology sections in the APA, BPS and APS, as well as in the International Association of Applied Psychology, is likely to turn the tide in this respect.

3.5. Specialised International Degree Programmes

ISSP has not run any international programmes or courses up to now, however, a Directory of Graduate Programs in Applied Sport Psychology (Burke, Sachs, & Smisson, 2004, 7th ed.) supplies details on academic programmes in the field, e.g., in Australia, Canada, Great Britain, South Africa, and the United States. The Directory contains many details about the requirements, length, history, and emphasis of each of the programmes. In 1996, FEPSAC initiated and established a PanEuropean master's degree in Exercise and Sport Psychology. Students can attend courses at 10 different universities in Europe (Belgium, Finland, France, Germany, Greece, Netherlands, Norway, Portugal, and United Kingdom) and be accredited via the European Credit Transfer System (ECTS). The programme consists of distance learning courses, intensive courses, dissertations, and levelling courses. According to Salmela (1992), all the Scandinavian countries, except Denmark, offer graduate programmes in sport and exercise psychology. Few programmes are offered in Africa (Nigeria), the Middle-East (Egypt), Asia (China, India, Japan), and South America (Brazil). Refer to Lidor et al. (2001) for more specific information, although this information is dating rapidly. A global review of training and selection processes for sport psychologists was instigated by ISSP in 1998. The results were reported during the 2001 ISSP Congress and were published in the IJSEP (Morris, Alfermann, Lintunen, & Hall, 2003). Updated information is provided in the *Handbook of Research in Applied Sport and Exercise Psychology: International Perspectives* by Hackfort, Duda, and Lidor (2005).

4. Appendix Material

4.1. Terminology

Terms used in sport and exercise psychology are principally derived from the parent domain of psychology. Some unique terms that relate psychological constructs to sport are also common. Sport psychologists frequently use terms from related sport and motor disciplines, such as sports medicine, exercise physiology, biomechanics, motor control, and sociology of sport.

4.2. Position Statement(s)

Position Stands/Statements are published by the ISSP (see ISSP homepage: www.issponline.org) and FEPSAC.
ISSP Position Stands:

- Physical Activity and Psychological Benefits. 1992. Published also in *The Physician and Sport Medicine*, 20, 179-184, 1992; The *International Journal of Sport Psychology*, 23, 86-91, 1992; Journal of Applied Sport Psychology, 4, 94-98,1992.
- The Use of Anabolic-Androgenic Steroids (AAS) in Sport and Physical Activity. 1993. Published also in *The Sport Psychologist*, 7, 4-7, 1993; *International Journal of Sport Psychology*, 24-27, 1997.
- Aggression and Violence in Sport. 1996. Published also in the *International Journal of Sport Psychology*, 27, 229-236, 1996; *The Sport Psychologist*, 11, 1-7, 1997; *International Journal of Sports Medicine and Physical Fitness*, 1, 1997.
- Use of the Internet in Sport Psychology. 2001. Published also in the *International Journal of Sport Psychology*, 32, 207-222, 2001.
- Ethical Principals.

- Competencies and their Accomplishment in Sport and Exercise Psychology. 2003.
- Training and Selection of Sport Psychologists: An International Review. 2004.

FEPSAC Position statements:

- Definition of Sport Psychology.
- Children in Sport.
- Sports Career Transitions.

FEPSAC released a more specific statement of "Sports Career Termination". Whereas Position Statement 3 referred to the transitions that occur at all ages and levels of sport, the career termination statement focuses on the issues of retirement from elite sport.

4.3. Varia

There are also Videos and DVDs available to support learning and training in sport and exercise psychology published/distributed by FIT (see www. fitinfotech.com/ video/videoresults.tpl).

In addition to ethical standards, internationally acceptable standards for education and training in sport and exercise psychology, and criteria to determine professional differentiations and specifications (certification) in sport and exercise psychology (e.g., Sport Psychologist, Mental Coach, Sport Psychology Consultant, Exercise Psychology Consultant) are needed. This is one of the missions of the ISSP and various position stands contribute to the fulfilment of this mission.

4.4. Free Statement

Sports can be regarded as a model for globalisation. Hence, intercultural and interindividual actions and relations are subject to the future theoretical, empirical, and applied work in sport and exercise psychology. Cross-cultural methodologies and concepts as well as interdisciplinary cooperation in Sport Science, Health Science, and Psychology are needed to cope with the challenges associated with the increasing complexity in research and practice – not only, but also in sport and exercise psychology.

Sport and Leisure Facilities

Klaus Meinel

Contact

Klaus Meinel
Managing Director
International Association for Sports and Leisure Facilities (IAKS)
Am Sportpark Müngersdorf 3
50933 Köln
Germany
Phone: +49 221 4912991
Fax: +49 221 4971280
Email: iaks@iaks.info
www.iaks.info

1. General Information

1.1. Historical Development

The ideas behind the design and construction of sports facilities in the modern age go back to the beginning of the 19th century. They were initially strongly influenced by sport in England, later by the gymnastics movement in Germany and Scandinavia, and finally by the European-wide propagation of swimming as a means of preventing drowning and staying healthy. The principles of facility planning were established at the beginning of the 20th century, to a large extent between 1920 and 1940 and, with greater intensity, after the Second World War. An exchange of experience on the European level was initiated in 1957. This culminated in the founding of the International Association for Sports and Leisure Facilities (IAKS) in 1965.

1.2. Function

With the exception of a few sports practised solely in the countryside or natural environment (e.g. skiing or mountain-biking), sports generally require an infrastructure built and maintained specifically for the particular sport. These have to meet the sport's functional needs as well as the needs of safety, economy and ecology. The elements of such an infrastructure can include competition, training and leisure facilities, as well as ancillary facilities for athletes, spectators, the media and administration. The goal is therefore to produce appropriate planning, construction and operating principles, which are regularly updated to bring them into line with the latest findings.

1.3. Body of Knowledge

Essential for compliance with the requirements of sport facilities, as outlined in section 1.2 above, is the networking of findings from the following sports-related fields of knowledge: education, sociology, psychology, medicine, biomechanics, accident prevention, architecture, landscape design, engineering, materials testing, economics and ecology.

1.4. Methodology

A comprehensive presentation of the scientific methodology for designing and evaluating facilities is not possible because of the diversity of approaches in the various scientific fields. The goals of the methods in their totality are:

- Definition of the demand for the various facility types in accordance with today's and tomorrow's needs of facility users
- Demand-oriented planning and realisation of the respective construction project in accordance with the requirements of sports function, economics, ecology and design
- A mode of operation respecting economic and ecological requirements and taking into account the needs of the respective sports disciplines and leisure activities

1.5. Relationship to Practice

Compliance with the requirements for sporting facilities, as outlined in points 1.2 to 1.4 above, is not possible without an ongoing and intensive exchange of information between scientists and practitioners and other institutions and bodies active internationally (such as IOC, IPC, GAISF, ANOC, IANOS, TAFISA, UIA and ICSSPE). Within IAKS, information is exchanged particularly in its multi-disciplinary congresses and working groups, in which internationally recognised experts discuss the effects of major sports and leisure activity trends on the associated infrastructure and prepare respective planning guidelines as well as other planning aids.

1.6. Future Perspectives

If one considers the life cycle of a sports facility from the point of view of costs, the sum required for construction amounts to only 20% to 25% according to various resources of Ministries of Finance and of facilities management groups. The remaining 75% to 80% is required for operation, maintenance, necessary replacement of technical components, demolition and disposal. In view of the fact that the funding available for sports facilities is becoming increasingly scarce, the goal should therefore be to lighten the overall financial burden on the client and operator by ascertaining the demand for the facility as accurately as possible, by designing it meticulously and by ensuring cost-effective construction and operation throughout the facility's life cycle. This is one of the main focuses of IAKS for the future, without neglecting its other goals or the monitoring of trends in sport and their effects on the facilities provided.

2. Information Sources

2.1. Journals

Available in the fields of the planning, realisation and operation of sports facilities on the international scene is the bi-monthly journal "Sports Facilities and Swimming Pools" (sb) from IAKS. As part of IAKS' services, it is provided free of charge to members.

2.2. Reference Books, Encyclopaedias, etc.

There are no comprehensive national or international encyclopaedias. This is due to the broad overall spectrum of information required and the changing nature of developments in many of the individual sectors.

2.3. Book Series

There are over 30 titles to choose from within the IAKS series of publications. Included within these publications, are the results of conferences, panels of experts and seminars.

2.4. Conference/Workshop Proceedings

Every two years, IAKS holds its international congress for the design, construction, modernization and management of sports and leisure facilities. The congress papers and presentations are published as a hardcopy and on the IAKS website.

2.5. Data Banks

There are two data banks with information and contact addresses of specialized planners and companies for sports and leisure facilities available on the IAKS website.

2.6. Internet Sources

More information about the IAKS, congresses and publications can be found at www.iaks.info.

3. Organisational Network

3.1. International Level

The IAKS is the international organization for the design, construction, modernization and management of sports and leisure facilities. IAKS is a non-profit organisation that has been active in the field of sports and leisure facilities for over 40 years. IAKS has about 1,000 members in over 100 countries throughout the world: These include:

- Ministries of sport and housing,
- Sports organizations such as Olympic Committees, sports councils and sports federations/associations,
- Universities, technical colleges, schools of engineering, colleges of physical education and sports institutes,
- Local government departments such as sports departments, housing departments and parks and gardens departments,
- Architects and engineers,
- Industrial companies and business federations.

Members benefit from the association's international outlook, its international exchange of ideas and its many services:

- Bimonthly international journal "Sports Facilities and Swimming Pools", which is concerned with the latest developments, trends and experience on key themes

such as stadiums and sports grounds, sports halls and arenas, pools and wellness, sports facilities for the Olympic Games, sport and the environment, sport in the urban setting and winter sports facilities,
- Use of the IAKS logo with a reference to membership,
- Basic entry in the Architects/Engineers or Sports Facility Industry database at the IAKS homepage. These databases enable potential customers to find specialists in the field of design, project management and expert opinions in the sports facility sector, and
- Advice from IAKS on technical issues.

The institutional bodies of IAKS are the General Assembly and the Executive Board. These are supported in their daily work by the office in Cologne, Germany, run by full-time staff.

The International Olympic Committee (IOC) has recognized the IAKS as the only international organization in the field of sports facility development.

The IAKS also cooperates with:

- The International Paralympic Committee (IPC),
- The General Association of Sports Federations (AGFIS/GAISF),
- The International Council of Sport Science and Physical Education (ICSSPE),
- The Sport and Leisure Programme of the International Union of Architects (UIA).

The IAKS is also included in the United Nations' list of NGOs with consultative status at the UN Economic and Social Council (UN ECOSOC).

3.2. Regional Level
See Section 3.3 National Level.

3.3. National Level
In countries or regions with a certain number of members, these members can form IAKS sections whose goals on the national level must concur with the goals

of IAKS. This must be evident from the section's rules and is anchored in an agreement between the respective section and IAKS. The national or regional sections generally participate in IAKS activities (events, work group activities) but also organise, in some cases with the involvement of IAKS, their own events.

3.4. Specialised Centres

Not applicable.

3.5. Specialised International Degree Programmes

Not applicable.

4. Appendix Material

4.1. Terminology

Not applicable.

4.2. Position Statement

The International Association for Sports and Leisure Facilities was founded in Cologne, Germany in 1965. The IAKS collects, evaluates and disseminates the experience acquired by its members or by other bodies during the planning, construction, equipping and management of all kinds of facilities for recreation, games and sport. As a result of these activities, the IAKS serves as an expert and consultant to:

- Ministries of sport, education and building,
- Sports organisations (Olympic committees, associations and federations),
- Universities, technical colleges, schools of engineering, and schools and institutes of physical education,
- Administrations of medium-size and large towns and cities (sports departments, construction departments, parks departments),
- Architects and engineers,
- Industrial companies and business associations.

IAKS Goals

In view of the growth in sports-oriented lifestyles, close examination of the space required for exercise in the urban environment and in the open countryside is more urgently needed today than ever before. In the past, the emphasis often used to be solely on meeting the quantitative needs for sports facilities. Today, high priority is accorded to quality. The goal is to develop a sports and leisure facility that is equally functional, well designed and environmentally compatible. In the planning, construction and operation of such a high-quality facility, the sensible use of new materials, technologies and methods is also essential from the point of view of economy. In this way the goals are in full agreement with Agenda 21 of the Olympic Movement.

Through its work, IAKS contributes to the realisation of such sports and leisure facilities. At the same time, it highlights the citizen's right to suitable sports facilities in sufficient numbers. Important goals for future-oriented sports facilities are their multi-functionality, integration into their surroundings and environmental compatibility. Consideration should therefore be given to the following aspects:

- Forms of facility with varied uses and an attractive appearance (fun or adventure quality) for a wide range of age and interest groups and standards of performance (sport for all),
- Promotion of health-related activities,
- Suitability for, or the possibility of, conversion to new forms of exercise or play,
- Integration of other leisure-oriented, social or cultural facilities,
- Spatial interconnection of sports facilities and integration into urban open-space schemes and into the residential environment, with easy access for less mobile users,
- As far as possible, unlimited access and low admission fees,
- Environmentally compatible construction, e.g. maximising soil permeability, avoidance of contaminated construction materials as well as material- and energy-intensive production processes, possibility of recycling building materials,
- Environmentally compatible operation, e.g. minimising consumption of energy and water and the use of alternative energy sources.

In the public relations sector, the IAKS attempts to combine the efforts of the associations and decision-makers responsible for sports facility construction and to develop argumentation aids for the battle for shrinking reserves of finance and space. To ensure satisfactory provision of sports facilities in the future, those in decision making positions must unite with sports organisations, leaders and active members in answering the questions, how to convince the community to meet the sports facilities requirements of future generations.

A central element of the IAKS' public relations work is the Award for Exemplary Sports and Leisure Facilities. This is a joint competition for operators and architects, which deliberately focuses on the special importance of facility quality. Held for the first time in 1987, it pursues the goal of raising world-wide awareness, not only for functionality, but also for well-designed buildings and facilities. The importance of this competition prompted the International Olympic Committee in 1999 to co-sponsor this award and it has since been called the IOC/IAKS Award. In 2005, the International Paralympic Committee joined this project which resulted in the additional awarding of a Distinction for Accessibility.

IAKS Activities

The tasks and activities of the IAKS are:

- Creation of an international and interdisciplinary platform for the design, construction, equipping, modernization, financing and management of sports and leisure facilities,
- Involvement in the drafting of standards and guidelines in Europe and Germany,
- Issuing publications, e.g. the bi-monthly journal "Sports Facilities and Swimming Pools" (sb),
- Holding congresses, seminars, conferences and exhibitions,
- Presentation of the IOC/IAKS Award, the international architecture award for exemplary sports and leisure facilities, and of the IPC/IAKS Distinction for Accessibility,
- Advising clients, designers, operators and users of sports facilities, from the individual project through to Olympic bids,
- Maintaining contacts with international organizations of sport, sports science and the construction industry.

The 20 congresses already held are an outstanding example of the IAKS' expertise in sports and leisure facilities. They have been held at two-yearly intervals since 1969 and cover a wide range of subjects, from facility planning and the construction of sports and leisure facilities to the operation of these indispensable facilities for school sports, sport for all and top-level sport. Over the last 39 years, a total of 708 experts from 35 countries have given talks and supplied technical information at the congresses. The congresses are accompanied by parallel events on allied subjects and by an international trade fair, the FSB (International Trade Fair for Amenity Areas, Sports and Pool Facilities).

4.3. Varia

Not applicable.

4.4. Free Statement

Not applicable.

Sport History

Gertrud Pfister

Contact

Prof. Dr. Gertrud Pfister
Institute of Exercise and Sport Science
University of Copenhagen
Norreallé 51
2200 Copenhagen
Denmark
Phone: +45 35320861
Fax: +45 35321747
Email: GPfister@ifi.ku.dk

1. General Information

1.1. Historical Development

Sport History, especially the tradition of Greek antiquity, was closely connected with the development of movement cultures in Western societies. Johann Christoph Friedrich GutsMuths in his famous book about "Gymnastic for youth" (1793) as well Friedrich Ludwig Jahn, the "father" of German "Turnen" (Gymnastics) referred to Greek athletics as their models. In the second part of the 19th century, history was part of the education program for physical education teachers and in the first half of the 20th century, a large and indepth work on the history of gymnastics, Turnen and Sport, was published. After the Second World War, sport history developed as an academic discipline, which is integrated in sport sciences and dependent on the "mother discipline", that is, the historical sciences.

The International Society for the History of Physical Education and Sport (ISHPES) is the umbrella organisation for sport historians from all over the world. ISHPES formed in 1989 through the merging of the International Committee for the History of Physical Education and Sport (ICOSH) and the International Association for the History of Physical Education and Sport (HISPA). ICOSH was founded as early as 1967 in Prague and HISPA was founded in 1976 in Zürich. ISHPES has been affiliated, since 1990, with the International Council of Sport Science and Physical Education (ICSSPE) and in 2000, ISHPES became a member of the International Committee of Historical Sciences (ICHA). With the establishment of these organisations, a network of professional sport historians developed.

1.2. Function

The aims of sport history are to:

- detect and describe developments of physical activities, physical education and sport in different historical periods and different areas;
- identify reasons, processes, connections and effects of historical developments in the area of physical activities, physical education and sport;
- interpret the interdependent influences between physical cultures and societies;
- describe and explain the history of physical activities before the backdrop of the political, economical and social history;

- investigate the developments and changes of different concepts of physical activities, different types of sport and performance levels, different organisations
  and institutions and different persons and groups involved in physical cultures;
- determine the influence of gender, class, race and religion on the opportunities
  and barriers for participation in physical activities;
- identify norms, values and ideologies connected with physical activities and
  sport;
- gather knowledge about local, regional, national and international developments;
- conduct intercultural comparisons;
- discover sport as a social and cultural phenomenon;
- support the understanding of the present situation of sport;
- contribute to reflections about the future of physical education and sport; and
- develop visions and perspectives for the new millennium.

1.3. Body of Knowledge

The history of games, gymnastics and sport has a long tradition. The "fathers" of
sport, gymnastics and "Turnen", among others Gutsmuths and Jahn, used historical sources in order to find and select exercises and activities, which they included
in their concept of physical culture. Since the middle of the 19th century, a large
number of articles and books has been published in Western industrialised societies focusing on the different areas of sport history in many parts of the world.
Additionally, there are publications about the sport history of a country, a region
or a city, and also many books on the history of a specific sport discipline. Finally,
biographies as well as histories of sport organisations contribute to the body of
knowledge.

The range of the body of knowledge is defined by the above-mentioned aims and
functions of sport history. In order to explain processes and developments, theoretical approaches must be included in the work of sport historians. In addition,
the results of the "mother" discipline, or better, of the many disciplines of historical sciences, must be integrated into the knowledge of sport history. There are
especially close relations between sport history and the history of health, leisure
and medicine. Moreover, there is a close connection with other disciplines of sport
sciences, including sport sociology and sport pedagogy.

1.4. Methodology

Sport history uses the same methodological approaches as history, depending on the questions and the accessible material of the research. The traditional methodology of history is the phenomenological-hermeneutical approach, meaning the collection, selection, critical evaluation and interpretation of sources. For certain historical periods and for specific problems, the interrogation of contemporary witnesses, i.e. those who experienced the event, may be useful. Possibilities and problems of oral history are discussed extensively in the textbooks about methods and auxiliary sciences of history. In addition, empirical analytical methods can be used, i.e. content analyses. Sport history aims to reconstruct the developments, processes and connections between physical activities and socio-cultural conditions. In order to explain results and determine insights into causes and effects, sport history employs different theoretical approaches. Many scholars understand sport history as part of social history and they describe the interconnectedness of the relations between the interests of social classes and groups, the economical and political developments and the physical culture of a given period. However, there are also sport historians who try to gain insights into the cultural-historical configuration with the help of the civilisation theory founded by Norbert Elias. In addition, the approaches of the French sociologists Michel Foucault and Pierre Bourdieu, or Eric Hobsbawm and Richard Sennet among others, play a role in the discussions of the sport history scientific community. In recent years, the work of Pierre Nora on the functions of history as "Lieux de Mémoire", places of remembering, has influenced the historical scientific community. Thereby, the role of the collective memory and political myths for constructing and strengthening a nation's states and for the identification with nations, regions or groups has been highlighted. Because sport, events, successes and heroes can be powerful myths and places of remembrance, this approach will be very useful for sport history.

1.5. Relationship to Practice

It is evident that sport history does not have an immediate influence on learning, training and practising sport. However, sport history can offer knowledge, which gives insight into developments, causes and effects and the backgrounds of physical culture. Thus, sport history can contribute to an understanding of the present situation and can provide knowledge that is necessary for making decisions, developing strategies and clarifying perspectives. Sport history also helps to detect myths and ideologies and to destroy the conviction of the self-evidence of sporting practices. For example, research in sport history can show that performance, competition and records are not an anthropological constant, but characteristics of the physical culture of modern industrialised countries. As a second example, investigations about the development of women's and men's sport demonstrate the change of gender roles in sport and society. These studies can be used to fight against stereotypes and can remove barriers for women (and men) in sport today.

1.6. Future Perspectives

Sport history is today in an ambivalent situation. On one hand, interest in and the need for historical approaches and knowledge has increased in recent years and the number of publications and the standard of research is higher than ever before. Conferences in sport history attract more and more colleagues, also from other fields. But on the other hand, sport history has lost its role as an integrative part of the curricula of sport and physical education studies. In many countries, decreasing resources of universities have led to a concentration on the so-called applied sciences. It is a great challenge for all involved in sport history to fight for the recognition of their subject and to transfer the positive attitude towards history in the sport institutes. Sport history will only flourish if it has its academic background and stronghold in universities. In order to get more public attention and support, it will also be necessary to discuss new ways of researching and teaching sport history.

2. Information Sources

2.1. Journals

There are several national and international journals in the area of sport history. Among them:

Canadian Journal of History of Sport and Physical Education
Idrætshistorisk Årbog
Journal of Olympic History

Journal of Sport History
Ludica
NIKEPHOROS. Zeitschrift fur Sport und Kultur im Altertum
NINE: a Journal of Baseball History and Social Policy Perspectives
Olympika
Skiing Heritage
Soccer and Society
Sportzeiten
Sportimonium
Sporting Traditions. Journal of the Australian Society for Sports History
Sport History Review
Sport in History (formerly The Sports Historian)
Sport und Gesellschaft. Zeitschrift für Sportsoziologie, Sportphilosophie,
Sportökonomie, Sportgeschichte
Stadion
The International Journal of the History of Sport

Articles on sport history are also published in interdisciplinary journals of sport sciences (in a broad sense), such as in the German journal *Sportwissenschaft* or in the United States' publication *Women in Sport and Physical Activity Journal*. Other interdisciplinary journals include Quest, the journal of The National Association for Kinesiology and *Physical Education in Higher Education and the Scandinavian Journal of Medicine and Science in Sports*.

2.2. Reference Books, Encyclopaedias, etc.

Levinson, D. and Christensen, K. (Eds.). (1996). *Encyclopaedia of World Sport. III Volumes.* Santa Barbara, Denver, Oxford: ABC-Clio.

Christensen, K., Guttmann, A. and Pfister, G. (Eds.). (in print). *International Encyclopaedia of Women and Sport.* Berkshire: ABC-Clio.

Cox, R, Jarvie, G and Vamplew, W (Eds.). (2000). *Encyclopedia of British Sport.* Oxford: ABC Clio.

Cox, R.W. (2003). *History of Sport: A Guide to Historiography*, Research Methodology and Sources of Information. London: Frank Cass.

Good references to contextualise sport history are:

Coakley, J. (2001). *Sport in Society: Issues and Controversies*, 7th ed, ch. 3. New York: McGraw Hill.
Horne, J., Tomlinson, A. and Whannel, G. (1999). *Understanding Sport: An Introduction to the Sociological and Cultural Analysis of Sport*, chs. 1-3. London: E & FN Spon.

2.3. Book Series

The International Society for the History of Physical Education and Sport (ISHPES) publishes a book series with Academia, Sankt Augustin. Editors are Thierry Terret, Gertrud Pfister and Michael Salter. 15 volumes have been published.
Sport in the Global Society Series edited by J.A. Mangan is published by Taylor and Francis.

2.4. Congresses/Workshop Proceedings

All of the above mentioned associations conduct conferences. Proceedings of sport historical seminars and congresses are published, among others, by ISHPES, European Committee for Sports History (CESH), ISOH and Australian Society for Sport History (ASSH).

2.5. Data Banks

Sport historical research is integrated into databases of sport and sport sciences. The most important databases are Vifa:Sport (Bundesinstitut für Sportwissenschaft, Bonn, Germany; Deutsche Vereinigung für Sportwissenschaft, Hamburg, et al.) and *Sportdiscus* (Sport Information Centre, SIRC, Canada).

2.6. Internet Sources

ISHPES organises a sport history Internet Listserv called *sporthist*. This network is supported by Richard Cox. ishpes.mcs-creations.com/

The best internet source for archives, associations, museums, publishers and many other types of information is the Scholarly Sport Sites organised by Gretchen Ghent www.ucalgary.ca/lib-old/ssportsite/

Other helpful webpages are:

Resource Guide to Sports History
    www.heacademy.ac.uk/hlst/resources/guides/guides_sport
Higher Academy Network for Hospitality, Leisure, Sport & Tourism Network
    www.hlst.heacademy.ac.uk/about/aboutus.html
Hickok Sports
    www.hickoksports.com/history.shtml
Intute: Social Sciences
    www.intute.ac.uk/socialsciences/altislost.html
How To Find Out History of Sport
    www.sprig.org.uk/htfo/htfohistory.html
Sportspages
    www.sportsbooksdirect.co.uk/
La84Foundation
    www.la84foundation.org/
Richard Cox's Sports History Bibliographical Service and Internet Gateway
    www.ishpes.org

3. Organisational Network

3.1. International Level

ISHPES, the International Society for the History of Physical Education and Sport, is the successor to HISPA (International Association for the History of Physical Education and Sport) and ICOSH (International Committee for History of Sport and Physical Education). ICOSH was founded in 1967 in Prague, HISPA in Zürich in 1973. At the 13th International HISPA Congress in Olympia (Greece) in 1989, the two former international sport history organisations decided to merge into one world-wide society, ISHPES.

ISHPES promotes research and teaching in sport history. Its purpose is to facilitate exchanges in sport history through international congresses and seminars and through the production and dissemination of appropriate publications. ISHPES organises seminars and congresses, publishes a bulletin (twice a year) and a book series. The Society gives two awards: the ISHPES award for the outstanding work of a scholar of high reputation; and an award for a young scholar for his/her research.

The International Society of Olympic Historians was formed in 1991, in London. The purpose of the organisation is to promote and study the Olympic Movement and the Olympic Games. This purpose is achieved primarily through research into the history of the Olympic Movement and the Olympic Games, the gathering of historical and statistical data and the publication of research via journals, monographs, etc.

There are several international associations focusing on a specific type of sport, such as the International Skiing History Association www.skiinghistory.org/

3.2. Regional Level

CESH (The European Committee for the History of Sport) was founded in 1995, in Bordeaux, France, to further the interest of cooperation between European scholars of different language backgrounds. It is supported by a college of fellows and organises annual conferences, which emphasise the work of young scholars, and a peer reviewed yearbook.

The Northeast Asian Society for the History of Physical Education and Sport was founded in 1994. It includes primarily scholars from China, Korea, Chinese Taipei and Japan. The first congress was held in 1995. Members organise congresses every two years and produce a publication twice a year.

3.3. National Level

The British Society of Sport History was founded by Richard Cox in 1982. Today, it has 250 members, publishes a journal, *Sport in History*, a Newsletter and occasional monographs through its imprint Sports History Publishing. It has a website, a listserv (see 2.6 Internet Sources) and an annual conference.

The North American Society for Sport History (NASSH) was founded in 1972 by a group of American and Canadian sports historians, most of whom were then working in departments of physical education. The first president was Marvin H. Eyler. NASSH meets once a year and publishes the *Journal of Sport History*, which has become one of America's most frequently cited scholarly journals.

Australian Society for Sports History (ASSH) was founded at the fourth Sporting Traditions conference in 1983. The Society's first president was Colin Tatz and the Society launched the first issue of its journal, *Sporting Traditions*, in November, 1984. Wray Vamplew was the first editor of *Sporting Traditions* and is credited with being the driving force behind ASSH in its early days.

Since its expansion, beginning in the early 1990s, sport history in Brazil has developed by means of a network of universities, rather than a national society. The embodiment of the network takes place during an annual national congress and its respective proceedings.

There are also very active sport history societies in Finland, Denmark and other countries. In Germany, sport historians form a section in the interdisciplinary German Association of Sport Sciences.

3.4. Specialised Centres

The International Centre for Sports History and Culture, De Montfort University, Leicester, United Kingdom was founded in 1995. It has full-time academics attached to it, as well as a host of visiting national, European and international academics, journalists and experts in the field of sport. Staff at the Centre supervise a wide range of PhD and other research students who are all pursuing their interest in the history, practice and importance of sport at the highest level. The Centre also runs a successful masters program called Sport, History and Culture.

The International Centre for Olympic Studies at the School of Kinesiology, The University of Western Ontario, London, Canada is a research, resource and service facility which aims to encourage, generate and disseminate scholarship on a broad range of social and cultural themes related to the Olympic Movement.

There are several centres in areas connected with sport history, specifically concerned with traditional sports and games. For example, the Vlaamse Volkssport Centrale, Sportmuseum Vlanderen, Leuven, Heverlee, Belgium which organises and promotes traditional games.

Information about sport museums is provided by the British Society of Sports History www.ishpes.org

Centre d'Estudis Olímpics (CEO-UAB) at the Universitat Autònoma de Barcelona collaborates with the International Olympic Movement and with the various national and international sports organisations, conducts research into Olympism and sport, organises courses, compiles information and documentation and organises conferences.

The LA84 Foundation is a private, non-profit institution endowed with Southern California's share of the surplus funds generated by the 1984 Olympic Games held there. Its purpose is to serve youth through sport. It has an excellent library and archive and also provides literature and materials online. www.la84foundation.org/

In addition, there are centres and/or working groups for sport history embedded in faculties/institutes of sport sciences, like the working group for current sport history at the University of Potsdam. In addition, there are museums, archives and documentation centres with a focus on sport history at the Institute for Sport History of Lower Saxionia in Hoya, Germany.

3.5. Specialised International Degree Programmes

Not applicable.

4. Appendix Material

4.1. Terminology

European and especially non-English speaking scholars have great problems with the terms sport and science. In Europe, sport is a very broad term for physical activities of all kinds and on all levels. Also, there is no differentiation between science and humanities with only one term for both approaches to knowledge. Clarification on these issues is therefore necessary in publications and international research or discussions.

4.2. Position Statement(s)

Not applicable.

4.3. Varia

Not applicable.

4.4. Free Statement

In many countries, sport history has many adherents but very little official support. Even if sport history is included in most of the curricula for physical education and sport students, it is very seldom represented by a Professor position. Among the many barriers and problems, sport history is challenged by the lack of financial resources, which tend to be concentrated on applied sciences. The discipline will need the joint efforts not only of sport historians, but also the scientific community of sport sciences (and humanities) to fight for the continuation, propagation and extension of sport history. In a time when developments for the future must be decided upon, sciences such as history, philosophy and sociology are necessary in order to provide the insights and knowledge, which form the basis for all decisions.

Sport Information

Gretchen Ghent

Contact

Gretchen Ghent, Librarian Emeritus
President, International Association for Sport Information and
Chair, North American Sport Library Network
C/o University of Calgary Library, 405A MLB
Calgary, Alberta T2N 1N4
CANADA
Phone: +403 220 6097
Email: gghent@ucalgary.ca

1. General Information

1.1. Historical Development

The development of sport information and documentation services parallels that of the larger world of information science, documentation and information management. These developments have been advocated, standardised and promoted by the International Federation of Library Associations and Institutions (IFLA) (www.ifla.org), the International Organisation for Standardization (ISO) (www.iso.ch), national library associations and more specifically, the International Association for Sport Information (IASI) (www.iasi.org). Sport documentation centres/libraries or academic sport sciences collections originated at universities, colleges and institutes where information resources were needed to support physical education and sport sciences programs and research. A more recent development is the government-supported sport documentation centre that provides resources for coaches, sport administrators, athletes, the media and sport sciences researchers. Specific tools have been developed to aid sport information and documentation, e.g. sport specific thesauri, dictionaries, encyclopedias and databases. The development of these tools follows similar guidelines and standards found in libraries in general and other disciplines. Recently, with the advent of the computer, increased data storage capacity and improved software, sport information and documentation services have greatly improved access and delivery of information for the enquirer or researcher.

1.2. Function

The sport information and documentation functions are to:

- Identify and organise sources of information and resources pertinent to the needs of sport information users, expressed or latent
- Facilitate access and delivery of that information in whatever format is necessary
- Actively promote and disseminate the availability of information on sport and related disciplines.
- Promote international cooperation in the field of sport information

1.3. Body of Knowledge

Sport information and documentation draws from all major disciplines including, library science, psychology, sociology, biology, education, medicine and technology and includes multi-media information, data, published and digital sources of information, both theoretical and applied.

The current sports information professional, documentalist or librarian's body of knowledge consists of an understanding of how the world's knowledge is organised and an appreciation of various classification systems, e.g. Universal Decimal System, Dewey Decimal System or Library of Congress Classification. The professional knows how to manage the registration, representation, cataloguing, classification and indexing of documents and is versed in bibliographical description including record structures, e.g. MARC (MAchine Readable Cataloguing) or the database field structure. The professional also acts as a consultant and analyst, interpreting and finding information to assist clients in their research. Management and administrative expertise facilitates the planning and organisation of library and documentation services, computer systems and technology for the intranet and for internet website content. Skills and knowledge of how to train and educate staff and users of library services are indispensable.

1.4. Methodology

In the early years of sport documentation work, focus was on the creation of thesauri, applying logic and specificity to the study of the hierarchical and inter-relational structure of sport terminology. The goal was to agree on a uniform, specialised terminology. This early work was necessary to provide the foundation for the building of databases, describing documents and promoting good storage and retrieval practices. Much of this work was done by IASI pioneers, Robert Timmer, Josef Recla, Karl Ringli, Siegfried Lachenicht and Gilles Chiasson. Their work resulted in the *Sport Thesaurus* (6th ed, 2002). Methods of various sciences are applied to information and documentation and can be used in a wide range of applied methodologies, e.g. psychology for user behaviour, computer science methodology to document analysis and searching tools.

1.5. Relationship to Practice

Sport information professionals work daily with clients of widely divergent educational, technical and professional backgrounds. Such diverse clients require the information professional to identify, acquire and make accessible, resource tools of all levels of complexity and scope. Sport information professionals are expected to ascertain the information needs of the secondary school student, the sport fan and the sport scientist equally.

1.6. Future Perspectives

The future is encouraging for the specialised sport database, greater access to fulltext documents and the educational use of the website. Sport organisation administrators and documentation managers can utilise software that enables the organisation to develop results, ranking and media information systems either for internal use, their national clientele or for website promotion. Websites can be the main communication vehicle for sports organisations to keep in touch with athletes, coaches and other administrators who need to coordinate activities. Websites can also be a powerful educational tool for organisations and researchers. Bibliographical databases, e.g. SPORTDiscus (www.sirc.ca/products/sportdiscus. cfm) provide enhanced delivery services, are indexing fulltext resources and providing hypertext links among other projects. Video and other audiovisual material is being delivered to monitors, workstations and remotely via the sport organisation's intranet. Faster document delivery can now be achieved utilising improved interlibrary loan software so that articles not available in-house or fulltext on a website, can be delivered to the individual's workstation in fulltext format.

2. Information Sources

2.1. Journals

There are many journals in the area of information and documentation with some containing sport documentation articles and bibliographies. The journal contents are indexed in *Library & Information Science Abstracts* (London: Bowker-Sauer) and *Library Literature and Information Science* (Bronx, NY: H.W. Wilson). Recently, two journals published special issues on sport information. Eleven papers can be found in *Revista General de Información y Documentación*, (2000) 10 (1). This issue is entitled, La Información deportiva en España. The second source with ten papers is in *North Carolina Libraries* (2001), 59 (2, Summer). Guest Editor is Suzanne Wise. The issue has the title, Sport: *the Liveliest Art*. Many other sport-specific or subject-based journals have occasional or frequent bibliographies. For example, in the journal, *New Studies in Athletics*, various issues contain a bibliography on topics of current interest. In addition, the semi-annual issues of NASLINE (www.naslin.org/nasline.html) have an extensive bibliography of new print and online publications selected and suggested for addition to the college, university or special sport library.

Journals utilised by sport scientists, athletes, administrators and others, number in the thousands with more titles being added each year, usually in electronic format. Academic libraries are converting their print subscriptions to electronic format for use by their campus faculty, students and staff. Book collections are now supported by electronic book services from vendors, e.g ebrary (www.ebrary.com) and Netlibrary (www.netlibrary.com).

2.2. Reference Books, Encyclopaedias, etc.

Sport librarians and information professionals utilise many sport specific reference works in the course of their responsibilities to provide reference service to library/documentation centre users. Dictionaries consulted include:

- Dictionary of the Sport and Exercise Sciences (1991);
- Oxford Dictionary of Sports Science and Medicine, 3rd ed. (2006);
- Dictionary of Sport and Exercise Science (2006);
- Wörterbuch der Sportwissenschaft (1992);
- Dictionnaire des Sports (1995);
- Dictionary: Sport, Physical Education, Sport Science (2003), ed. by Herbert Haag.

Four recently published encyclopedias provide essential starting points for many library users:

- Encyclopedia of Sports Science, 2v. (1997);
- International Encyclopedia of Women and Sports, 3v. (2001);

- Encyclopedia of World Sport: from Ancient Times to the Present, 3v. (1996); and
- Berkshire Encyclopedia of World Sport, 4v. (2005).

Other guides to sport information sources include:

Scarrott, Martin (Ed.). (1999). *Sport, Leisure, and Tourism Information Sources: a Guide for Researchers*. Oxford: Butterworth Heinemann.

To assist in the planning and organisation of sport documentation centres, members of IASI have collaborated in publishing a manual in four languages:

Clarke, Nerida et al (Eds.). (2000). *Manual for a Sports Information Centre*. Lausanne, International Olympic Committee.
This manual is also available in Spanish, *Manual del Centro de Información Deportiva Manual* and French, *Manuel du Centre d'Information Sportive* and Portuguese available in PDF format at www.iasi.org/publications/monographs.html

2.3. Book Series

Volume 4 of ICSSPE's Perspectives series contains 12 papers on Sport and Information Technology (Meyer and Meyer, 2002). For a listing of the contents see the Publications section on the ICSSPE website www.icsspe.org

2.4. Conference/Workshop Proceedings

The International Association for Sports Information (IASI), the leading professional body for sport information and documentation, holds a World Congress every four years and papers from these congresses are indexed in the international sport database, SPORTDiscus and are available fulltext on the IASI website.

The latest four IASI Congress proceedings are:

- The Value of Sports Information: Toward Beijing 2008: Proceedings of the 12th IASI World Congress, 19-21 May, 2005, Beijing, Beijing Sport University, 2005.

- Sports Information in the Third Millennium: Proceedings of the 11th IASI World Congress, Lausanne, 25th – 27th April 2001, Lausanne: Olympic Museum and Studies Centre, 2001.
- Scientific Congress of the International Association for Sports Information (10th: 1997: Paris). Actes = Papers, Paris: INSEP Publications, 1997.
- International Association for Sports Information, 9th Scientific Congress: Sports Information in the Nineties: Roma, 7-10 June, 1993. Roma: CONI, Scuola dello Sport, 1993.

Bibliographic information for earlier congresses in Graz, 1975 and Duisburg, 1977 are listed on the IASI website www.iasi.org

2.5. Data Banks

Sport documentation centres and libraries serve a wide range of clientele, from the academic student or faculty member to sport administrators, government officials, independent researchers, journalists, athletes and sports fans. In addition to the library's in-house collection, the database is the most important tool.

The most comprehensive bibliographic sport database is the SPORTDiscus. The majority of the indexing is undertaken by the Sport Information Resource Centre (SIRC) in Ottawa (www.sirc.ca/products/sportdiscus.cfm), with additional citations and fulltext material added by EBSCO, the current owner of the database. A database of over 700,000 records, SPORTDiscus includes references to periodical articles, books, book chapters and essays, conference papers, reports and videotapes. The topics covered comprise all aspects of sports sciences, psychology, administration, sociology, coaching, training, physical education, physical fitness and recreation. The University of Oregon, International Institute for Sport and Human Performance, Kinesiology Publications, theses records (from 1949 to date) are also incorporated. Other major additions are the sociology of sport records from the discontinued SIRLS database, records from the Canadian sport history project 1900-1995 and from other sport specific projects. The major indexing partners contribute their records to SPORTDiscus. They include:

Australia's National Sport Information Centre (see also the NSIC catalogue at www.ausport.gov.au/information/nsic/catalogue);

Catalogue du Musee Olympique, Lausaune, Switzerland (see their library catalogue at www.olympic.org/uk/passion/studies/library/index_uk.asp);

LA84 Foundation Sports Library (formerly the Amateur Athletic Foundation of Los Angeles) (www.la84foundation.org/4sl/over_frmst.htm); and

Heracles, the French database produced by the INSEP in Paris (ceased publication in 2005).

Other important sport sciences databases include SPOLIT, produced by the Bundesinstitut for Sportwissenschaft/Federal Institute of Sport Science (BisP), Bonn, Germany. This database has over 130,000 advanced-level records (40,000 on sports medicine) from 1970 to date and includes periodical articles, books, dissertations and conference papers. Approximately 85% of the records are in German or English. SPOLIT and two other databases are freely available at www. bisp-datenbanken.de

The China Sport Information Center in Beijing maintains the *Chinese Sports Database* from 1985 to date. The database contains over 103,500 records (as of Sept 2001) indexed from 75 Chinese sport sciences periodicals. This subscription-based database also contains records of conference papers, theses and reports. The book references are listed in the union catalogue, China Joint Sport Books Catalogue, which has the holdings from 16 of China's physical education colleges and sport research institutes.

In the past five years, many sport sciences and medical journals have made their periodical articles available in fulltext via the internet. The database *SPONET* was created in 2000 by the Universitat Leipzig, Institut fuer Angewandte Trainingswissenschaft, Abteilung Dokumentation Sport / University of Leipzig, Institute for Applied Training Science, Information and Documentation Centre. This is a database of over 14,000 references to fulltext periodical articles and essays on all aspects of coaching and training sciences. A search engine on the website provides an author, title, subject approach to finding information. It is freely available via the web at www.sponet.de

The comprehensive sport documentation centre and library will also have access to one of two large medical databases to serve sport sciences/medicine clients. Both *SPORTDiscus* and *SPOLIT* contain references to sports medicine literature, but a subscription to *EMBASE/Excerpta Medica* or *Medline* (also known as *PubMed* or *Medlars*) is essential to provide the best reference service. Both databases contain millions of records from worldwide medical journals and are updated weekly. *Medline's PubMed* database is freely available on the web at www.ncbi.nlm.nih.gov/sites/entrez

Other databases that the academic sport/kinesiology librarian will have access to include those produced for the major disciplines. Each of these databases has some sport-related references. Consult: *BIOSIS Previews, PsycINFO, Sociological Abstracts, ERIC* and *Proquest Digital Dissertations (Dissertation Abstracts), Philosophers Index* and *ABI INFORM Global* (on sport business/management topics).

One other type of fulltext, online database important to sport history is the digital library of the LA84 Foundation (formerly the Amateur Athletic Foundation of Los Angeles). After obtaining permission from appropriate bodies, LA84 has completed the OCR digitization of the Olympic movement's *Official Reports, Olympic Review* and *Revue Olympique*. In addition, other academic sport history and management journal back runs are available fulltext along with selected sport magazines. An in-house search engine provides access to author, title or keywords in the fulltext or to each article citation.

2.6. Internet Sources

Many sport documentation centres and libraries have embraced the information delivery and educational aspects of the internet and have created websites that provide information on their country's national sport structure. These websites also include information on the national and local sport organisations, competition/events calendars, training opportunities, coaching education, available sport facilities, career information, sport news and statistics, library services (and sometime access to their online library catalogue), fulltext of important policy documents and many other topics of interest to their clientele. Some outstanding examples are the:

Australian Sports Commission
    www.ausport.gov.au
INSEP
    www.insep.jeunesse-sports.fr

Sport and Recreation South Africa
www.srsa.gov.za; and
Spain's Consell Català de l'Esport
www16.gencat.net/esport.

Other specialised websites to consult include the Olympic Studies Centre CEO-UAB (Barcelona) http://olympicstudies.uab.es/eng/index.asp that maintains comprehensive listings, fulltext documents, contacts and links on the Olympic-related topics.

A number of sport directories have also been created to assist the web user in finding website appropriate to their interests. *Scholarly Sport Sites: a Subject Directory* www.ucalgary.ca/lib-old/ssportsite/ provides subject-based links of interest to the sports sciences researcher. Key links are found in the subject-arranged sections on archives, associations, databases/directories, museums and halls of fame, sport charters, codes, declarations, etc and serials. The *SPORTQuest* website www.sirc.ca/online_resources/sportquest.cfm has thousands of sport-specific links and includes a comprehensive section on sport sciences/kinesiology/physical education departments and faculties and an international conference calendar.

Australia's National Sport Information Centre's website www.ausport.gov.au/information/nsic is noted for their four directories to print and online sport sciences periodicals, many fulltext documents and comprehensive directories to Australian sport organisations and libraries.

There are thousands of sport-specific websites available on the Internet. Keyword searches may be done on the various search engines (Google, Yahoo) and as well, these services and others have organised sport specific links by topics. They include:

- The Virtual Library of Sport (sportsvl.com);
- Google Directory: Sports (www.google.com/Top/Sports);
- Infosport.org (France) (www.infosport.org);
- Open Directory Project: Sports (dmoz.org/Sports);
- Yahoo! Sports (dir.yahoo.com/recreation/sports/index.html); and

- Intute: Social Sciences: Sport and Leisure Practice (www.intute.ac.uk/socials-
ciences/sport).

3. Organisational Network

3.1. International Level

The International Association for Sports Information (IASI) is the major international group whose members represent sport information centres and sports libraries
and information professionals. IASI has a membership of over 100, representing
over 60 countries from all continents. IASI is recognised by the United Nations
Educational, Scientific and Cultural Organisation (UNESCO), the International Olympic Committee and has partnerships with ICSSPE, International Association of
Computer Science in Sport (IACSS), the European Association for Sport Management (EASM) and the European Network of Sport Science, Education & Employment (ENSSEE). The International Council on Archives has recently created the
Provisional Section on Sport Archives www.ica.org and is actively collaborating
with IASI administrators.

3.2. Regional Level

There are regional associations who are also IASI members and work in conjunction with IASI. Their websites have information on their membership, activities and
publications. These include:

- Australasian Sport Information Network (AUSPIN) (www.ausport.gov.au/infor-
mation/nsic/memberships);
- Nordic Committee for Sport Libraries (NORSIB) (kirjasto.jyu.fi/showpage.
php?lang=eng&keyword=norsib-frontpage); and
- North American Sport Library Network (NASLIN) (www.naslin.org).

3.3. National Level

At the national level the United Kingdom's Sport and Recreation Information Group,
SPRIG (www.sprig.org.uk), provides a focus for sport, leisure and tourism librarians. Many countries have documentation and special library organisations where

sport professionals also find colleagues. For example in the United States, there is the American Society for Information Sciences (www.asis.org) and Special Libraries Association (www.sla.org).

3.4. Specialised Centres

Most countries have one or more sport library, information or documentation centre that are attached to academic institutions, government departments or are run independently. Outstanding examples of the academic institutions include:

- Norges Idrettshogskoles Bibliotek/Norwegian University of Physical Education and Sport Library (www.nih.no);
- HPER Library, Indiana University (www.libraries.iub.edu/index.php?pageId=83);
- Zentralbibliothek der Sportwissenschaft, Deutsche Sporthochschule Koln (http://zb-sport.dshs-koeln.de); and
- Institut für Angewandte Trainingwissenschaft Leipzig (www.iat.uni-leipzig.de).

Of the many government supported sport documentation centres, those top-ranked and active institutions include:

- France's Service d'Information et de Documentation, Institut National du Sport et de l'Education Physique (INSEP) (mediatheque.insep.info/cgi-bin/prog/index.cgi?langue=fr);
- La Biblioteca de l'Esport (Generalitat de Catalunya, Consell Català de l'Esport, Barcelona) (www16.gencat.net/esport/biblio);
- China Sport Information Center (www.sport.gov.cn/sport_zixun/csic);
- Australia's National Sport Information Centre (www.ausport.gov.au/information/nsic); and
- Japan Institute of Sport Science. Dept of Sports Information (www.jiss.naash.go.jp/english/shisetsu/research.html).

Independent sport foundation libraries and Olympic studies libraries also play an important part in sport documentation and information delivery. Examples are the:

Paul Ziffren Sports Resource Centre of the LA84 Foundation
    www.la84foundation.org;
Canada's Sport Information Resource Centre (SIRC)
    www.sirc.ca;
IOC's Library
    www.olympics.org; and
Olympic Studies Centre CEO-UAB (Barcelona)
    http://olympicstudies.uab.es/eng/index.asp.

3.5. Specialised International Degree Programmes

Library and Information Science Degree Programs
Education in the art and science of library, information and archival studies varies from country to country. There are many un iversity programs that offer a
Bachelor's degree in library and information science, or a Master's degree. For
the latter degree, the candidate is usually required to obtain a bachelor degree in
another discipline prior to taking the Master's degree in library science. Library
technician education is usually delivered by junior colleges or as a two-year
program in technical schools and institutes. Programs are accredited through a
country's library association. For the United States and Canada, the American
Library Association accredits library science programs for both countries (www.
ala.org). The Chartered Institute of Library and Information Professionals (www.
cilip.org.uk/default.cilip) is the overall body that accredits library programs in
the UK.

A World List of Departments and Schools of Information Studies, Information Management, and Information Systems lists departments and schools of information
studies, information management and systems from many countries of the world
(informationR.net/wl/). The UNESCO Libraries Portal also provides a world-wide
list of training and educational opportunities. See www.unesco.org/cgi-bin/web-
world/portal_bib2/cgi/page.cgi?d=1.

In addition, consult the American Library Association's website section called Education that has a Directory of Institutions Offering Accredited Master's Programs
in Library and Information Studies (US and Canada). Web address is www.ala.org/
education

The Chartered Institute of Library and Information Professionals has a *List of Graduate Training Opportunities* section on their website at www.cilip.org.uk/qualificationschartership/GraduateTrainingOpportunities

4. Appendix Material

4.1. Terminology

A number of dictionaries used by sport librarians and information professionals were mentioned in section 2.2 Reference books. In addition, sport and physical education thesauri play (or used to play) an important role in the indexing of journal articles, books and theses for databases. Key titles include:

- *SPORT Thesaurus*: the Thesaurus of Terminology Used in the SPORTDiscus, 6th ed. (2002) Ottawa: Sport Information Resource Centre. (on CD ROM, Note: this thesaurus is no longer available through the subscription to SPORTDiscus as of 2006. EBSCO seems to be using a more general thesaurus. Many searchers, looking for very specialised sport information, now have to use keywords/phrases when searching)
- *Thésaurus Héraclès* (2001). Paris: INSEP. (No longer available online)
- *Sportdokumentation*: die Deskriptoren der Datenbank SPOLIT ; SPOLIT data base descriptors / Jürgen Schiffer. - Schorndorf : Hofmann. - Bd. 1. Deutsch - Englisch. - 1990. - Vol. 2. English - German. - 1992

Library and Information science definitions may be found online at lu.com/odlis/ (ODLIS: the online dictionary for library and information science).

4.2. Position Statement(s)

Not applicable.

4.3. Varia

Not applicable.

4.4. Free Statement

Not applicable.

5. References

Olsen, A. Morgan (1984). International Sport Information and Documentation: principles for further development. *International Bulletin of Sport Information* 6 (3), 6-14.

Ghent, Gretchen, Kluka, Darlene and Jones, Denise (Eds.). (2002). *Sport and Information Technology*, Oxford/Aachen: Meyer & Meyer Sport (Perspectives, vol. 4), 186p.

Powell, Ronald R and Creth, Sheila D. (1986). Knowledge Bases and Library Education. *College and Research Libraries*, 47 (1), 16-27.

Yesterday, Today and Tomorrow: Better Sport Documentation Through International Cooperation (1994) Brussels, International Association for Sport Information www.iasi.org/publications/monographs.html in pdf format

Sports Law

Jochen Fritzweiler

Contact

Dr. Jochen Fritzweiler
Advocate
President International Sport Lawyers Association
Marktler Str. 19
84489 Burghausen
GERMANY
Phone: +49 8677 3034
Fax: +49 8677 62093
Email: Dr.fritzweiler@t-online.de
www.fritzweiler-niebler.de
www.fritzweiler-sportrecht.de

1. General Information

1.1. Historical Development

In the past fifty years, sport has developed from play to work, from entertainment to marketing and money has become an important issue. Sport can no longer exist outside of the law systems and without legal regulations, whether at the national or international level.

In direct connection with the increasing role of money in professional and leisure sports, is the increasing number of conflicts.

Legal experts, scientists, judges and advocates world-wide, but particularly in those countries where the infrastructure of leisure sports and commercialisation in professional sports is the furthest advanced, have more often dealt with questions regarding law in sports. Since approximately 1980, an increasing number of scientific-oriented sports law organisations have formed, e.g. the International Association of Sports Law (IASL), the International Sport Lawyers Association (ISLA) or in Germany the Deutsche Vereinigung für Sportrecht (DVSR). As the courts of law in different countries have had to face cases from the field of sport, especially accidents, questions concerning labour law and more recently, marketing matters, sports law has became a specialised scientific discipline at universities. As a consequence, dissertations in the field of sports law have begun to be written with the first special publications and commentaries on sports law published in 1985. Now, sports law has become an established area of law.

1.2. Function

The working area of sports law results from the development of sport itself. Sport practice has organised itself in different countries fairly independently and has developed its own rules. This is the first pillar of sports law, namely the self-created corpus of legislation and legal texts of sports federations. Through the increasing power of national and international federations, the second pillar of sports law, namely the national law of the countries, has gained more importance. The responsibility of the countries is to keep the balance between national laws and the legal texts of federations. This task of sports law has been further developed through science. Constitutions of different countries/states grant sporting federations a level of autonomy, which is limited through constitutional and federal/state laws. The commercialisation of sport, concerns for physical safety of participants in sport as well as forms of abuse in sport e.g. corruption and doping, call for intervention by the state. Therefore, it is the task of sports lawyers to clarify in constitutions and public laws which state interventions against sporting federations are justified, and to which extent federations can resist interventions. In certain cases, the state needs to intervene in order to maintain public order, e.g. in the case of riots at large sport events, damage caused by environmental pollution and emissions from sport facilities that impact on neighbouring areas.

The primary task of sports law is to consider the distinctive characteristics of sport in the different areas of law. Employment contracts of professional athletes need to be considered, as do sponsorship agreements or the marketing of TV and radio retransmission rights. Liability issues in the context of sport accidents must also take into account this distinctive character of sport, which in this case implies voluntary exposure to specific dangers and risks.

Finally, one of the main functions of sports law is to understand and govern the international relevance of sport and sport practice. Professional, high-performance sport takes place on an international level, is organised world-wide and is dominated by sovereign multi-sport international sport federations such as the International Olympic Committee when it comes to the Olympic Games, and by the international Sport Federations as far as World Championships are concerned. Many conflicts of interest arise between national and international sport federations, as well as between athletes and their federations when it comes to the enforcement of their right to participation, and when athletes that have engaged in manipulation and doping are sanctioned. In this context, sports law has to maintain the order of the sports system through international legal panels of the federations, as well as through decisions made by public courts. The Court of Arbitration for Sport (CAS) in Lausanne, Switzerland, has increased, even doubled, its activities since 1990. The Swiss Federal Court has recognised the jurisdiction of this court, which is the highest at the international level in the area of sport.

1.3. Body of Knowledge

Sports law results from the rules and statutes of national and international federations, e.g. the Olympic Charter, the World Anti-Doping Code and, in addition, of the written law of countries.

1.3.1. Constitutional Law, Administrative Law

Constitutions contain the basic rights of single athletes to participate in sport, or the basic right to free choice of profession. Constitutions also contain the provision that clubs and federations may govern their affairs on their own i.e. they are autonomous. Furthermore, constitutional regulations of countries define the rights of the state to restrict certain constitutional rights, as well as the tasks and rights of the state in the realm of sport.

In all these cases, the term "sport" is not mentioned explicitly in the law, however the jurisdiction has declared the general legal regulations to be applicable for sport as well, which is also the case concerning legal regulations of public administrations. By means of such regulations, government organs are allowed to intervene in sport in case of any looming dangers and risks, thereby considering the specific character of sport. Administrative regulations for the protection of the environment and nature consider the distinctive character of sport as well.

1.3.2. Organisational Law (Clubs and Federations)

As a consequence of the constitutionally granted autonomy, regulations of private law govern the details of the legislative process, how for example, statutes and by-laws arise and how they are handled in detail. The limitations of the federations' legislative powers are governed by law and formed by jurisdiction. The legislative powers of the federations allow them to issue rules concerning their organisational forms, their economic undertakings, their financing and the issuing of licences in the context of sport activities in leagues.

In the same way, private law provides the reference framework within which the federation is entitled to take regulatory measures, i.e. declare sanctions in case of violations of the federation's rules and regulations. The federations' jurisdiction is in most cases predetermined by the regulations of the proceedings of the lawsuit.

The jurisdiction has in many cases decided how the tribunals of federations have to keep constitutional principles and has often overruled their decisions in cases where violations have occurred.

1.3.3. Labour Law, Commercial Law

The legal regulations of labour law are being applied in sports as well, especially professional sports. For example, the employment contracts of professional athletes in team sports are of an importance, including items such as which athletic performances the athlete is covenanted to his employer for, how long the employment contract is going to last and how the transfer to another employer will be carried out.

In commercial law, when it comes to marketing a sport, the prerequisites for the individual contract result from civil law, competition law and copyright laws.

1.3.4. Media Law

The special relevance of sports reporting through media is characterised by different contracts between the states and the broadcasting corporations, especially through TV guidelines, conventions and press law. Sport-typical situations in sport advertising and sponsorship have developed in relation to the guaranteed freedom of the press and broadcasting in most countries, based on newer jurisdiction. In addition, specific details for sports arise from competition laws, and those laws surrounding copyright and trademarks for logos, emblems, brands and names that are often used in sports.

1.3.5. International Sports Law

The international character of sport implies that the international private law and international lawsuit procedures are applied to sport as well. The respective international legal regulations or agreements define how international sport federations are legally codified and how specific sport-typical contracts such as employment, sponsoring or marketing, betting and equipment contracts, are to be interpreted.

1.3.6. Sport Accidents and Sport Injuries

In cases of sports accidents and injuries, most countries refer to the general legal regulations that govern entitlements to damages. The responsibility for accidents in sports is especially characterised by the high risk that each athlete accepts when practicing his or her sport. The existing legal regulations are being interpreted by the courts when judging the faults and responsibilities of the injured and/ or injuring protagonist. This applies especially for sports with a high risk such as motor sport, aviation and equine sports. In the case of sports accidents for which the organisers of the sport are responsible, it is necessary to consider the type of sport and the specific responsibility of the organisers such as sport schools, mountain rails and ski-lifts, producers of sports equipment as well as public administration that provides sports facilities.

1.3.7. Sanctions

Sport practice often leads to assault and battery as well as to cases of death for which penal sanctions are defined; here the specific type of sport is also considered. With the increasing commercialisation of sport, the amount of manipulation and corruption is growing – again. For these activities, which are in a sense alien and contradictory to sports, the penal laws of the individual states are to be applied.

1.4. Methodology

The scientific approaches for specific methods and doctrines for sports law differ, depending on the legal system of the country, such as the Common Law, the Case Law or the Codified Law. The "specificity of sport", however, is relevant for all systems. Whenever the state acknowledges and supports sport such as it is practised in the federations, or even recognises sport in its constitution, the "system of values" that has been developed by sport (namely the "specificity of sport") has to be acknowledged as socially adequate. If, for example, boxing is permitted and recognised as a competitive sport, then an injury that is typical for this sport may not be sanctioned as illegal. Instead, the state has to form a compromise between sport-specific necessities and state law. What matters is an exact consideration of each single case. A well-known and acknowledged specificity of sport is the monopolistic, hierarchic organisational structure of sports federations, which would, outside of sport, constitute a violation of anti-trust laws.

1.5. Relationship to Practice

Since approximately 1970, the number of conflicts in leisure and professional sport taken to court has increased, as well as the number of advocates that have been engaged in cases. Professional athletes, entrepreneurs and event organisers are seeking legal advice from advocates and legal advisers.

Science and practice have developed in parallel over recent years, which means that more attorneys, especially advocates, have specialised in sport, labour and economy. In almost every cabinet, sports law has become a working area, especially in the large cabinets where there are specialists who have gained specific knowledge at universities or from the recorded literature. These advocates are known to be used by professional athletes and in many aspects of sport business, but there is no special education or title for them.

1.6. Future Perspectives

In the future, sports law will continue to become more specialised with additional universities and institutes developing courses to focus on the topic. Current scientific projects and teaching areas will continue to grow and the jurisdiction of state courts and courts of arbitration, especially the Court of Arbitration of Sport (CAS), will also further develop. Among the many problems that sports law has to tackle, the fight against doping and corruption in sport and the many issues connected to marketing rights are key.

2. Information Sources

2.1. Journals

A.N.Z.S.L.A. Newsletter – published by the Australian and New Zealand Sport Law Association.

Entertainment and Sports Lawyer - publication of the American Bar Association, 750 N. Lake Shore Drive, Chicago, IL 60611, USA.

Exercise Standards and Malpractice Reporter - PRC Publishing, 3976 Fulton Drive NW, Canton, OH44718, USA.

Gym to the Jury - publication of the Center for Sports Law and Risk Management, 6917 Wildglen Drive, Dallas, TX 75230 USA.

Journal of Legal Aspects of Sport - official publication of the Society for the Study of the Legal Aspects of Sport and Physical Activity, published by Marquette University Law Institute, Marquette University, 1103 W. Wisconsin Avenue, Milwaukee, WI 53233, USA.

Outdoor Education & Recreation Law Quarterly - 2336 Pearl St, Boulder, CO 80302, USA.

Pandektis International Sports Law Review - the official organ of IASL, published by ION Publishing Group, Hellin Publications.

Recreation & Parks Law Reporter - publication of the National Recreation and Park Association, 22377 Belmont Ridge Road, Ashburn, VA 20148, USA.

Revue Juridique et Economique du Sport - published by the Comité national olympique et sportif français.

Rivista di Diritto Sportivo - published by the National Olympic Committee of Italy.

Seton Hall Journal of Sport Law - publication of Seton Hall University School of Law, 1111 Raymond Blvd, Newark, NJ 07102, USA.

Sports & the Courts - P.O. Box 2836, Winston-Salem, NC 27102, USA.

Sport und Recht (SpuRt) - 1994, C.H.BECK, 80801 München, Germany. www.spurt.de

Sports Facility Law Reporter - Sport Administration, Garrison Gym, University of Houston, Houston, TX 77204, USA.

Sports Lawyers Journal - publication of the Sports Lawyers Association, 11250 Roger Bacon Drive, Suite 8, Reston, VA 22090, USA.

Sports, Parks & Recreation Law Reporter - PRC Publishing, 3976 Fulton Drive NW, Canton, OH 44718, USA.

University of Miami Entertainment and Sport Law Review - P.O. Box 248087, Coral Gables, Florida 33124, USA.

Virginia Journal of Sport & the Law - publication of the University of Virginia School of Law, 580 Massie Road, Charlottesville, VA 22903, USA.

World Sports Law Report - London, UK. www.e-comlaw.com

2.2. Reference Books, Encyclopaedias, etc.

Anderson, P.M. (1999). *Sports law: A desktop handbook*. Milwaukee, WI: National Sports Law Institute, Marquette University Law School.

Adophsen, Jens (2003). *Internationale Dopingstrafen*. Tübingen.

Appenzeller, H. (Ed.). (1998). *Risk management in sport: Issues and strategies*. Durham, NC: Carolina Academic Press.

Appenzeller, T. (2000). *Youth sport and the law*. Durham, NC: Carolina Academic Press.

Baddeley, Margareta (1996). Le Sportif, sujet ou objet? La protection de la personnalité du sportif. *ZSchwR Bd. 115*.

Beloff, M. et al. (1999). *Sports law*. Oxford, England: Hart.

Blackshaw, Ian S. and Siekmann Robert, C.R. (Eds.). (2005). *Sports Image Rights in Europe*.

Boyes (2001). Regulation Sport after the Human Rights Act 1998. *New Law Journal*.

Carpenter, L.J. (2000). *Legal concepts in sport: A primer*. Champaign, IL: Sagamore.

Champion, W.T. Jr. (2000). *Sports law in a nutshell*. St. Paul, MN: West Group.

Clement, A. (1998). *Law in sport and physical activity*. Cape Canaveral, FL: Sport and Law Press.

Clement, A. (1997). *Legal responsibility in aquatics*. Cape Canaveral, FL: Sport and Law Press.

Cloutier, R. (2000). *Legal liability and risk management in adventure tourism*. Kanloops, BC, Canada: Bhudak Consultants.

Coopers and Lybrand (1993). *Der Einfluß der Tätigkeit der Europäischen Gemeinschaft auf den Sport*.

Cotton, D., Wolohan, J.T. and Wilde, T.J. (Eds.). (2001). *Sport law for recreation and sport managers*. Dubuque, IA: Kendall/Hunt.

Davis, T., Mathewson, A.D. and Shropshire, K.L. (Eds.). (1999). *Sports and the law, a modern anthology*. Durham, NC: Carolina Academic Press.

Dougherty, N., Auxter, D., Goldberger, A. and Heinzmann, G. (1993). *Sport, physical activity, and the law*. Champaign, IL: Human Kinetics.

Fritzweiler, Pfister and Summer (2007). *Praxishandbuch des Sportrechts*, 2. Auflage, München.

Gallup, E.M. (1995). *Law and the team physician*. Champaign, IL: Human Kinetics.

Gardiner, Simon, et al. (1998). Sports law. London: Cavendish.

Grayson, E. (1998). *Sport and the law*. London: Butterworth.

Grayson, E. (1999). *Sports medicine and the law*. London: Butterworth.

Greenberg, M.J. and Gray, J.T. (1998). *Sports law practice*. Charlottesville, VA: Lexis Law Publishing.

Griffith-Jones, D. (1997). *Law and the business of sport*. London: Buttersworth.

Haas, Haug and Reschke (2006). *Handbuch des Sportrechts, Lose Blattsammlung mit Erläuterungen*.

Herbert, D. and Herbert, W. (1993). *Legal aspects of preventive, rehabilitative and recreational exercise programs*. Canton, OH: Professional Reports Corporation.

Hylton, J.G. and Anderson, P.E. (Eds.). (1999). Sport law and regulation. Milwaukee, WI: Marquette University Law School.

Jarvis, R.M. and Coleman, P. (1999). *Sports law, cases and materials*. St. Paul, MN: West Group.

Karaquillo, J.P. (Ed.). (1995). *L'Activite sportive dans les balances de la justice*. Paris: Tom, II.

Kelly, G.M. (1987). *Sport and the law: An Australian perspective*. North Ryde, NSW, Australia: The Law Book Company.

Koeberle, B.E. (1998). *Legal aspects of personal fitness training*. Canton, OH: Professional Reports Corporation.

Nafziger, J.A.R. (1988). *International sports law*. Irvington, NY: Transnational Publishers.

Osborn, G. and Greenfield, S. (Eds.). (2000). *The sport and law reader*. London: Frank Cass.

Panagiotopoulos, D. (1990). *Doping – legal confrontations*. Athens: Ant Sakkoulas.

Panagiotopoulos, D. (1990). *Sport law theory*. Athens: Ant Sakkoulas.

Panagiotopoulos, D. (1991). *Olympic games law*. Athens: Ant Sakkoulas.

Panagiotopoulos, D. (1997). *Sport law code*. Athens: Ant Sakkoulas.

Raupach, Arndt (1995). *Structure follows strategy – Grundfragen der Organisation, des Zivil- und Steuerrechts im Sport, dargestellt am Thema Profigesellschaften*. SpuRt.

Riffer, J.K. (1985 and annual supplement). *Sports and recreational injuries*. Colorado Spring, CO: Shepard's McGraw Hill.

Romano, Santo (1975). *L`Ordinamento Giuridico (Die Rechtsordnung), deutsche Übersetzung*. Berlin.

Ruxin, R.H. (1993). *An athlete's guide to agents*. Boston: Jones and Bartlett.

Sawyer, T.H. (2001). *Golf and the law*. Durham, NC: Carolina Academic Press.

Silance, L. et al. (1997). *Les sports et le droit.* Bruxelles: DeBoeck Universite.

Steiner, U. (2003). *Die Autonomie des Sports.*

Uberstine, G.A. and Stratos, K. (Ed.). (1998). *Law of professional and amateur sport.* St. Paul, MN: West Group.

Van der Smissen, B. (1990). *Legal liability and risk management for public and private entities – sports and physical education, leisure services, recreation and parks, camping and adventure activities.* Cincinnati, OH: Anderson.

Vieweg, Claus (1990). Nomsetzung und – Anwendung deutscher und internationaler Verbände.

Weistart, J.C. and Lowell, C.H. (1985). *The Law of Sports.* Charlottesville, 1979, supplement 1985.

Wise, Aron N. and Meyer, Bruce S. (1997). *International Sports Law and Business, 3 Bände.* Cambridge (USA).

Wittenberg, J.D. (1985 with regular supplements). *Product liability: Recreation and sport equipment.* New York, NY: Law Journal Seminar.

Wong, G.M. (2000). *Essentials of sport law.* Westport, CT: Praeger.

Wong, Glenn M. (1988). *Essentials of Amateur Sports Law.* Dover.

Yasser, R., McCurdy, J.R. and Goplerud, C.P. (1997). *Sports law, cases and materials.* Cincinnati: Anderson.

2.3. Book Series

Not applicable.

2.4. Conferences/Workshops Proceedings

Not applicable.

2.5. Data Banks

Lexis/Nexis, a private corporation, provides results of court decisions, constitutions and statutes as well as standard library publications. WestLaw, a private corporation, provides results of court decisions, constitutions and statues and standard library publications relating only to law.

2.6. Internet Sources

American Bar Association Forum on the Entertainment and Sport Industries
    www.abanet.org/forums/entsports/home.html
Australian and New Zealand Sports Lawyers Association
    www.anzsla.com.au
International Association of Sports Law (IASL)
    www.iasl.org
Marquette University Law School National Sports Law Institute
    www.mu.edu/law/sports/index.html
Sport and Recreation Law Association
    http://srlaweb.org/
Sports Lawyers Association
    www.sportslaw.org

3. Organisational Networks

3.1. International Levels

Association Internationale des Juristes du Sport, Rouen, France.
Centre International D'Etude Du Sport, Neuchâtel, Switzerland.
International Association of Sports Law (IASL), Athens, Greece.
International Sport Lawyers Association (ISLA), Burghausen, Germany.

3.2. Regional Level

Australia and New Zealand Sports Law Association (ANZSLA), Melbourne, Australia.

3.3. National Level

Centre de Droit et d'Economie du Sport, Limoges, France.
Hellenic Center of Research on Sports Law, Athen, Greece.
National Sports Law Institute, Milwaukee, WI, USA.
Sport and Recreation Law Association, Wichita State University, Wichita, USA.

Journal of Legal Aspects of Sport is their official publication, published by Marquette University Law Institute, Marquette University, 1103 W. Wisconsin Avenue, Milwaukee, WI 53233, USA.
Deutsche Vereinigung für Sportrecht (DVSR), Germany.

3.4. Specialised Centres

- British Association for Sport and Law, Manchester Metropolitan University, England.
- Centre de Droit du Sport de Nice, Nice, France.
- Institute of Sports Law, Athen, Greece.
- Sport Law Study, University of Athens, Athens, Greece.
- South African Sports Law Association, Durban, South Africa.
- National Sport Law Institute, Marquette University, Milwaukee, WI, USA.
- Center for Sports Law and Risk Management, 6917 Wildglen Drive, Dallas, TX 75230 USA.

3.5. Specialised Degree Programs

Law Degree:

- Marquette University, Milwaukee, WI, USA.

Law with an emphasis on Sport Management:

- University of Massachusetts, Amherst, MA, USA.
- Florida State University, Tallahassee, FL, USA.

4. Appendix Materials

4.1. Terminology

Not applicable.

4.2. Position Statement(s)

Not applicable.

4.3. Varia

Not applicable.

4.4. Free Statement

Not applicable.

Sport Management

Dianna Gray, Alberto Madella, Kari Puronaho,
David Shilbury and Berit Skirstad

Contact

Europe
Associate Prof. Berit Skirstad
President
European Association for
Sport Management (EASM)
Postboks 4014 - Ullevaall Sta-
dion
0806 Oslo, Norway
Phone: +47 23 26 24 28
Email: berit.skirstad@nih.no

Dr. Kari Puronaho
Secretary General
European Association for
Sport Management (EASM)
Kaskelantie 10
19120 Vierumäki, Finland
Phone: +358-40-5867493
Fax: +358-3-84245755
Email: secretariat@easm.net or
kari.puronaho@vierumaki.fi
www.easm.net

Oceania
Prof. David Shilbury
Director of Sport Management
Deakin University
221 Burwood Highway
Burwood 3125, Australia
Phone: +61 3 9244 6164
Fax: +61 3 9251 7083
Email: shilbury@deakin.edu.au

North America
Dr. Diana Gray
University of Northern Colorado
School of SES, Gunter Hall
501 20th St.
Greeley, CO 80639, USA
Email: dianna.gray@unco.edu

1. General Information

Sport management is a growing field, world-wide, however, it is not represented by a single, international organisation. In order to present the most comprehensive selection of information available, this chapter includes contributions from the four main continental sport management associations: the Sport Management Association of Australia and New Zealand (SMANZ), the European Association for Sport Management (EASM), the Asian Association for Sport Management (AASM) and the North American Society for Sport Management (NASSM). Some similarities exist between the four regional perspectives and many resources and organisations that are specific to the geographic region are referred to in each section. The four organisations are members of the International Alliance (See section 3.1 International Level).

1.1. Historical Development

During the last century, development of the sports industry has increased the complexity of the tasks of sports managers. This has facilitated the professionalisation of sports organisations and the development of specific academic curricula. Since 1966, when the first Master's program in sport management was established at Ohio University (USA), there has been an enormous increase in academic activities in this regard. Today, there are over 400 colleges and universities offering a curriculum in sport management in the United States, Europe, Australia and Asia.

1.2. Function

Sport management is concerned with all kind of sports services, regardless of the level of athletes and teams or the competitive or non-competitive nature of the participants. Functions of sport management are to analyse the activities, to co-ordinate the technologies and to provide support units that facilitate both the production and marketing of sports services, as well as the context for where the production and marketing of services take place (Chelladurai, 1994, EJSM: 1,1).

Its principle focus as an academic discipline is to prepare managers with the skills necessary to provide leadership and direction for sporting organisations and to craft strategies at an organisation-wide level. These include marketing and sponsorship-related strategies, financial strategies and strategies designed to increase spectator attendance and sports consumption, as well as facilitate participation in the community, and devise the pathways into and through sports to semi-professional and elite levels.

The sports manager is therefore the conduit between the sport, the activity and the structured systems that are designed to promote sports participation in the community. The sports manager is also the conduit between a plethora of academic disciplines and bodies of knowledge relevant to managing sporting organisations.

1.3. Body of Knowledge

Sport management is essentially an interdisciplinary body of knowledge, with programs in universities or higher education institutions world-wide either linked to business education or physical education that emphasises management, marketing and financial knowledge. Increasingly, programs are grounded within business education. The bodies of knowledge contributing to sport management are many and complex in their level of detail and application. Sport management education typically draws upon a solid sociological analysis of sport and its role in society, as well as management, which is rooted in sociological, socio-psychological and psychological theory and is applicable to leadership, motivation and general human resources theories.

Marketing theory also draws upon many facets of psychological theory in relation to understanding consumer behaviour, the research process generally and, like management, the strategy literature. Much of the knowledge applied to facility and event management in particular, draws upon the combined theories present in the study of management and marketing in sport, cognisant of the socio-cultural role of sport and its history in society. Such education also relies upon a fundamental study of sport and the law and an understanding of financial transactions as they relate to the management of organisations, facilities and events.

Also embedded in the study of sport management is an examination of sport governance at a micro- and macro-level. Increasingly, sporting organisations are re-examining the most appropriate way to manage their sport, given the mix of pressures for corporate reform and the community aspects of sports participation and involvement in general. Theoretical underpinnings of governance issues can be found within the management literature and specifically within organisation theory. Overall, the body of knowledge evident in sport management is a composite of traditional disciplines with an increasing focus on theories that specifically address the unique aspects of sport business as they relate to sport management practice in community and club, state and national associations and professional levels.

1.4. Methodology

Quantitative and qualitative methods are used in management sciences. The study of sport management has typically used methodologies consistent with the most sophisticated quantitative analysis techniques, and in the case of phenomenological methodologies, includes case studies, ethnographic investigations, feminist perspectives, grounded theory and action research. Surveys dominated early research efforts in the field and were used to describe and better understand the practices of what was a relatively new field of endeavour, at least in the paid professional sense. Marketing research in particular, employs both survey methods and focus group work to better understand consumer buying patterns and behaviours. Contemporary sport management research studies employ rigorous designs, sophisticated modelling techniques and analyses.

1.5. Relationship to Practice

As sport management practices evolved prior to any recognised needs for a theoretical basis, there has always been an obvious link between practice and theory. The majority of members in sport management organisations are sport management educators who teach, engage in research, consult in the sports industry and serve their individual institutions. As a discipline, sport management provides a conceptual background for practitioners, teachers, trainers and developers of curricula on sport management. Sport management programs seek to maintain relationships with sports organisations for the purposes of analysing the industry, co-operating in field-based needs, such as research and analysis of problems, as well as cultivating field experience/education sites for student's practical performances in the form of practice and internships. The multitude of sports and events and the range of complex problems confronting these organisations in an evolving area facilitates a rich learning environment, suited to the link between theory construction and testing. As a consequence, sport management graduates are required to apply their knowledge to evolving and complex problems.

The objectives of the completed European 'Aligning a European Higher Educational Structure In Sport Science (AEHESIS) Thematic Network Project' were to firstly integrate programs and time frames of educational structures and secondly, to ensure that the identified structures relate to the needs of the labour market in the four main areas of sport management, physical education, health and fitness and sport coaching. In the sport management area, a lot of research material was collected from employers to support curriculum development work and to create a direct link between the labour market and education. More information about this project can be found at www.aehesis.de or at the new Sport Education Information Platform. The free platform is named 'Sophelia' www.eseip.eu/index. php?option=com_frontpage&Itemid=44 and presents general information on the Bologna process, the development of curricula, quality assurance and an extensive database for researching programs of study within sport education.
Strongly connected with the outcomes of the AEHESIS project, a new book, Higher Education in Sport in Europe, will be published by Meyer and Meyer in Spring, 2008.

1.6. Future Perspectives

As the sports management area continues to grow and evolve, so too will the direction of action and research on topical issues. Because of the diverse nature of sports management in the different regions of the world, there are unique problems and thus goals and directions for each of the organisations. For example, in Europe, sports managers and scholars are currently confronted with the following major issues: the cohabitation of professionals and volunteers in most of the sports organisations; the impact of European Union policy and legislation on sport management and sport management education; and the impact of new technologies on management and administration of sport, as well as, the search for new funding strategies.

In North America, the most recent estimate of the size and impact of the sports industry in the United States (US) was reported in Street and Smith's Sports Business Journal (SBJ) (December 20-26, 1999) as having an economic value of $213 billion dollars. When consumer spending on mainstream and alternative participatory sports is added to SBJ's figure, the total amount of annual expenditure exceeds $250 billion (Howard and Crompton, 2004). This places the sports industry as sixth among industries in the US. At least four studies on the economic value of the sports industry in the US have been conducted since the early 1990s, each documenting the continuing growth of the sport sector in industrial markets. This growth trend can be expected to continue, as sport, hospitality and tourism work more in concert in delivering sport products and services to a mounting number of consumers.

2. Information Sources

2.1. Journals

European Sport Management Quarterly (formerly European Journal for Sport Management) - Published by the European Association for Sport Management. Taylor and Francis.

International Journal of Sport Management - American Press Publishers.

Journal of Sport Management - North American Society for Sport Management - Human Kinetics.

Revue Européenne de Management du Sport - Presses Universitaires du Sport.

Sport Management Education Journal - New joint refereed publication of the North American Society for Sport Management (NASSM) and the National Association for Sport and Physical Education (NASPE).

Sport Management Review - Sport and Management Association of Australia and New Zealand.

Sport Marketing Europe - Sportfacilities and Media. www.sportfacilities.com

Sport Marketing Quarterly - Fitness Information Technology.

Sporting Traditions - Australian Society for Sports History.

Sports Marketing and Sponsorship - Winthrop Publications Limited.

2.2. Reference Books, Encyclopaedias, etc.

Adair, D. and Vamplew, W. (1997). *Sport in Australian History*. Melbourne: Oxford University Press.

Cameron, J. (1996). Trail blazers: *Women who Manage New Zealand Sport.* Sports Inclined: Christchurch, N.Z.

Comfort, P.G. (2005). *The Directory of Academic Programs in Sport Management,* www.fitinfotech.com

Comfort, P.G. (2005). *Directory of Undergraduate Programs in Sport Management,* www.fitinfotech.com

Chelladurai, P. (2006). *Human Resources in Sport and Recreation.* Champaign II: Human Kinetics.

Chelladurai, P. (2005). *Managing organizations for sport and physical activity: A systems perspective,* 2nd ed. Scottsdale, AZ: Holcomb Hathaway.

Hernandes, R.A. (2002). *Managing Sport Organizations.* Champaign II: Human Kinetics.

Howard, D.R. and Crompton, J.L. (2004). *Financing Sport.* Morgantown, W.V.: Fitness Information Technology.

Miller, L.K. (1997). *Sport Business Management.* Gaithersburg Maryland: Aspen Publishers. Inc.

Masteralexis, L.P., Barr, C.A. and Hums, M.A. (1998). *Principles and Practice of Sport Management.* Gaithersburg Maryland: Aspen Publishers. Inc.

National Association for Sport and Physical Education/North American Society for Sport Management. (2000). *Sport management program standards and review protocol.* Reston, VA: Author.

National Association for Sport and Physical Education/North American Society for Sport Management. (2000). *Sport management program review registry.* Reston, VA: Author.

Parks, J.B., Quarterman, J. and Thibault, L. (2007). *Contemporary Sport Management,* 3rd. ed. Champaign II: Human Kinetics.

Masteralexis, Pike, L., Barr, C.A. and Hums, M.A. (2005) *Principles and practice of sport management* Boston : Jones and Bartlett

Sharp, L.A., Moorman, A.M. and Claussen, C.L. (2007). *Sport Law: A Managerial Approach.* Scottsdale, AZ: Holcomb Hathaway, Publishers.

Shilbury, D. and Deane, J. (2001). *Sport Management in Australia.* An Organisational Overview. Melbourne: Strategic Sport Management.

Shilbury, D., Quick, S. and Westerbeek, H. (1998). *Strategic Sport Marketing.* Sydney: Allen & Unwin.

Slack, T. and Parent, M.M. (2006). *Understanding Sport Organisations.* Champaign II: Human Kinetics.

Smith, A. and Stewart, B. (1999). *Sports Management. A Guide to Professional Practice.*Sydney: Allen & Unwin.

Trenberth, L. and Collins, C. (1999). *Sport Business Management in New Zealand.* Palmerston North, New Zealand: The Dunmore Press.

Vamplew, W., Moore, K., O'Hara, J. Cashman, R. and Jobling, I. (Eds.). (1994). *The Oxford Companion to Australian Sport,* 2nd ed. Melbourne: Oxford University Press.

Watt, D.C. (2003). *Sports Management and Administration.* London: E & FN Spon.

2.3. Book Series

Not applicable.

2.4. Conference/Workshop Proceedings

For proceedings of all European Congresses on Sport Management, contact EASM directly for information on how to purchase these publications.

1993 Groningen (The Netherlands)
1994 Florence (Italy)
1995 Budapest (Hungary)
1996 Montpellier (France)
1997 Glasgow (United Kingdom)
1998 Madeira (Portugal)
1999 Thessaloniki (Greece)
2000 San Marino (Republic of San Marino)
2001 Vitoria-Gasteiz (Basque Countries, Spain)
2002 Jyvaskyla (Finland)
2003 Stockholm (Sweden)
2004 Ghent (Belgium)
2005 Newcastle-Gateshead (UK)
2006 Nicosia (Cyprus)
2007 Turin (Italy)
2008 Heidelberg-Bayreuth (Germany)
2009 Amsterdam (The Netherlands)

Conference abstracts of the North American Society of Sport Management (NASSM) are published in printed form and on the internet. Typically, the publication contains over 100 abstracts of research and scholarly presentations delivered in oral, poster, symposium, roundtable or workshop format. A distinguished scholar lecture is delivered regularly at the NASSM conference by the Earle F. Zeigler Lecture Award recipient. The lecture is subsequently published in the Journal of Sport Management.

2.5. Data Banks

Sport management professionals refer students and colleagues to the following data banks, among others:

The Laboratory for Leisure, Tourism and Sport, based at the University of Connecticut, provides online bibliographies and is divided into a number of sub-categories. Search for Laboratory for Leisure, Tourism and Sport on www.uconn.edu/schools/business.php

Sport Information Resource Centre for information/collections on sports, fitness and related fields www.sirc.ca/

SBRNet - Sports Business Research Network. Combines market research from the National Sporting Goods Association; the US Department of Commerce; various sports governing bodies. www.sbrnet.com. Subjects covered include sporting goods industry, sports sponsorship and marketing, broadcasting, facilities, broadcasting and new media.

Databank of European Sport Management programs, Universities and Higher Education Institutions www.aehesis.de

2.6. Internet Sources

European Association for Sport Management
    www.easm.net
North American Society for Sport Management
    www.nassm.com
The Faculty of Kinesiology at the University of New Brunswick (Canada) maintains a NASSM Listserve.

Please contact: SPORTMGT@UNB.CA

Sport Management Association of Australia and New Zealand

http://smaanz.cadability.com.au/

Australian Sports Commission

www.ausport.gov.au/

SPARC (Sport and Recreation New Zealand), NZ

www.sparc.org.nz/

Asian Association for Sport Management

http://english.peopledaily.com.cn/200202/02/eng20020202_89801.shtml

3. Organisational Network

3.1. International Level

There is a formal International Alliance between the major sports management associations world-wide (European Association for Sport Management, North American Society for Sport Management, Sport Management Association of Australia and New Zealand and the Asian Association for Sport Management). The Alliance designates one of the regional conferences of its members as the Conference of the International Alliance on a two year basis. The Alliance has the goal to promote international collaboration in the areas of research, teaching and to facilitate the circulation of information and the diffusion of effective management practices.

3.2. Regional Level

North American Society for Sport Management (NASSM)

The North American Society for Sport Management (NASSM) was one of the first scholarly organisations formed to meet the unique interests of sport managment professionals in the academy for scientific study and professional preparation. Founded in 1985, the organisation's first conference was held at Kent State University (Ohio). Since its founding, the organisation has held an annual conference during the first week of June each year and rotates sites between Canada and the US. NASSM is actively involved in supporting and assisting professionals working in the fields of sport, leisure and recreation. "The purpose of the North American Society for Sport Management is to promote, stimulate and encourage study, research, scholarly writing, and professional development in the area of sport management - both theoretical and applied aspects. Topics of interest to NASSM members include sport marketing, future directions in management, employment perspectives, management competencies, leadership, sport and the law, personnel management, facility management, organisational structures, fund raising and conflict resolution". NASSM endeavours to support and cooperate with local, regional, national and international organisations, which have similar purposes. In addition, it organises and administers meetings to promote its purposes. The first issue of the Journal of Sport Management was published in 1987 and subsequently has increased in stature and distribution. The Journal is now published quarterly and boasts a global distribution. In 2007, the organisation will introduce a second journal. The Sport Management Education Journal is a joint refereed publication of the North American Society for Sport Management (NASSM) and the National Association for Sport and Physical Education (NASPE). The Sport Management Education Journal will advance the body of knowledge in pedagogy as it relates to sport management education and disseminate knowledge about sport management courses, curricula and teaching. NASSM's membership, conference delegates and scholarly programs, continue to grow and the organisation has been used as a model for other sport management professional organisations.

European Association for Sport Management (EASM)

The aims of EASM are to promote, stimulate and encourage studies, research, scholarly writing and professional development in the field of sport management. It was founded in Groningen, the Netherlands in 1993, and its journal commenced publication in the same year. The office of the Association is based at the Sport Institute of Finland, Vierumäki. It provides services and information to current and potential members, keeps and develops contacts with sport organisations and universities in Europe, manages and carries out various sport management related projects (e.g. AEHESIS and Sport Coordinator) and develops relationships with international organisations.

The main outputs of the Association are:

- the European Sport Management Quarterly, which covers a wide range of sport management topics, ensuring proper balance of practical application and theory.

- a website (www.easm.net) with the latest information on the activities of the Association and the general development in the field in Europe.
- EASM E-News, published every second month on the EASM website.
- an annual Conference hosted in different European countries. The Conference offers scientists and professionals the best opportunity to discuss the basic issues and the latest topics in the field. During the Conference, special attention is paid to the main trends at the European level and to the specific local aspects of sport, culture and industry in the host city and country.
- an annual Sport Management Student Seminar, which offers students the chance to work in an international context, to work in international groups and to deal with international sport management tasks.

The European Association for Sport Management has developed strong connections with the sports industry and sports organisations. With this in mind, the Annual European Congress on Sport Management has, over the last few years, increased in participant numbers. This has facilitated the exchange of information and the identification of outstanding practices and new problem areas.

Sport Management Association of Australia and New Zealand (SMAANZ)

The Sport Management Association of Australia and New Zealand is the regional association dedicated to enhancing and facilitating sport management research throughout the two countries comprising its membership. It was formally established at its first conference hosted by Deakin University in Melbourne, Australia in 1995 by members of sport management related faculties from universities offering sport management programs in Australia and New Zealand. Since then, an annual conference has been held (usually in November), which typically attracts academics and practitioners with the emphasis on presenting the latest research emerging from within Australia and New Zealand. Increasingly, the conference is attracting international delegates from a variety of universities and countries. SMAANZ has, since 1998, produced the journal, *Sport Management Review*, which is currently published twice a year in May and November. *Sport Management Review* has published manuscripts from local academics as well as a number from international researchers.

SMAANZ aims to continue pursuing its charter by promoting scholarly enquiry in sport management and by providing an outlet for the communication of sport management related research. Over time, this is likely to see the activities of SMAANZ expand and move beyond the boundaries of its member universities.

Asian Association for Sport Management

The Asian Association for Sport Management was founded in 2002 and other known associations can be found in India, Japan and Korea.

3.3. National Level

China: The Chinese Association for Sport Management
France: Societé Française de Management du Sport
Greece: Hellenic Association of Sports Management
Italy: SIMS (Società Italiana di Management dello Sport)
Japan: The Japanese Society of Management for Physical Education and Sport
Korea: Korean Institute for Sports Marketing
Portugal: APOGESD: Associacao Portuguesa de Gestion de Desporto www.ag-esport.org
South Africa: South Africa Society for Sport Management
Spain: KAIT
Sweden: Swedish Association for Sport Management
Switzerland: Association Suisse des managers du sport
United Kingdom : British Institute of Sports Administration

3.4. Specialised Centres

The Laboratory for Leisure, Tourism and Sport (University of Connecticut) is a training centre for doctoral students and provides a working lab where both academic and market forms of research are conducted. It also promotes a think-tank environment in which to pursue and exchange ideas and viewpoints. Search for Laboratory for Leisure, Tourism and Sport on
    www.uconn.edu/schools/business.php

The Sport Marketing Research Team of the Bureau of Tourism and Recreation Research (Illinois State University) conducts research for sporting events and organisations, hosts a sport tourism marketing conference and delivers course instruction via the internet.

www.kinrec.ilstu.edu/graduate/sport_management.shtml

The International Institute for Tourism Studies (The George Washington University) conducts research in the area of sport tourism and organises an annual international conference on sport tourism (TEAMS: Travel, Events and Management in Sports).

www.gwu.edu/~iits

The Sport Marketing Research Institute (University of Northern Colorado) conducts research in sport sponsorship and marketing, provides sports business consultation, conducts seminars and publishes technical reports.

www.unco.edu/smri/

The Sport Alliance of Ontario provides business support services to Ontario sport associations, leadership training to sport volunteers and coordinates delivery of Canada's National Coaching Certification Program in Ontario.

www.sportalliance.com

The Warsaw Sports Marketing Centre (Lundquist College of Business, University of Oregon) is dedicated to the study of sports marketing (i.e. sponsorship, licensing, communications, sports marketing law, consumer behaviour) from both a theoretical and applied perspective.

www.warsawcenter.com/index2.htm

The Bureau of Sport and Leisure Commerce (University of Memphis) provides: professional assistance to sport and leisure organisations in educational, experiential and analytical needs of the field; delivers professional seminars; designs, executes and analyses research studies; develops and executes market plans; and gives event and facility management support.

http://umdrive.memphis.edu/g-HSS/Bureau/

The Centre for Sport Management (Stillman School of Business, Seton Hall University) hosts lectures, continuing education and professional development opportunities, and provides research expertise for the sport industry in the New York, New Jersey metropolitan area.

http://business.shu.edu/sports/programdescription.html

The major universities in Australia and New Zealand offering sport management study at either undergraduate or postgraduate level include:

- Ballarat University – Ballarat, Victoria
- Deakin University – Melbourne, Victoria
- Curtin University – Perth, Western Australia
- Edith Cowan University – Perth, Western Australia
- Griffith University – Gold Coast, Queensland
- Massey University - Auckland, New Zealand.
- Southern Cross University – Lismore, NSW
- University of Canberra – Canberra, ACT
- University of Technology Sydney – Sydney, NSW
- University of Otago - Otago, New Zealand
- UNITEC Institute of Technology – Auckland, New Zealand
- Victoria University of Technology - Melbourne, Victoria

3.5. Specialised International Degree Programmes

- MEMOS (Master Européen de Management des Organisations Sportives), contact Jean Loup Chappelet, IDHEAP, Route de Chavannes, Lausanne, Switzerland.
- Currently, in Europe there are more than 200 universities offering specialised or combined education, focused on sport management. At the same time, many national sport governing bodies in Europe are offering courses and training for those volunteers who are leaders in sport federations and clubs. See more at www.aehesis.de

4. Appendix Material

4.1. Terminology

An international project on terminology is now being developed by EASM, in co-operation with the European Network of Sport Science, Education and Employment and the International Association for Sports Information (IASI) Europe.

4.2. Position Statement(s)

1. Statement of the European Association For Sport Management on Fair Play
The European Association for Sport Management called for ethical policies and practices regarding sport, at the close of its 7th Annual Congress in Thessaloniki, Greece. The Board of the Association issued the following statement on behalf of the Congress delegates:
"The process of globalisation, Europeanisation, nationalism and economic growth must be addressed by all organisations and individuals involved in sport because these processes have a profound impact on social issues.
First, governments throughout Europe are stressing the social inclusion agenda as we approach the next millennium. At the same time, sadly, governments are reducing the funds at the local level available to deal with the exclusion from sport of disadvantaged people.
Second, governments and sport organisations at international, national and local levels include the issue of equal opportunities for women and ethnic minorities within their policy statements; however, constitutional and rule changes are not enough to allow women and ethnic minorities to play an equal part in leading sport and participating in sport.
The adoption and practice of ethical values of sport policies, consistent with the Olympic ideal and adequate funding are needed urgently to turn fine words into the deeds and ensure fair play for all".

2. NASSM operates according to the following canons or principles:

   a) That sport managers shall hold paramount the safety, health and welfare of individuals; perform services only in areas of competence; issue public statements in an objective and truthful manner; seek employment only where a need for service exists; maintain high standards of personal conduct; strive to become and remain proficient in professional practice and in the performance of professional functions; and act in accordance with the highest standards of professional integrity.

b) That professionals shall (a) hold as primary their obligations and responsibilities to students/clients; be a faithful agent or trustee when acting in a professional matter; (b) make every effort to foster maximum self-determination on the part of students/clients; (c) respect the privacy of students/clients and hold in confidence all information obtained in the course of professional service; and (d) ensure that private or commercial service fees are fair, reasonable, considerate and commensurate with the service performed and with due respect to the student/clients to pay.

c) That professionals' ethical responsibilities to employers/employing organisations are characterised by fairness, non-malfeasance and truthfulness.

d) That professionals (a) treat colleagues with respect, courtesy, fairness and good faith; (b) relate to the students/clients of colleagues with full professional consideration; (c) uphold and advance the values and ethical standards, the knowledge and the mission of the profession; (d) take responsibility for identifying, developing and fully utilising established knowledge for professional practice; and (e) shall, when engaged in study and/or research, be guided by the accepted convention of scholarly inquiry.

e) That professionals shall (a) promote the general welfare of society; (b) regard professional service to others as primary; and (c) report minor and major infractions by colleagues to the appropriate committee of the professional society when and where such a mechanism exists.

4.3. Varia

1. Sport Management Program Approval

As of June 2006, 40 American undergraduate sport management programs have received NASPE/NASSM curriculum approval. 29 programs have received master's curriculum approval, and 6 programs have received doctoral curriculum approval.

2. NASSM Archives and Business Office

NASSM maintains comprehensive archives within the Archival Collections at the Bowling Green State University (Ohio, USA) Libraries. The NASSM Archives contain the presidential papers of all past NASSM executives as well as a number of oral (video) histories (e.g. conversation among the NASSM founders, term summaries from past-presidents, historical perspectives delivered by the Journal of Sport Management Editors). The business office of the Association is based at Slippery Rock University, West Gym 014, Slippery Rock University, Slippery Rock, PA 16057, USA, and manages the Association's business, provides services and information to current and potential members, keeps and develops contacts with sport organisations and universities in North America and carries out various NASSM projects.

4.4. Free Statement

P. Chelladurai, in various literature, defined sport management as a field concerned with the coordination of the production and marketing of sport services (which have been broadly classified as participant services and spectator services). Such coordination is achieved through the managerial functions of planning, organising, leading and evaluating. Thus, sport management, as an area of scientific study and professional preparation, addresses these functions within the specific environment of sport enterprise.

Sports Medicine

David A. Parker

Contact

Dr. David Parker
FRACS
Sydney Orthopaedic Research Institute
Email: dparker@sydneyortho.com.au

1. General Information

1.1. Function

Sports Medicine, in general terms, is the study of medicine relevant to sporting activities, including injury prevention, diagnosis and treatment. The field of sports medicine encompasses a large number of areas of science and practice, such that its definition can vary depending on the context of its use. It can however be broadly divided into sports science and clinical sports medicine. Sports science involves the scientific study of areas relevant to sporting participation, and includes fields such as exercise physiology, biomechanics and sports psychology, whereas clinical sports medicine focuses more on management of conditions related to sports, and includes practitioners such as orthopaedic surgeons, sports physicians, physiotherapists and exercise physiologists. There is clearly a complementary relationship between sports science and clinical sports medicine, with the basic science study providing a sound basis for clinical practice.

For the purpose of this publication, the focus will be on the clinical aspect of sports medicine in the management of conditions relevant to sporting participation. Whilst there is often a focus on musculoskeletal conditions and therefore orthopaedic surgery, sports medicine also involves management of any medical condition relevant to sport, hence the relatively recent evolution of the sports physician as the primary care person in sports medicine, usually with secondary involvement of a surgeon if necessary. In an increasing number of countries, Sports Medicine is recognised as a specialty in its own right.

The history of sports medicine can be traced back to the civilisations of Ancient Greece and Rome, where athletic contests were a part of everyday life, and there appeared to be an appreciation of the health benefits of physical activity. The first use of therapeutic exercise is credited to Herodicus, one of Hippocrate's teachers. Until the 2nd century, when the first "team doctor", Galen, was appointed to the gladiators, physicians were only involved in the case of injury. In the 5th century, care of athletes was largely the responsibility of coaches who were thought to be experts, not just on sporting techniques but also diet and physical therapy. However, more formal recognition and controlled study of the science of sports medicine did not occur until the first half of the 20th century. In 1912, the German National Sports Medicine organisation was established, the first sports medicine association, followed by the Italian body in 1929. In 1928, at the Olympics in St Moritz, a committee was formed to plan the First International Congress of Sports Medicine. As a result, the Association Internationale Medico-Sportive (AIMS) was formed, which subsequently changed its name to the Federation Internationale Medecine Sportive (FIMS). FIMS exists today as the main international sports medicine association, and is discussed in more detail in a later section of this chapter.

In 1958, in Koln, Germany, the first modern definition of sports medicine was offered by the Institute for Cardiology and Sports Medicine as: "sports medicine incudes those theoretical and practical branches of medicine which investigate the influence of exercise, training, and sport on healthy and ill people, as well as the effects of lack of exercise, to produce useful results for prevention, therapy, rehabilitation, and the athlete". The main aspects of sports medicine were described by the director of this Institute, Prof. Wildor Hollmann, in 1988, as:

1. Medical treatment of injuries and illnesses.
2. Medical examination before starting a sport to detect any damage, which could be worsened by the sport.
3. Medical performance investigation to assess the performance capacity of heart, circulation, respiration, metabolism and the skeletal musculature.
4. Performance diagnosis specific to the type of sport.
5. Medical advice on lifestyle and nutrition.
6. Medical assistance in developing optimal training methods.
7. Scientifically based control of training.

What has become clear throughout the history and evolution of sports medicine is that it has always been a multidisciplinary specialty, involved in injury prevention and treatment, as well as providing athletes with optimum preparation for competition. Modern sports medicine has continued this practice, involving a broad range of medical speciality areas, including orthopaedics, cardiology, rehabilitation, paediatrics, general practitioners and specialist sports medicine physicians.

1.2. Body of Knowledge

There is a large and ever expanding volume of knowledge regarding sports medicine that is readily available. This exists in the form of original research articles in refereed journals and in textbooks and reference books. Original research studies published in journals must be peer-reviewed and pass appropriate, stringent criteria for scientific quality regarding methodology and conclusions. Case studies may also be published, usually describing an interesting case affecting one patient. These studies have less scientific validity and are usually published on the basis that they present an unusual case that adds to the current body of knowledge on a subject. Textbooks usually present a compilation of current expert opinions around a specific area of sports medicine, with individual chapters written by acknowledged experts on specific topics, and edited by leading sports medicine practitioners. There are also a number of book series, representing comprehensive publication that address general areas of sports medicine, usually updated at regular intervals. A listing of the international research journals of sports medicine is presented in section 2.1 Journals, section 2.2 Reference Books, Encyclopaedias, etc. and section 2.3 Book Series.

1.3. Methodology

Research into the many areas of sports medicine encompasses a wide range of methodologies and can be divided into basic science studies and more clinically oriented work. Basic science studies can be in either animal or cadaveric models and look to examine the response of tissues to conditions relevant to sporting activity, as well as investigate the validity of newer treatment methods prior to the introduction of human use. These studies encompass many areas including exercise physiology and biomechanics.

Clinical studies include studies of the epidemiology of sporting injuries, to better understand injury patterns and guide preventative measures, as well as clinical trials of treatment methods. These can include trials comparing surgical techniques with other non-surgical interventions, although at the elite level it can be difficult to enrol athletes in such trials when a perception of the "best" treatment may already exist, often without valid evidence. Therefore, although prospective, randomised, controlled trials are considered Level 1 or best available evidence for clinical studies and are increasing in number, they are often not feasible. For this reason, many clinical studies will simply follow subjects over a defined period of time to examine the natural history of conditions and treatments in cohort studies, whereas cross-sectional studies will compare different groups at a specific point in time. Case control studies will compare a group of subjects with a certain condition and/or treatment to a matched group of controls, but are often subject to a number of possible biases.

Meta-analyses involve combining data from carefully selected, high-quality studies to increase validity, often providing a valuable summary of the best available evidence on a topic. Case studies can provide interesting insight into a condition by virtue of individual examples, although they lack the scientific power or validity of larger controlled trials. Journals often also solicit review articles on specific topics, in which recognised authorities will review the literature on a specific topic and provide additional expert opinion. Proceedings from conferences are published by many journals, including abstracts of original research and conclusions from symposia.

1.4. Relationship to Practice

Sports Medicine is practised in many locations, from universities to hospitals, to sporting venues themselves. Universities are the usual location for basic science research and are often affiliated with sports medicine clinics and hospitals. At the sporting venues, trainers and physical therapists will assist in preparation for competition and in the initial management of injuries. Team doctors will be involved in initial assessment and management of injuries, in conjunction with the physical therapist, and will usually facilitate subsequent management. The background of the team doctor varies in different countries, but has largely moved towards the specialist sports physician being the primary care doctor for the athlete. The role of the sports physician usually involves on-field coverage, as well as running clinics between competition to assess and follow up athletes' progress and guide management, including referral when necessary. The most common referral would be to an orthopaedic specialist, for assessment and management of conditions likely to require surgical intervention.

All of the above professionals should have complementary roles in sports medicine, with both basic science and clinical research providing a sound basis for management of conditions by these professionals. Ongoing study at all levels should lead to a greater understanding of the benefits of sporting participation, as well as the prevention and management of sporting injury for all athletes from recreational to elite level. Improved knowledge can enhance these areas, as well as often leading to health policy change at government level and rule changes at competitive level. The end result of these collaborations should be a healthier population with reduced sports related injury.

2. Information Sources

2.1. Journals

English Language
Acta Orthopaedica (Sweden)
American Journal of Sports Medicine (USA)
Archives of Physical Medicine and Rehabilitation (USA)
British Journal of Sports Medicine (United Kingdom)
Cardiopulmonary Physical Therapy Journal
Chinese Journal of Physical Therapy
Clinics in Sports Medicine (USA)
Clinical Journal of Sports Medicine (USA)
Human Movement Science (Holland)
International Journal of Sports Medicine (Germany)
International SportMed Journal (International)
Journal of Orthopaedic and Sports Physical Therapy (USA)
Journal of Science and Medicine in Sport (Australia)
Journal of Sports Sciences (England)
Journal of Sports Sciences supplements (England)
Journal of Sports Science and Medicine (Turkey)
Journal of Sport Rehabilitation (USA)
Journal of Sports Science and Medicine (International online journal)
Knee Surgery, Sports Traumatology, Arthroscopy (Europe)
Medicine and Science in Sport and Exercise Sports Medicine (USA)
Operative Techniques in Sports Medicine (USA)

Physical Therapy in Sport (UK)
The American Journal of Sports Medicine (USA)
The Journal of Sports Medicine and Physical Fitness (Italy)
The Physician and Sports Medicine (USA)
The Swedish Medical Journal (Sweden)
Scandinavian Journal of Medicine and Science in Sports (Denmark)
South African Journal of Sports Medicine (South Africa)
Sports Medicine (New Zealand)
Sports Medicine Reports (USA)
Sports Medicine and Arthroscopy Review (USA)

Non-English language

Annales de Réadaptation et de Médicine Physique (France)
Archivos de Medicina del Deporte (Spain)
Deutsche Zeitschrift für Sportmedizin (Germany)
Journal de Traumatoligie du Sport (France)
Medicina dello Sport (Italy)
Medicina del Ejercicio (Spain)
Medecine du Sport (France) (in English, French, Italian and Portuguese)
Östereichisches Journal für Sportsmedizin (Austria)
Schweizerische Zeitshrift für Sportsmedizin / Revue Suisse de Médecine et Trau-
matologie du Sport (Switzerland)
Sport-Orthopaedie und Traumatologie (Germany)

2.2. Reference Books, Encyclopaedias, etc.

Sports medicine and training

Ahonen, J. et al. (1994). *Sportmedizin und Trainingslehre*. Stuttgart: Schatthaue.
    ISBN: 3794510275
Almekinders, L.C. (1996). *Soft Tissue Injuries in Sports Medicine*. Blackwell. ISBN:
    0865423822
American College of Sports Medicine (1996). *ACSM's Handbook for the Team
    Physician*. Baltimore: Williams and Wilkins. ISBN: 0683000284
Badtke, G. (1995). *Lehrbuch der Sportmedizin*. Leipzig: Hüthig-Barth-Verlag. ISBN:
    3825280985

Billat, V. (1998). *Physiologie et Méthodologie de l'Entraînement: de la Théorie à la Practique.* De Boeck Université. ISBN: 2744500364

Birrer, R.B. and O'Connor, F.G. (2004.) *Sports Medicine for the Primary Care Physician.* Boca Raton: CRC Press, ISBN: 084931464X

Bloomfield, J., Fricker, P.A. and Fitch, K.D. (1995). *Science and Medicine in Sport* (2nd Ed). ISBN: 0867933216

Brukner, P. and Khan, K. (2002). *Clinical sports medicine 2nd Ed.*, Sydney; London: McGraw-Hill, ISBN: 0074711083

Chan, K.M. et al., (1998). *Controversies in Orthopaedic Sports Medicine.* Williams & Wilkins. ISBN: 9623560257

Chan, KM et al (2006). *FIMS Team Physician Manual, 2nd Ed.* Hong Kong, ISBN 962-356-029-X

DeLee, J.C. and Drez, D. (2003). *Orthopaedic Sports Medicine: Principles and Practice 2nd Ed.*, Philadelphia. W.B. Saunders Company. ISBN: 0721688454.

Fields, K.B. and Fricker, P.A. (1997). *Medical Problems in Athletes.* Oxford. Blackwell Science. ISBN: 0865424802

Garrett, Jr., W.E., Speer, K.P. and Kirkendall, D.T. (Eds.) (2001). *Principles and practice of orthopaedic sports medicine*, Philadelphia: Lippincott Williams & Wilkins, ISBN: 078172578X

Hackney, R. and Wallace, A. (1999). *Sports Medicine Handbook.* BMJ Publishing Group. ISBN: 0727910310

Higgins, R., English, B. and Brukner, P. (Eds.) (2005). *Essential sports medicine*, Malden: Blackwell Publishing, ISBN: 140511438X

Hollmann, W. (Ed.) (1995). *Lexicon der Sportmedizin.* Leipzig: Barth-Verlag. ISBN: 3335004116

Irvin, I. (1998). *Sports Medicine.* Prentice Hall. ISBN: 0130374660

Johnson, D.L. and Mair, S.D. (2006). *Clinical sports medicine*, Philadelphia, PA: Mosby Elsevier, ISBN: 0323025889

Kent, M. (2006). *The Oxford Dictionary of Sports Science and Medicine, 3rd Ed.* Oxford University Press. ISBN: 0198568509

Knight, K.L. (1995). *Cryotherapy in Sport Injury Management.* Champaign, IL: Human Kinetics. ISBN: 0873227719

MacAuley, D. (1999). *Sports Medicine: Practical Guidelines for General Practice.* Butterworth-Heinemann. ISBN: 0750637307

Marcos Becerro, J.F. and Santonia Gomez, R. (1996). *Olimpismo y Medicina.* Madrid: Rafael Santonja. ISBN: 8489355045

Marcos Becerro, J.F. (1992). *Medicina del Deporte*. Guia Practica. Editado por el Comité Olimpico Español. ISBN: 8487114349

Maughn, R.J. (1999). *Basic Science for Sports Medicine*. Butterworth-Heinemann. ISBN: 0750634669

Miller, M.D., Cooper, D.E. and Warner, J.J.P. (2002). *Review of sports medicine and arthroscopy, 2nd Ed*. Philadelphia: W.B. Saunders ISBN: 0721694209

Miller, M.D. and Sekiya, J.K. (2006). *Sports medicine: core knowledge in orthopaedics*, Philadelphia, PA: Mosby Elsevier, ISBN: 0323031382

Monod, H. and Kahn, J. (1995.) *La Médicine du Sport*. Paris: Autropsi. ISBN: 2717828974

Monod, H. and Kahn, J. (1995). *La Médicine du Sport pour la Practique*. Paris: SIMEP ISBN: 2225841152

Narvani, A.A., Thomas, P. and Lynn, B. (Eds.) (2006). *Key topics in sports medicine*, Abingdon, Oxon ; New York : Routledge, ISBN: 041541122X

Neumann, G. and Schüler, K.P. (1994). *Sportmedizinische Funktionsdiagnostik*. Leipzig: Hüthig-Barth-Verlag. ISBN: 33350035207

O'Conner, F.G. (Ed.) (2006). *Sports Medicine: Just the Facts*, New York: McGraw-Hill, Medical Pub. Division, ISBN: 0071421513

Oakes, E.H. (2006). *The A to Z of sports medicine*, New York: Checkmark Books, ISBN: 0816066922

Ortega Sánchez Pinilla, R. (1992). *Medicina del Ejercicio Físico y del Deporte para la Atención de la Salud*. Edita Librería Díaz de Santos. ISBN: 8479782533

Perrin, D.H. (2005). *Athletic Taping and Bracing*. Champaign IL: Human Kinetics. ISBN: 0736048111

Pilardeau, P. (1998). *Dictionaire Encyclopaedie Pedicatrique en Médicine du Sport*. French and European Publications, Inc. ISBN: 0320003922

Saillert, G. (1996). *Pathologie Chirurgicale du Genou du Sportif*. Paris: Expansion Scientifique Francaise. ISBN: 270461525X

Scuderi, G.R. and McCann, P.D. (2005). *Sports Medicine: A Comprehensive Approach, 2nd Ed.*, Philadelphia: Mosby-Elsevier ISBN: 0323023452

Sherry, E. (1997). *Sports Medicine: Common Problems and Practical Management*. New York: Oxford University Press. ISBN: 1900151553

Pellicia, A., Caselli, G. and Bellotti, P. (1997). *Advances in Sports Cardiology*. Milano: Springer-Verlag Italia. ISBN: 3540750363

Whyte, G.P., Harries, M. and Williams, C. (2005). *ABC of sports and exercise medicine, 3rd Ed*. Mass.: Blackwell Pub, ISBN: 0727918133

Zeppilli, P. (nd). *Cardiologica Dello Sport*. Roma: C.E.S.I> - Via Cremona, 19-00161. ISBN: 8886062176

Sports injuries

Caine, D.J., Caine, C.G. and Lindner, K.J. (1996). *Epidemiology of Sports Injuries*. Oxford: Blackwell Science. ISBN: 0873224663

Ciullo, J. (1996). *Shoulder Injuries in Sport*. Champaign, IL. Human Kinetics. ISBN: 087326518

Danowski, R.G. and Chanussot, J.C. (1999). *Traumatologie du Sport*. Masson. ISBN: 2225837473

Engelhardt, M., Huitormann, B. and Segesse, B. (1997). *GOTS Manual Sport Traumatologie*. Bern. Haus Hube. ISBN: 345682792X

Ferretti, A. (1996). *Traumatologia Dello Sport*. Rome. Casa Editrice Scientifica Internazionale. ISBN: 8886062214

Karageanes, S.J. (2005). *Principles of manual sports medicine*, Philadelphia: Lippincott Williams & Wilkins, ISBN: 0781741890

Norris, Christopher M. (2004). *Sports injuries: diagnosis and management, 3rd Ed.*, Edinburgh ; New York : Butterworth Heinemann, ISBN: 0750652233

O'Connor, R., Budgett, R., Wells, C. and Lewis, J. (1998). *Sports Injuries and Illnesses: Their Prevention and Treatment*. Crowood Press, Ltd. ISBN: 1861261071

Potparic, O. and Gibson, J. (1996). *A Dictionary of Sports Injuries and Disorders*. New York/ London: Parthenon Publishing Group. ISBN:1850706867

Simonian, P.T., Cole, B.J. and Bach, B.R. (2006). *Sports Injuries of the Knee: Surgical Approaches*, New York: Thieme, ISBN: 3131402911

Specific sports

Cousteau, J.P. (1999). *Médicine du Tennis*. Masson. ISBN: 2225828938

Ekstrand, J., Karlsson, J. and Hodson, A. (2003). *Football Medicine*, London: Martin Dunitz, ISBN: 1841841641

Irvin, R., Iversen, D. and Roy, S. (1998). *Sports medicine: prevention, assessment, management, and rehabilitation of athletic injuries*, Boston: Allyn & Bacon, ISBN: 0130374660

Sands, W. (2003). *Scientific Aspects of Women's Gymnastics*, Basel; New York: Karger, ISBN: 3805574762

Troup, J.P. et al., (1996). Biomechanics and Medicine in Swimming VII, London; New York: E & FN Spon, ISBN: 0419204806

Rehabilitation

Kumbhare, D.A. and Basmajian, J.V. (2000). *Decision making and outcomes in sports rehabilitation*, Philadelphia: Churchill Livingstone, ISBN: 0443065462

Prentice, W.E. (2004). *Rehabilitation techniques for sports medicine and athletic training 4th Ed.*, Boston: McGraw-Hill, ISBN: 0072462108

Tippet, S.R. and Voight, M.L. (1995). *Functional Progressions for Sport Rehabilitation*. Champaign, IL: Human Kinetics. ISBN: 0873226607

Zachazewski, J.E. Magee, D.J. and Quillen, W.S. (1996). *Athletic Injuries and Rehabilitation*. Philadelphia: W.B. Saunders Company. ISBN: 0721649467

Nutrition

Brouns, F. (1994). *Les Besoins Nutritionels des Athlètes*. Paris: Masson. ISBN: 2225843139

Brouns, F. (1995). *Necesidades Nutricionales de los Atletas*. Barcelona: Editorial Paidotribo. ISBN: 8480192089.

Burke, L. and Deakin, V. (Eds.) (2006). *Clinical Sports Nutrition 3rd Ed*. Sydney, McGraw Hill, ISBN: 0074716026.

Pujol-Amat, P. (1997). *Nutrición, Salud y Reudièuto Deportivo (2nd Ed.)*. Barcelona: Expas, Publicaciones Médicas. ISBN: 847179277X

Ryan, M. (2002). *Sports Nutrition for Endurance Athletes*. Boulder, CO: VeloPress, ISBN: 1931382158

2.3. Book Series

The Encyclopaedia of Sports Medicine – produced by the IOC Medical Commission in collaboration with the Fédération Internationale de Médicine du Sport (FIMS).

Vol I. Dirix, A. Knuttgen, H.G. and Tittel, K. (1998). *The Olympic Book of Sports Medicine*. Oxford: Blackwell Science, Ltd. ISBN: 0632019638; Soft cover ISBN: 0632030844

Vol II. Shephard, R. and Astrand, P. (1991). *Endurance in Sport.* Oxford: Blackwell Science, Ltd. ISBN: 0632030356; Soft cover ISBN: 0632037075

Vol II. Shephard, R. and Astrand, P. (2000). *Endurance in Sport 2nd Ed.* Oxford: Blackwell Science, Ltd. ISBN: 0632019638. Soft cover ISBN: 0632030844

Vol III. Komi, P. (1993). *Strength and Power in Sport.* Oxford: Blackwell Science, Ltd. ISBN: 0632030313; Soft cover ISBN: 063203863

Vol IV. Renström, P. (1993). *Sports Injuries: Basic Principles of Prevention and Care.* Oxford: Blackwell Science, Ltd. ISBN: 0632033312

Vol V. Renström, P. (1994). *Clinical Practice of Sports Injury Prevention and Care.* Oxford: Blackwell Science, Ltd. ISBN: 0632037857

Vol VI. Bar-Or, O. (1996). *The Child and Adolescent Athlete.* Oxford: Blackwell Science, Ltd. ISBN: 0865429049

Vol VII. Maughan, R. (2000). *Nutrition in Sport.* Oxford: Blackwell Science, Ltd. ISBN: 0632050942

Vol VIII. Drinkwater, B. (2000). *Women in Sport.* Oxford: Blackwell Science, Ltd. ISBN: 0632050845

Vol IX. Zatisiorsky (2000). *Biomechanics in Sport: Performance Improvement and Injury Prevention.* Oxford: Blackwell Science, Ltd. ISBN: 0632053925

Vol X. Frontera, W.R. (2002). *Rehabilitation of Sports Injuries: Scientific Basis.* Oxford: Blackwell Science, Ltd. ISBN: 9780632058136

Vol XI. Kraemer, W. and Rogol, A. (2005). *The Endocrine System in Sports and Exercise.* Oxford: Blackwell Science, Ltd. ISBN: 97814051301

Vol XII. Woo, S., Renström, P. and Arnoczky, S. (2007). *Tendinopathy in Athletes.* Oxford: Blackwell Science, Ltd. ISBN: 97814051567

Vol XIII. Hebestreit, H. and Bar-Or, O. (2007). *The Young Athlete.* Oxford: Blackwell Science, Ltd. ISBN: 9781405156479

Handbooks of sports medicine

Costill, D.L., Maglischo, E.W. and Richardson, A.B. (1991). *Swimming.* Oxford: Blackwell Science, Ltd. ISBN: 0632030275

Stager, J. and Tanner, D. (2004). *Swimming 2nd Ed.* Oxford: Blackwell Science, Ltd. ISBN: 0632059141

Leach, R., Fritschy, D. and Steadman, J.R. (1994). *Alpine Skiing.* Oxford: Blackwell Science, Ltd. ISBN: 063203033X

Ekblom, B. (1994). *Football.* Oxford: Blackwell Science, Ltd. ISBN: 0632033282

Hawley, J. (2000). *Running*. Oxford: Blackwell Science, Ltd. ISBN: 0632053917

Gregor, R. and Conconi, J. (2000). *Road Cycling*. Oxford: Blackwell Science, Ltd. ISBN: 086542912X

Kraemer, W. and Hakkinen, K. (2001). *Strength Training for Sport*. Oxford: Blackwell Science, Ltd. ISBN: 0632055685

Renström, P. (2002). *Tennis*. Oxford: Blackwell Science, Ltd. ISBN: 0632050349

Rusko, H. (2002). *Cross Country Skiing*. Oxford: Blackwell Science, Ltd. ISBN: 0632055715

Maughan, R. and Burke, L. (2002). *Sports Nutrition*. Oxford: Blackwell Science, Ltd. ISBN: 0632058145

McKeag, D. (2003). *Basketball*. Oxford: Blackwell Science, Ltd. ISBN: 0632059125

Reeser, J. (2003). *Volleyball*. Oxford: Blackwell Science, Ltd. ISBN: 0632059133

Year Books of Sports Medicine

Shephard, R. et al (Eds.) (2000). *Year Book of Sports Medicine*. Mosby. ISBN: 0323007309

Shephard, R. et al (Eds.) (2001). *Year Book of Sports Medicine*. Mosby. ISBN: 0323007317

Shephard, R. et al (Eds.) (2002). *Year Book of Sports Medicine*. Mosby. ISBN: 0323015735

Shephard, R. et al (Eds.) (2003). *Year Book of Sports Medicine*. Mosby. ISBN: 0323020569

Shephard, R. et al (Eds.) (2004). Y*Year Book of Sports Medicine*. Mosby. ISBN: 0323020577

Shephard, R. et al (Eds.) (2005). *Year Book of Sports Medicine*. Mosby. ISBN: 0323021174

Shephard, R. et al (Eds.) (2006). *Year Book of Sports Medicine*. Mosby. ISBN: 1416033017

Shephard, R. et al (Eds.) (2007). *Year Book of Sports Medicine*. Mosby. ISBN: 9780323046473

Perspectives in Exercise Science and Sports Medicine

Vol 1. Lamb, D.R. and Murray, R. (Eds.) (1988). *Prolonged Exercise.* Benchmark Press, Inc. ISBN: 0936157348

Vol 2. Gisolfi, C.V. and Lamb, D.R. (Eds.) (1989). *Youth, Exercise and Sport.* Benchmark Press, Inc. ISBN: 0936157321

Vol 3. Gisolfi, C.V. and Lamb, D.R. (Eds.) (1990). *Fluid Homeostasis During Exercise.* Benchmark Press, Inc. ISBN: 0697148165

Vol 4. Lamb, D.R. and Williams, M.H. (Eds.) (1991). *Ergonomics: Enhancement of Performance in Exercise and Sport.* William C. Brown Publishers. ISBN: 0697149773

Vol 5. Gisolfi, C.V. and Lamb, D.R. (Eds.) (1992). *Energy Metabolism in Exercise and Sport.* William C. Brown Publishers. ISBN: 0697162753

Vol 6. Gisolfi, C.V., Lamb, D.R. and Nadel, E. (Eds.) (1993). *Exercise, Heat and Thermoregulation.* William C. Brown Publishers. ISBN: 0697204928

Vol 7. Lamb, D.R., Knuttgen, H.G. and Murray, R. (Eds.) (1994). *Physiology and Nutrition for Competitive Sport.* Cooper Publishing Group. ISBN: 1884125093

Vol 8. Gisolfi, C.V., Lamb, D.R. and Nadel, E. (Eds.) (1995). *Exercise in Older Adults.* Cooper Publishing Group. ISBN: 18841252204

Vol 9. Bar-Or, O., Lamb, D.R. and Clarkeson, P.M. (Eds.) (1996). *Exercise and the Female: A Life Span Approach.* Cooper Publishing Group. ISBN: 188412528X

Vol 10. Lamb, D.R. and Murray, R. (Eds.) (1997). *Optimizing Sport Performance.* Cooper Publishing Group. ISBN: 1884125638

Vol 11. Lamb, D.R. and Murray, R. (Eds.) (1998). *Exercise, Nutrition and Weight Control.* Cooper Publishing Group. ISBN: 1884125700

Vol 12. Lamb, D.R. and Murray, R. (Eds.) (1999). *The Metabolic Basis of Performance in Sport and Exercise.* Cooper Publishing Group. ISBN: 1884125735

Sports Medicine for Coaches and Athletes

Vol 1. Shamoo, A.E., Baugher, W.H. and Germeroth, R.M. (1995). *Soccer,* Luxembourg: Harwood Academic, ISBN: 3718606011

Vol 2. Shamoo, A.E., Silberstein, C.E. and Germeroth, R.M. (2000). *Baseball,* Amsterdam, Netherlands: Harwood Academic, ISBN: 9057026112

Vol 3. Rogers, M.A., Wernicki, P.G. and Shamoo, A.E. (2000). *Older individuals and athletes over 50,* Australia: Harwood Academic, ISBN: 9057026007

Vol. 4. Ekland, A., Crocket, L.J. and Shamoo, A.E. (2000). *Skiing*, Amsterdam, Netherlands: Harwood Academic, ISBN: 9057025825

2.4. Congress/Workshop Proceedings

Proceedings of conferences are printed and distributed to delegates by the various host organisations. Frequently, abstracts of the presented papers are published in associated journals.

2.5. Data Banks

There are a number of information databases, which are useful for locating further information sources such as original research articles in scientific journals and reference articles.

Pubmed
      www.ncbi.nlm.nih.gov/sites/entrez
Sportdiscus
      www.sirc.ca/products/sportdiscus.cfm
Cinahl
      www.cinahl.com/index.html
AusSportMed
      www.informit.com.au/browse.asp?itemID=AUSPORTMED&ContainerID
      =info_product_indexes_bytitles
The NLM Gateway
      http://gateway.nlm.nih.gov
ScienceDirect
      www.sciencedirect.com

2.6. Internet Sources

Many sports medicine organisations have websites offering information for practitioners and patients about sports medicine, as well as links to other useful websites.
American College of Sports Medicine
      www.acsm.org

American Medical Society for Sports Medicine
www.amssm.org/
American Medical Athletic Association
www.amaasportsmed.org/
American Orthopaedic Society for Sports Medicine
www.sportsmed.org/
Asian Federation of Sports Medicine
www.afsmonline.com
Association of Chartered Physiotherapists in Sports Medicine
www.acpsm.org/index.asp
Australasian Academy of Podiatric Sports Medicine Inc.
www.aapsm.org.au/
Australian Association for Exercise and Sport Science
www.aaess.com.au/
Australian Institute of Sport (AIS)
www.ausport.gov.au/ais
Australian Orthopaedic Association
www.aoa.org.au
British Association for Sport and Exercise Medicine
www.basem.co.uk/
British Association of Sport and Exercise Sciences
www.bases.org.uk/newsite/home.asp
Canadian Academy of Sports Medicine
www.casm-acms.org/
European Federation of Sports Medicine Associations
www.efsma.net/
Fédération International de Médicine du Sport
www.fims.org
Hong Kong Association of Sports Medicine and Sports Science
www.fmshk.com.hk/hkasmss/home.htm
International Society for Arthroscopy, Knee Surgery and Orthopaedic Sports Medicine
www.isakos.com/
Japanese Society of Physical Fitness and Sports Medicine
www.jspfsm.umin.ne.jp/
National Academy of Sports Medicine
www.nasm.org/

Orthopaedic Research Societ
    www.ors.org/web/index.asp
Sports Doctors Australia
    www.sportsdoctors.com.au/
Sports Medicine Australia
    www.sma.org.au/
Sports Medicine New Zealand (SMNZ)
    www.sportsmedicine.co.nz/
The American Academy of Orthopaedic Surgeons
    www.aaos.org/
World Anti-Doping Agency
    www.wada-ama.org/en/index.ch2
Worldortho (Orthopaedics, Trauma, Sports Medicine: educating the world)
    www.worldortho.com/

3. Organisational Network

3.1. International Level

International Federation of Sports Medicine (FIMS)

The first sports medicine event to ever be held was the 1st International Hygiene Exhibition in Dresden, 1911. A sport laboratory was opened to enable a six month period of anthropometrical, physiological, cardio-vascular, metabolic and x-ray-investigations of trained and untrained people before and after physical exercise – issues in sports medicine that are still being investigated today.

Following the success of this conference the First Congress for the scientific investigation of Sports and Physical Exercises took place in Germany in 1912, under the chairmanship of Ferdinand-August Schmidt, with 60-70 participants. Topics covered included "Exaggeration in sports' (F. Kraus), 'The importance of physical education from a hygienic point of view" (F-A. Schmidt) and 'Women and physical fitness' (R. Hirsch). It was at this time that the German Committee for Scientific Investigation of Sports and Physical Exercise was founded with F. Kraus being the first elected president. This Committee was the first Sports Medicine Association to be founded anywhere in the world. Two conferences were held in 1913

(French International Conference of Physical Education and International Olympic Congress), however, the development of the Committee was interrupted by the First World War.

It was not until 1921 when the sports medicine forum recommenced, with the French Society of Sports Medicine being founded (President Jean-Paul Langlois). Just one year later, the first sports medicine journal was published Bulletin de la Société Médicale d'Éducation Physique et du Sport. The Netherlands and Switzerland also formed sport medicine organisations in 1921-1922. The first conference since the First World War (2nd German Sports Medicine) was held in 1924. At this time, the Committee changed its' name to German Federation of Physicians for the Promotion of Physical Exercises and from here, the German Federation conducted annual sports medical congresses.

The 4th Congress of the German Federation of Physicians for the Promotion of Physical Exercises was held in Berlin, where 12 countries were invited to elect a new President of the German Federation, Walter Schnell. The foundation of an International Association was discussed, as the International Olympic Committee (IOC) would not agree to the formation of a Medical Commission. The discussions were based on the common desire to share knowledge about the influence of regular physical training on physiological processes, to improve and standardise methods of sports medical examinations for sports performance used in different countries and to support the International Olympic Committee in providing the best possible sports medical supervision of athletes taking part in the Olympic Summer and Winter Games.

The Association Internationale Medico Sportive (AIMS) was formed in 1928, after the International Conference on the Medical Control in Sports, which was approved for the first time by the IOC President. The objective of the organisation was to co-operate with and support the International Sports Federations, increase provision of support for the further development of National Sports Medicine Associations, exchange information on a regular basis about the basic science and practical aspects of sports medicine. The 1st Congress of the Association Internationale Medico Supportive took place in 1928, in Amsterdam during the Olympic Summer Games. In the first official AIMS statement, three objectives of the Association were listed:

1. the promotion of scientific research in the areas of biology, psychology and sociology in connection with sports;
2. the promotion of investigations about medical problems in connection with training and sports; and
3. the organisation of international sports medical congresses.

Frederick Buytendijk was elected as second President of the AIMS. The 2nd International Congress took place in Italy where the name was changed to 'Fédération Internationale Médico – Sportive et Scientific' and Andre Latarjet was elected as President. The 3rd International Congress in France saw the renaming of the organisation to "Fédération Internationale de Médecine Sportive' (FIMS). New recommendations were also developed:

1. the demand for physical education in schools;
2. the recommendation to the Member Associations to develop a sports physician diploma after especially prescribed aspects;
3. the introduction of obligatory medical examinations for athletes; and
4. the standardisation of Sports Medicine examination methods.

The annual congresses took place without interruption until the Second World War, occurring in Germany (4th), France (5th) and Belgium (6th). The Second World War interrupted the congress until 1948 when the 7th International FIMS Congress took place in Czechoslovakia.

The IOC officially announced its cooperation in 1952 at the 9th International FIMS Congress in France, describing FIMS 'as the designed competent international organisation for biological and medical research related to medicine and sport and the medical care of athletes'. Meetings every two years continued to be held and at the 13th International FIMS Congress, in Austria (1960) the first edition of the FIMS' *Journal of Sports Medicine and Physical Fitness* was developed. This journal addressed the urgent need for international information to be dispersed, and strengthened the work of the Federation.

Conferences continued to be held every two years until 1974 and thereafter every four years (due to the rising costs involved).

In Luxembourg in 1962, the possible development of International Sports Medicine FIMS courses was raised, and in Germany, 1965 the first course was held. This served as a prototype for the subsequent courses with a particular focus on developing nations. Olympic Solidarity provided financial support. The purpose and themes of international FIMS courses were:

- Value, dangers and limits of competitive sport as well as sport for health purposes from the medical point of view;
- Preservation of good health as well as prevention and cure of internal, surgical and nervous disorders of our time;
- Standardisation of examination methods and medical functional tests;
- Necessity and possibilities of medical co-operation in Sport Federations and Associations;
- Medical problematic and sport-technical regulations; and
- Questions of international documentation and evaluation of results.

The course was intended to be a world-wide, two week comprehensive productive talk for the benefit of all sportsmen and sportswomen.

The Basic Book of Sports Medicine was developed in 1976, as an adjunct to the FIMS International Sports Medicine Course and was prepared for physicians and professionals enrolled in the course. The text was developed by more than 40 international well-known sports physicians and scientists.

In 1986, Australia (Brisbane) held the 23rd FIMS World Congress where the Council of Delegates elected a new Executive Committee and President, W Hollmann (Germany). Hollmann declared that the FIMS existed as the only world organisation in sports medicine and that two streams should be developed, one for foreign policy and one for internal policy.

The main points of foreign policy should be:

1. The relations to the World Council of Sports must be intensified in order to advance research, education and practice in all fields of sport science;
2. The relations to the World Health Organisation (WHO) must be intensified;
3. The relations to the International Olympic Committee (IOC) must be improved and

4. Direct influence should be exerted on national policy concerning sports medicine, whereever this seems to be necessary,

For internal policy, the main aim of FIMS was to give assistance to the Member National Association concerning all questions of:

1. Research (basic and applied research);
2. Education in sports medicine (at universities as well as in post-graduate courses with and without examination of medical education and training);
3. Practice of sports medicine; and
4. Representation of sports medicine in other medical associations. Today, FIMS serves as the sole sport medicine organisation at the international level and their meetings and congresses continue to be held every 4 years (Council of Delegates every 2 years).

IOC Medical Commission

www.olympic.org/uk/organisation/commissions/medical
The IOC Medical Commission was created in 1967 to manage the increasing problem of doping in sports. Three main principles were subsequently identified:

- Protection of the health of athletes;
- Respect for both medical and sports ethics;
- Equality for all competing athletes.

Through study in the anti-doping area, the Commission has developed alternative methods to assist athletes, including sports medicine, biomechanics, exercise physiology and nutrition. Since creation of the World Anti-Doping Agency (WADA), the Commission's scope has expanded to include all of the main medical issues, which may occur in sports. The Commission will facilitate consensus meetings on all of these issues whilst giving priority to protection of the health of anyone participating in sport at any level.

The Commission has established sub-commissions:
- Doping and Biochemistry of Sports Medicine;
- Biomechanics and Physiology of Sports;

- Sports Medicine and Coordination with the National Olympic Committees;
- Publications in the Sports Sciences.

3.2. Regional Level

Within FIMS, various continental and multinational groups exist that are made up of the FIMS Member National Associations.

Continental groups:

- African Union of Sports Medicine / Union Africaine de Medecine du Sport – founded in 1982 (15 nations)
- Asian Federation of Sports Medicine – founded in 1990 (22 nations)
- Confederación PanAmericana de Medicina del Deporte – founded in 1975 (all 28 nations of the Western Hemisphere)
- European Federation of Sports Medicine – founded in 1997 (36 nations)

Multinational groups:

- Arab Federation of Sports Medicine – founded in 1982 (9 Arabic speaking countries)
- Association of Sports Medicine of the Balkans – founded in 1967 (9 nations of the Balkan area)
- Caribbean Association of Sports Medicine – founded in 1986 (7 nations)
- Confederación Centroamericana de Medicina del Deporte – founded 1986 (6 nations)
- Confederación Sudamericana de Medicina del Deporte – founded in 1986 (10 nations)
- Federation Magrebine de Medecine du Sport – founded in 1990 (5 nations of North Africa)
- Groupement Latin et Mediterraneen de Medecine du Sport GLMMS (LMGSM) – founded in 1956 (16 Latin based language nations of the Mediterranean)
- Northwest European Chapter of Sports Medicine (12 nations)

3.3. National Level

Germany

The German Federation of Sports Medicine (Deutscher Sportärztebund) was founded in 1912. In 1920, the Federation participated in the founding of the first German Academy for Sports in Berlin and in 1928 participated in the founding of FIMS. In 1955, a Board of Trustees for sports medicine research was founded in Köln. In 1970, the additional title of "Sports Physician" was introduced for medical doctors who successfully completed 240 hours of specialised education.

The Federation conducts approximately 200 sports medicine courses per year and funds are provided on an annual basis for research by ministries of the Federal government. The Federation organises annual meetings and sports medicine courses and also publishes the research journal, Zeitschrift für Sportmedizin.

Italy

The Italian Federation of Sports Medicine (Federanzione Medico Sportiva Italiana) is an independent association with its own technical, organisational and management structure. It was founded in 1929. The Federation is one of the first Member National Associations of FIMS.

The by-laws of the Italian Federation identify the responsibilities of the association, which include the following areas: training of its members; dissemination of information about the practice of sports medicine; the promotion of research; protecting the well-being of all competitive athletes by carrying out activities of prevention, consultation and medical supervision of athletes; performing doping control; providing medical services during competitive events; and carrying out other promotional and educational activities aimed at increasing the awareness of sport as a means towards physical and moral improvement of the population.

Over the years, the Federation has organised a number of refresher courses in sports medicine and similar events of both international and national scope. Every three years, the Federation organises a National Congress of Sports Medicine for its member associations that work on a provincial basis throughout Italy.

Japan

The Japanese Society of Physical Fitness and Sports Medicine was established in 1949 by Toshiro Azuma, MD, PhD, to satisfy the desire for an organisation consisting of researchers in physical fitness, fatigue and occupational health. It is dedicated to the promotion of medicine and science to improve physical fitness, health and disease prevention for the benefit of both athletes and non-athletes.

The four main activities of the Society are research, publication of an annual journal, provision of an annual meeting and education. Since 1988, the Society has been involved in a number of projects in both basic exercise research and practical exercise research for health promotion and rehabilitation.

Each year, the Society provides a conference, which highlights this research work as well as other new ideas in sports medicine. In addition to the annual meeting, the seven regional sections of the Society each hold their own meetings, symposia and other forums, which disseminate and highlight Japanese research in sports medicine.

The Japanese Society of Physical Fitness and Sports Medicine collaborates nationally with the Japanese Society of Physiology, the Japanese Society of Clinical Sports Medicine, as well as the Japanese Council of Science.

United States of America

The American College of Sports Medicine (ACSM) was founded by a small group of physicians, scientists and educators in 1954. It was incorporated in 1955 and currently has over 20 000 members. The membership covers 40 professional fields allowing for interdisciplinary interaction dedicated to improving health at all levels of the community.

The ACSM promotes and provides opportunities to integrate scientific research, education and practical application of sports medicine and exercise science, to maintain and enhance physical performance, fitness, health and quality of life.

Through the ACSM Foundation and the Office of Research Administration and Programs, the College offers research grants to ACSM student members, as well as new and senior investigators. Funding areas include, but are not limited to: injury prevention, weightlessness and space physiology, exercise and aging, exercise and cardiovascular disease risk factors and exercise and heart rate response.

The annual meeting of ACSM takes place in a different location in the United States each year. There are also several other specialty conferences and expositions that are organised throughout the year. Team Physician Sports Medicine Courses are conducted each year to aid physicians in enhancing medical skills required for work with sports teams and athletes.

There are also 12 regional chapters located throughout the United States that hold scientific meetings and educational activities throughout the year. The chapters are organised by states or clusters of states: Alaska Chapter, Central States Chapter, Greater New York Chapter, Mid-Atlantic Chapter, New England Chapter, Northland Chapter, Northwest Chapter, Rocky Mountain Chapter, Southeast Chapter, Southwest Chapter and Texas Chapter.

ACSM publishes a scientific journal (Medicine and Science in Sport and Exercise) containing original articles in the clinical sciences, basic sciences, regulatory physiology and applied sciences. The journal also publishes proceedings of symposia and clinical case studies.

Australia

The Australian Sports Medicine Federation, now commonly referred to as Sports Medicine Australia (SMA), was founded in 1963. SMA has a broad membership of sports medicine and health professionals, sports trainers, sporting clubs and community members. Members of Sports Medicine Australia can be found at every level of physical activity and sport from elite competition to grass-roots participation.

Sports Medicine Australia (SMA) prides itself on being a truly multidisciplinary organisation and while the affiliated organisations are independent, SMA provides the forum that brings together the various sports science, medicine and allied health discipline and professional groups in Australia under a common umbrella. These groups include exercise physiologists, sports scientists, sports dietitians, sports physicians, sports doctors, physiotherapists, podiatrists and psychologists.

The association organises an annual multidisciplinary meeting (Australian Conference of Science and Medicine in Sport) directed towards the groups previously identified and health professionals with an interest or specialisation in sport and physical activity. The Conference program is interdisciplinary, and its purpose is to 'promote knowledge and practice in sports science and sports medicine by providing an interactive educational forum of the highest standard, so that the participation, performance and well-being of Australians engaged in sport and physical activity may be ultimately enhanced'.

Sports Medicine Australia offer a variety of injury prevention, first aid and sports medicine courses designed to educate people involved in sport and recreation. The Safer Sport Program is a community based education program aimed at:

- all participants engaged in exercise and activity, of all age groups and of all abilities; and
- those supporting the exercising community. eg parents, teachers, coaches and sports trainers.

This very successful program is increasing the number of volunteers at local sporting events with knowledge about injury prevention and management. Information can be found at www.sma.org.au/information/safer_sport_program.asp

Canada

The Canadian Academy of Sports Medicine (CASM) was founded in 1970, and is a not-for-profit organisation of physicians committed to excellence in the practice of sports medicine. Much of the inspiration for the organisation came from Dr J.C. Kennedy, an orthopaedic surgeon based in London, Ontario who, after watching his daughter compete at the 1968 Summer Olympics in Mexico City, decided that athletic teams from Canada should be accompanied by a qualified and well organised medical team. He subsequently became a founding father of CASM, and the chief medical officer of the medical team at the 1972 Summer Olympics. Membership is limited to medical doctors and the organisation is based in Ottawa.

The organisation has an annual meeting and also publishes the Clinical Journal of Sports Medicine, which is an international refereed journal that publishes original research and review articles. The organisation also offers a Diploma in Sports Medicine.

Asia

The idea of organising an Asian society of sports medicine begin in 1987, but it was not until 1990 during the International Federation of Sports Medicine (FIMS) congress in Amsterdam that a preparatory meeting was held by Asian participants. The Asian Federation of Sports Medicine (AFSM) was founded during the 11th Asian Games in Beijing in 1990.

It is one of the multi-national continental federations within FIMS and has more than 30 National Sports Medicine Association members and is still growing. The first AFSM congress was held in Hong Kong in 1992 and meetings have been held regularly since this time. The next Congress will be held in Iran in 2008.

The main objective of the AFSM is to foster the promotion of sports medicine in Asia. Its constitution, by-laws, executive and commission structures are built in accordance with those of FIMS.

The AFSM Forum for Innovation and Leadership (AFIL) was established in 2007 in Bangkok, to provide an opportunity for young scholars and professionals in sports medicine to develop their leadership skills. AFIL is committed, through partnership of national members, academia and industry, to building the capacity of the Asian Federation in areas of research, education and training. The forum will explore and implement new solutions to existing and emerging issues in sports medicine, with sensitivity to the unique background and opportunities provided by the members of the Asian Federation.

South Africa

The establishment of the South Africa Sports Medicine Association (SASMA) was preceded by sports medicine activities, including an International Sports Medicine Conference in Johannesburg in 1977 and a growing interest in the clinical practice of sports medicine.

Following this first successful congress, a group of doctors met on a monthly basis and decided in 1982 to form a sports medicine association. A steering committee was formed and a constitution developed, which was discussed at the first Association Congress held in Johannesburg in 1985. The SASMA congresses are now held every two years.

The Association numbers have varied over the years from between 150 to 800 active members. There are a number of International Members and a nationally presence, covering all areas of the country with seven regional branches. The South African Journal of Sports Medicine is a peer-reviewed journal published on average 3 times per annum. It is accredited with the Department of Education of South Africa, and will shortly become internationally accredited.

Scandinavia

Membership of the Scandinavian Foundation of Medicine and Science in Sports comprises the Danish Association of Sports Medicine (Dansk Idrætsmedicinsk Selskab), the Finnish Society of Sports Medicine (Suomen Urheilulääkäriyhdistys), the Finnish Society for Research in Sport and Physical Education (Liikuntatieteellinen Seurary), the Norwegian Society of Sports Medicine (Norsk Idrettsmedisinsk Forening) and the Swedish Society of Sports Medicine (Svensk Idrottsmedicinsk Förening).

Each of these national groups has made, and continues to make, significant contributions to sports medicine in areas of research, education and clinical practice.

The foundation has its own internationally recognised journal, the *Scandinavian Journal of Medicine and Science in Sport*,s which is the official publication representing the Scandinavian sports medicine and science associations. The content of the journal is multidisciplinary, encompassing all elements of research in sport.

Articles generally represent original research in areas of traumatology and ortho-paedics, physiology, biomechanics, injury prevention and rehabilitation, sociology, psychology, pedagogy, history and philosophy of sport.

United Kingdom

The British Association of Sports and Exercise Medicine (BASEM) was founded in 1953 and is the oldest Sport and Exercise Medicine Association in the United Kingdom. The organisation is dedicated to the promotion of good health through physical activity and the provision of sports medicine expertise to optimise ath-letic performance at all levels.

Currently, the BASEM works closely with the National Sports Medicine Institute of the United Kingdom, which was formed in 1992 out of the former London Sports Medicine Institute.

BASEM has its own internationally recognised journal, The British Journal of Sports Medicine (BJSM), which was first published in 1974. BJSM is an international, peer-reviewed journal covering the latest advances in clinical practice and research. Topics include all aspects of sports medicine, such as the management of sports injury, exercise physiology, sports psychology, physiotherapy and the epidemiol-ogy of exercise and health. To ensure international coverage, BJSM has dedicated North American editors, as well as Editorial Board members from countries includ-ing Canada, Australia, Scandinavia and South Africa.

3.4. Specialised Centres

Numerous centres of sports medicine, which provide clinical care for athletes, are found in all parts of the world. A large number of research centres for sports medicine, usually functioning as part of a hospital complex, are common in many major cities.

3.5. Specialised International Degree Programmes

Because of the complexities of licensing for the practice of medicine, which are different from one nation to another and, within nations from one state to another, there have been few attempts to establish international degree programs for either undergraduate or postgraduate medical education.

4. Appendix Material

4.1. Terminology

The terminology employed in sports medicine, as regards human anatomy, physiology, pathology and rehabilitation, is the same as for all other aspects of medical practice

4.2. Position Statement

Position statements on topics of interest can be found on the websites of the international and national organisations listed above in section 2.6 Internet Sources.

4.3. Varia

Not applicable.

4.4. Free Statement

Not applicable.

Acknowledgements

The work of Professor Howard Knuttgen, who provided content for the chapter related to sports medicine in the first edition of Vade Mecum, is gratefully acknowledged.

Sport Pedagogy

Maria Dinold and Michael Kolb

Contact

Univ. Ass. Mag. Dr. Maria Dinold
Prof. Dr. Michael Kolb
Universität Wien
Zentrum für Sportwissenschaft und Universitätssport
Abteilung Bewegungs- und Sportpädagogik
Auf der Schmelz 6, 1150 Wien
Phone: +43-1-427748811
Fax: +43-1-427748819
Email: maria.dinold@univie.ac.at

1. General Information

The chapter on Sport Pedagogy in the first edition of the Directory of Sport Science (Vade Mecum, 1998) was prepared by Roland Naul (Germany), Ken Hardman (England) and Risto Telama (Finland). The second edition (Vade Mecum, 2000) was revised by Ronald Feingold (USA) and Bart Crum (Netherlands) and the third edition was prepared by Ronald Feingold (USA), Bart Crum (Netherlands), Mary O'Sullivan (USA) and Roland Naul (Germany). In both revisions, the authors both updated the information from the preceding edition and expanded the perspective of sport pedagogy by utilising sport pedagogy specialists from additional countries. In addition to the authors, valuable contributions have been made by Doune Macdonald (Australia), Wolf Brettschneider (Germany), Jean-Francis Gréhaigne (France) and Francisco Carreiro da Costa (Portugal). This edition was revised by Maria Dinold and Michael Kolb (Austria) by just updating some information.

1.1. Historical Development

Sport pedagogy has its origins dating back to the 19th century when significant debates about teaching and practices within physical education as gymnastics, exercise and rehabilitation, games and sports took place. At this time, several European countries such as Denmark, Sweden, Germany and Great Britain each established their own practices within school physical education (cf. Naul, 1994; Renson, 1999). Diffusion of these systems travelled to all parts of the world and was established in the latter half of the century. (cf. Van Dalen and Bennett, 1971). The term "sport pedagogy", not readily used in the 19th century, was first noted in Pierre de Coubertin's famous book, *Pédagogie Sportive* in 1925. Indeed the early pioneers of the Olympic Movement (de Coubertin, Gebhardt, Guth and Kemeny) already highlighted physical education as "character building" through Olympic Principles which often had been translated as "sport education". However the term "pédagogie sportive" was not readily assimilated in France by the physical educators, since they preferred the Lingian approach to gymnastics and were opposed to sports or became supporters of their own system "la methode française".

In the German translation of de Coubertin's book (1928), the term *"Sportpädagogik"* was introduced. However, it wasn't until the end of the 1960s that the term *"Sportpädagogik"* became common in German discourses. This was due to the paradigmatic shift from "teaching physical education" to "sport instruction", partially launched by cold war politics of competitive sport (Hardman and Naul, 2002). At the time, the name of the school subject in Germany changed from "Physical Education" to "Sport" and the name of the scientific foundation or theory of physical education to "sport pedagogy" (Grupe, 1984).

Within the aforementioned socio-historical context, the implementation of the term "sport pedagogy" in Germany was felt necessary in order to delineate a scholarly field, that focused on more than just school physical education, namely on all educational interventions in different settings and for specific target groups in the domain of human movement and sport. Over the past decade, the focus of sport pedagogy has expanded from just children to all ages and abilities (pre-school through to the elderly, disabled through to elite), and from the school environment to other institutions in the community and work place.

In other regions, such as the rest of Europe and North America, the term was not readily recognised in the 1970s. It has only become more widely known and used during the last twenty years in Europe and is still not widely recognised in North America or some Asian countries.

However, today it is more commonly suggested to use the label "sport pedagogy" for a complex body of knowledge which is neither restricted on sports matters nor linked only to school physical education.

1.2. Function

Currently, sport pedagogy describes the disciplined inquiry of physical education teaching and sport coaching from different theoretical perspectives. This inquiry aims to inform and improve the educational practices in the domain of physical activity, movement and sport (Piéron, Cheffers and Barrette, 1990, p. 24). In order to accomplish this function, sport pedagogy carries out three interdependent research tasks: (1) a hermeneutic research task, dealing with the ideological clarification of the relationships between fundamental conceptions, objectives and quality criteria of sport educational practices; (2) a descriptive-explanatory research task, dealing with the description and explanation of empirical relationships between presage, process, product and context variables of sport educational practices; and (3) an evaluative research task, dealing with design, controlled implementation and evaluation of sport educational programs (Crum, 1986).

1.3. Body of Knowledge

In recent years, the body of knowledge produced in sport pedagogy has grown rapidly. For the most part, the early focus of sport pedagogy was on school physical education and thus reflected the available body of knowledge. More recently, knowledge about teaching and coaching in non-school settings has emerged. Three general areas of focus and investigation can be distinguished: (1) teachers, teaching and coaching; (2) teacher and coach education; and (3) school curriculum.

While research in German speaking countries generally dealt more with anthropological assumptions, justification and objectives, and has generally followed a hermeneutic paradigm (cf. Grupe and Krüger, 1996; Beckers, 1996), the North American sport pedagogy research focused more on an empirical paradigm with description and explanation of the content and delivery on students' perspectives toward the physical education subject (cf. Crum, 1988, 1996; Pieron and Cheffers, 1988; Schempp, 1996; Van der Mars, 1996). However, a number of issues have been at the centre of both research traditions. These issues include: 'curriculum' (Cothran and Ennis, 1998; JTPE Monograph, 1987; Stibbe and Aschebrock, 2007); 'teaching styles' (Bielefelder Sportpädagogen, 2007; Mosston and Ashworth, 1990; Wolters et al., 2000); 'teacher cognition' (Bräutigam, 1986, 2006; Griffey and Housner, 1996; Maraun, 1981; Miethling, 1986; Miethling and Gieß-Stüber, 2007); 'teacher socialisation' (Baur, 1981; Templin and Schempp, 1989, Fejgin et al., 1995); 'teacher-student interactions' (Martinek, et al., 1982; Martinek, 1996; Hanke, 1991; Miethling and Krieger, 2004); 'students' knowledge' (JTPE Monograph, 2001); 'students' perspectives' (JTPE Monograph, 1995), 'social learning' (Pühse, 1990; Ungerer-Röhrich, 1984), 'teacher education' (Bain, 1990; JTPE Monograph, 1984), and 'teacher induction' (Stroot, 1996).

In France, "didactics of physical education" has become an important research thrust. Didactics investigate the nature of instructional delivery (Amade-Escot, 2000; Gréhaigne and Godbout, 1995; Gréhaigne, Godbout and Bouthier, 2001).

It should be emphasised that sport pedagogy research is carried out in many countries throughout Europe, including University of Liège and University of Brussels (Belgium), University of Jyväskylä (Finland), Technical University Lisbon (Portugal) and University of Loughborough (UK).

The body of knowledge in sport pedagogy has expanded in the 1990s to include national and international studies on the state and status of school physical education (cf. Hardman Marschall, 2000, 2006), children's health-related fitness (cf. Feingold, 2000), active lifestyles and youth sport (cf. Armstrong and Welsman, 1997; de Knop, Engström, Skirstad and Weiss, 1996; Baur and Brettschneider, 1994; Brettschneider and Bräutigam, 1990; Brettschneider and Kleine, 2001), determinants of physical and psycho-social health behaviours (Dale, Corbin and Dale; McKenzie, Marshall, Sallis and Conway, 2000; Fox, 1997; Biddle and Mutrie, 2001), as well as socio-cultural development (cf. Hellison 1995; DeBusk and Hellison, 1989; Locke, 1987; O'Sullivan, 1994; Silverman and Ennis, 1996).

Recently, sports pedagogy allocated some new topics. Among them the interest in research for violence prevention through sport (Brettschneider, Brandl-Bredenbeck and Hofmann, 2005) and inter cultural learning in the field of physical activity and sport (Gieß-Stüber, 2005). In the area of school physical education, the development of school profiles related to sport and with respect to full-time schools (Hummel and Schierz, 2006) has become an important issue.

In addition, a greater recognition of areas of study, include adapted physical education/ activity (DePauw, 1990; DePauw and Doll-Tepper, 1989; DePauw and Gavron, 1995, Steadward, Wheeler and Watkinson, 2003; Sherrill, 2004; Winnick, 2005), gender (Gieß-Stüber, 2006; Hartmann-Tews, 2006), cross-cultural studies about sport and movement (cf. Brettschneider and Brandl-Bredenbeck, 1997; Telama, Naul, Nupponen, Rychtecky and Vuolle, 2001), and "Olympic Education" as a focus of teaching to moral standards like fair play and fairness at school and sport clubs as well as doping prevention (cf. Shields and Bredemeier, 1995; Kidd, 1996; Müller, 1998; Parry, 1998; Singler and Treutlein, 2001).

1.4. Methodology

To account for the differences between basic theoretical assumptions in paradigms of sport pedagogy, there are a variety of research methodologies used, both quantitative and qualitative. While in the German tradition the 'geisteswissenschaftliches' paradigm, with its hermeneutic-phenomenological and interpretive (qualitative) methods, still dominates (cp. Dietrich and Landau, 1990; Prohl, 2006), in North America, a strong emphasis on a behavioural paradigm with an empirical-analytical methodology has developed. In the past ten years, however, there appears to be a greater increase in qualitative research. Both qualitative and quantitative research approaches thrive equally well and provide insight into determining behavioural and attitudinal changes as a result of pedagogical or curricular interventions, e.g., utilisation of diary review, essay analysis, case study, etc., which have become more prevalent (Melograno, 1994; Kirk, et al., 1997). Many researchers recognise that there is great complexity when studying the human being when they encounter physical education and the movement environment. In the UK, Australia, Germany and North America, reflective methodologies of a more critical, feminist and post-structuralist perspective have emerged. (Bain, 1992; Nilges, 2000; Tsangaridou and O'Sullivan, 1997; Fernández-Balboa, 1997; Wright, 1995; Kirk, Macdonald and Tinning, 1997).

1.5. Relationship to Practice

In the past, sport pedagogy was limited to a prescriptive theory (theory of practice), which could be applied within the school physical education and coaching context. Today, sport pedagogy not only consists of descriptive–explanatory theories, but also relates to all forms of physical activity for all abilities, genders and age in both formal and informal settings. Therefore, sport pedagogy is related, not only to the school physical education and elite athlete performance, but also to community recreation, work-site centres and sports clubs, senior activities as well as programs for people with disabilities and "sport for all".

Aside from the extensive connection of sport pedagogy to practice, sport pedagogy as a field of scholarly inquiry includes studies not directly related to practice, such as curriculum histories and lifestyle analyses (Piéron, et al, 1996), teaching culture analyses (O'Sullivan, 1994), normative and ethical standards (cf. Scherler,

1992), values in education and quality management of physical education in autonomous local school programs (cf. EADS, 2001).

1.6. Future Perspectives

Over the past decade, an extensive body of knowledge relative to the function of sport pedagogy (to inform and improve practice) has pointed to the complexity of the field and complexity in study of the interaction between educator and learner within the variety of contextual factors. Also as noted above, the focus of sport pedagogy's expansion beyond the school-aged population to all ages and abilities has brought forth additional issues and research methodologies that affect one throughout one's lifestyle.

Through the increased research methodologies (qualitative and quantitative), more reflective practice and the complexity in teaching and curricular development, in addition to the expansion of inquiry throughout the life-span, there is evidence of an increased body of knowledge and increased importance to ones self, the community and to society.

References:

Amade-Escot, C. (2000). Pedagogical Content Knowledge and Didactics of Physical Education. *Journal of Teaching in Physical Education*, 20, 78-101.

Armstrong, N. and Welsman, J. (1997). *Young People and Physical Activity*. Oxford: University Press.

Bain, L. (1990). Physical Education Teacher Education. In W.R. Houston (Ed.), *Handbook of Research on Teacher Education*. New York: Macmillan.

Bain, L. (1992). Research in Sport Pedagogy: Past, Present and Future. In T. Williams, L. Almond and A. Sparkes (Eds.), *Sport and Physical Activity*. London: Spon.

Baur, J. (1981). *Zur beruflichen Sozialisation von Sportlehrern*. Schorndorf: Hofmann.

Baur, J. and Brettschneider, W.D. (1994). *Der Sportverein und seine Jugendlichen*. Aachen: Meyer and Meyer.

Beckers, E. (1996). Hermeneutic and Sport Pedagogy. In P. Schempp (Ed.), *Scientific Development of Sport Pedagogy* (pp. 203-222). Münster: Waxmann.

Biddle, S. and Mutrie, N. (2001). *Psychology of Physical Activity*. London, New York: Routledge.

Bielefelder Sportpädagogen. (2007). *Methoden im Sportunterricht* (5. Aufl.). Schorndorf: Hofmann.

Brettschneider, W.-D., Brandl-Bredenbeck, H. P. and Hofmann, J. (2005). *Sportpartizipation und Gewaltbereitschaft bei Jugendlichen.* Aachen: Meyer and Meyer.

Bräutigam, M. (1986). *Unterrichtsplanung and Lehrplanrezeption von Sportlehrern.* Schorndorf: Hofmann.

Bräutigam, M. (2006). *Sportdidaktik.* Ein Lehrbuch in 12 Lektionen (2. Aufl.). Aachen: Meyer and Meyer.

Brettschneider, W.D. and Bräutigam, M. (1990). *Sport in der Alltagswelt von Jugendlichen.* Frechen: Rittersbach.

Brettschneider, W.D. and Kleine, T. (2001). *Jugendarbeit in Sportvereinen - Anspruch und Wirklichkeit.* Düsseldorf: MWSKS.

Brettschneider, W.D. and Brandl-Bredenbeck, H.P. (1997). *Sportkultur und jugendliches Selbstkonzept - Eine interkulturell-vergleichende Studie über Deutschland und die USA.* Weinheim, München: Juventa.

Cothran, D. and Ennis, C. (1998). Curricula of Mutual Worth: Comparisons of Students' and Teachers; Curricular Goals. *Journal of Teaching in Physical Education,* 17, 307-327.

Crum, B. (1986). Concerning the Quality of the Development of Knowledge in Sport Pedagogy. Journal of Teaching in Physical Education, 5, 211-220.

Crum, B. (1988). Zur Entwicklung sportpädagogischer Forschung – ein Vergleich zweier maßgebender Subkulturen. *Sportwissenschaft,* 2, 176-184.

Crum, B. (1996). In search of Paradigmatic Identities: General Comparison and Commentary. In P. Schempp (Ed.), *Scientific Development of Sport Pedagogy* (pp. 239-258). Münster: Waxmann.

Dale, D., Corbin, C. and Dale, K. (2000). Restricting Opportunities to be Active During School Time: Do Children Compensate by Increasing Physical Activity Levels After School. *Research Quarterly for Exercise and Sport,* 71, 240-248.

De Coubertin, P. (1928). *Sportliche Erziehung.* Stuttgart: Dieck & Co.

DePauw, K. (1990). Sport, Society, and Individuals with Disabilities: Research Opportunities. In G. Reid (Ed.), *Problems in Motor Control* (pp. 319-337). New York: Elsevier Science.

DePauw, K. and Doll-Tepper, G. (1989). European Perspectives on Adapted Physical Activity. *Adpated Physical Activity Quarterly,* 6, 95-99.

DePauw, K. and Gavron, S. (1995). *Sport and Disability.* Champaign, IL: Human Kinetics.

Dietrich, K. and Landau, G. (1990). *Sportpädagogik*. Reinbek: Rowohlt.

European Academy of Sport (EADS). (2001). *Physical Education: From Central Governmental Regulation to Local School Autonomy*. Velen: EADS.

Feingold, R. (2000). Health and Physical Education: Partners for the Future. In M. Piéron and M.A.G. Valeiro (Eds.), *Ten Years of "José Maria Cagigal" Scholar Lectures* (pp. 149-166). La Coruna: Universidade da Coruña.

Fejgin, N., Ephraty, N. and Ben-Sira, D. (1995). Work Environment and Burnout of Physical Education Teachers. *Journal of Teaching in Physical Education, 15*, 64-78.

Fernández-Balboa, J.M. (Ed.). (1997). *Critical Postmodernism in Human Movement, Physical Education and Sport*. NY: SUNY Press.

Fox, K.R. (Ed.). (1997). The Physical Self. From Motivation to Well-Being. Champaign, IL: Human Kinetics.

Gieß-Stüber, P. (Hrsg.). (2005). *Interkulturelle Erziehung im und durch Sport*. Münster: Lit Verlag.

Gieß-Stüber, P. (Hrsg.). (2006). *Gleichheit und Differenz in Bewegung: Entwicklungen und Perspektiven für die Geschlechterforschung in der Sportwissenschaft*. Hamburg: Czwalina.

Gréhaigne, J.F. and Godbout, P. (1995). Tactical Knowledge in Team Sports from a Constructivist and Cognitivist Perspective. *QUEST, 47*, 490-505.

Gréhaigne, J.F., Godbout, P. and Bouthier, D. (2001). The Teaching and Learning of Decision Making in Team Sports. *QUEST, 53*, 59-76.

Griffey, D. and Housner, L. (1996). The Study of Teacher Cognition in Sport Pedagogy. In P. Schempp (Ed.), *Scientific Development of Sport Pedagogy* (pp. 103-122). Münster: Waxmann.

Grupe, O. (1984). *Grundlagen der Sportpädagogik* (3. überarb. Aufl.). Schorndorf: Hofmann.

Grupe, O. and Krüger, M. (1996). Sport Pedagogy: Anthropological Perspectives and Traditions. In P. Schempp (Ed.), *Scientific development of Sport Pedagogy* (pp. 103-122). Münster: Waxmann.

Hanke, U. (1991). *Analyse und Modifikation des Sportlehrer- und Trainerhandelns*. Göttingen: Hogrefe.

Hardman, K. and Marshall, J.J. (2000). *World-wide Survey of the State and Status of School Physical Education. Final Report*. Manchester, University of Manchester.

Hardman, K. and Marshall, J.J. (2006). Update on Current Situation of Physical Education in Schools. *ICSSPE Bulletin, 47*, May 2006.

Hardman, K. and Naul, R. (2002). Development of Physical Education and Sport in the Two Germanies 1945-1990. In R. Naul and K. Hardman (Eds.), *Sport and Physical Education in Germany*. London, New York: Routledge.

Hartmann-Tews, I. (Hrsg.). (2006). *Handbuch Sport und Geschlecht*. Schorndorf: Hofmann.

Hellison, D. (1995). *Teaching Personal and Social Responsibility Through Physical Activity*. Champaign, IL: Human Kinetics.

Hummel, A. and Schierz, M. (Hrsg.). (2006). *Studien zur Sportentwicklung in Deutschland*. Schorndorf: Meyer and Meyer.

JTPE Monograph, 1983 (Time to learn in PE)

JTPE Monograph, 1984 (PE Teacher Education)

JTPE Monograph, 1987 (Curriculum)

JTPE Monograph, 1995 (Students' Perspectives)

JTPE Monograph, 2001 (Understanding and Development of Learners' Domain Specific Knowledge)

Kidd, B. (1996). Taking the Rhetoric Seriously: Proposals for Olympic Education. *QUEST, 48*, 82-92.

Kirk, D., Macdonald, D. and Tinning, R. (1997). The Social Construction of Pedagogic Discourse in Physical Education Teacher Education in Australia. *The Curriculum Journal, 8*(2), 271-298.

Knop, P. de, Engström, L.M., Skirstad, B. and Weiss, M.R. (Eds.). (1996). *Worldwide Trends in Youth Sport*. Champaign, IL: Human Kinetics.

Locke, L. (1987). The Future of Research on Pedagogy: Balancing on the Cutting Edge. *American Academy of Physical Education papers, 20* (pp. 83-95). Champaign, IL: Human Kinetics.

Maraun, H.K. (1981). *Analysieren und Planen als Handlungsprobleme des Sportlehrers*. Schorndorf: Hofmann.

Martinek, T., Crowe, P. and Rajeski, W. (1982). *Pygmalion in the Gym. Causes and Effects of Expectations in Teaching and Coaching*. West Point, NY: Leisure Press.

Martinek, T. (1996). *Psycho-social Aspects of Student Differences in Physical Education*. In P. Schempp (Ed.), Scientific Development of Sports Pedagogy. Münster: Waxmann.

McKenzie, T., Marshall, S., Sallis, J. and Conway, T. (2000). Student Activity Levels, Lesson Context, and Teacher Behavior During Middle School Physical Education. *Research Quarterly for Exercise and Sport, 71*, 260-266.

Melograno, V. (1994). Portfolio Assessment: Documenting Authentic Student Learning. *JOHPERD, 65*(8), 50-55, 58-61.

Miethling, W.D. (1986). *Belastungssituationen im Selbstverständnis junger Sportlehrer.* Schorndorf: Hofmann.

Miethling, W-D. and Krieger, C. (2004). *Schüler im Sportunterricht: Die Rekonstruktion relevanter Themen und Situationen des Sportunterrichts aus Schülersicht (RETHESIS).* Schorndorf: Hofmann.

Miethling, W.-D. and Gieß-Stüber, P. (Hrsg.). (2007). *Beruf Sportlehrer/in. Über Persönlichkeit, Kompetenzen und professionelles Selbst von Sport- und Bewegungslehrern.* Hohengehren: Schneider.

Mosston, M. and Ashworth, S. (1990). *The Spectrum of Teaching Styles.* New York: Longman.

Müller, N. (1998). Olympische Erziehung. In O. Grupe & D. Mieth (Hrsg.), *Lexikon der Ethik im Sport* (S. 385-395). Schorndorf: Hofmann.

Naul, R. (1994). Physical Education Teacher Training – Historical Perspectives. In J. Mester (Ed.), *Sport Science in Europe 1993 – Current and Future Perspectives* (pp. 588-612). Aachen: Meyer and Meyer.

Naul, R. and Hardman, K. (Eds.). (2002). *Sport and Physical Education in Germany.* London, New York: Routledge.

Nilges, L. (2000). A Nonverbal Discourse Analysis of Gender in Undergraduate Educational Gymnastics Sequences Using Laban-Effort Analysis. *Journal of Teaching in Physical Education, 19,* 287-310.

O'Sullivan, M. (Ed.). (1994). High School Physical Education Teachers: Their World of Work. *Journal of Teaching in Physical Education. Monograph.*

Parry, J. (1998). Physical Education as Olympic Education. *European Review of Physical Education, 4*(2), 153-167.

Piéron, M. and Cheffers, J. (1988). *Research in Sport Pedagogy – Empirical Analytical Perspective.* Schorndorf: Hofmann.

Piéron, M., Cheffers, J. and Barrette, G. (1990). *An Introduction to the Terminology of Sport Pedagogy.* Liège: ICSP.

Piéron, M., et al. (1996). *Comparative Analysis of Young Lifestyles in Selected European Countries: Research report.* Liège: University of Liège.

Prohl, R. (2006). *Grundriss der Sportpädagogik* (2. Aufl.). Wiebelsheim: Limpert.

Pühse, U. (1990). *Soziales Lernen im Sport.* Bad Heilbrunn: Klinkhardt.

Renson, R. (1999). Physical Education in Europe from Cross-Cultural Perspective. In J.C. Bussard and F. Roth (Eds.), *Quelle Education Physique pour quelle*

Ecole? Berne: ASEP.

Schempp, P.G. (Ed.). (1996). *Scientific Development of Sport Pedagogy*. Münster: Waxmann.

Scherler, K. (Hrsg.). (1990). *Normative Sportpädagogik*. Clausthal-Zellerfeld: Geinert.

Sherrill, C. (2004). *Adapted Physical Activity, Recreation, and Sport. Crossdisciplinary and Lifespan* (6th ed.). New York: McGraw-Hill.

Shields, D. and Bredemeier, B. (1995). *Character Development and Physical Activity*. Champaign, IL: Human Kinetics.

Silverman, S. and Ennis, C. (1996). *Student Learning in Physical Education*. Champaign, IL: Human Kinetics.

Singler, A. and Treutlein, G. (2001). *Doping – von der Analyse zur Prävention*. Aachen: Meyer and Meyer.

Steadward, R.D., Wheeler, G.D. and Watkinson, E.J. (Eds.). (2003). *Adapted Physical Activity*. Alberta: University of Alberta Press.

Stibbe, G. and Aschebrock, H. (2007). *Lehrpläne Sport. Grundzüge der sportdidaktischen Lehrplanforschung*. Hohengehren: Schneider.

Stroot, S. (1996). Organizational Socialization: Factors Impacting Beginning Teachers. In S. Silverman and C. Ennis (Eds.), *Student Learning in Physical Education*. Champaign, IL: Human Kinetics.

Telama, R., Naul, R., Nupponen, H., Rychtecky, A. and Vuolle, P. (2001). *Physical Fitness, Sporting Lifestyle and Olympic Ideals: Studies on Youth Sport in Europe*. Schorndorf: Hofmann.

Templin, T. and Schempp. P. (Eds.). (1989). *Learning to Teach – Socialization into Physical Education*. Indianapolis: Benchmark Press.

Tsangaridou, N. and O'Sullivan, M. (1997). The Role of Reflection in Shaping Physical Education Teachers' Educational Values and Practices. *Journal of Teaching in Physical Education*, 17, 2-25.

Ungerer-Röhrich, U. (1984). *Eine Konzeption zum Sozialen Lernen im Sportunterricht und ihre empirische Überprüfung*. Darmstadt.

Van Dalen, D.D. and Bennett, B.L. (Eds.). (1971). *A World History of Physical Education* (2nd ed.). Englewood Cliffs NJ: Prentice Hall.

Van der Mars, H. (1996). Behavioral Traditions in Sport Pedagogy. In P. Schempp (Ed.), *Scientific Development of Sport Pedagogy* (pp. 41-62). Münster: Waxmann.

Weiss, M.R., & Gould, D. (Eds.). (1986). *Sport for Children and Youth. 1984 Olympic Scientific Congress Proceedings*, Vol 10. Champaign, IL: Human Kinetics.

Winnick, J.P. (Ed.). (2005). *Adapted Physical Education and Sport* (4th ed.). Champaign, IL: Human Kinetics.

Wolters et al. (2000). *Didaktik des Schulsports*. Schorndorf: Hoffmann.

Wright, J. (1995). A Feminist Poststructuralist Methodology for the Study of Gender Construction in Physical Education: Description of a Study. *Journal of Teaching in Physical Education, 11*, 177-194.

2. Information Sources

2.1. Journals

A variety of journals exist across all continents, focusing on sport pedagogical research and theory and also on the practical aspects of teaching and coaching strategies. Compared with other disciplines of the sport sciences (e.g. sport sociology, sport psychology), specific journals of sport pedagogy on an international level have a shorter tradition, yet, there is a number of research and theory oriented journals (in English language) that serve the international community of sport pedagogy scholars. Examples include:

Journal of Teaching in Physical Education and QUEST, (published by HK, USA);
Sport, Education & Society, Routledge, UK;
European Physical Education Review, Sage, UK;
Physical Education and Sport Pedagogy (until 2004 this was called the European Journal of Physical Education), Routledge, UK.

In the German language the general sport scientific journals Sportwissenschaft (German Journal of Sports Science), published by the German Society of Sport Science / Deutsche Vereinigung für Sportwissenschaft (DVS) and Spectrum der Sportwissenschaften (Journal of the Austrian Sports Science Society), published by the Austrian Sports Science Society (ÖSG) are important outlets for the publication of sport pedagogical research and theorising.

In France, the Science et Motricité (by De Boeck Université) has been published since 1987.

The International Journal of Physical Education is a review journal of sport pedagogy which addresses the following subtopics: instructional theory; historical foundations; curriculum theory; health foundations; physical education teacher and coach education; historical, psychological and sociological foundations; comparative sport pedagogy; and the nature and function of sport pedagogy. Additionally, there are a number of other journals (more or less continental based) in which sport pedagogy articles occasionally appear: e.g., African Journal of Physical Education, Recreation and Dance and the Asian Journal of Physical Education.

Another journal is that of the International Society of Comparative Physical Education and Sport, called International Sport Studies, which provides a cross-continental comparative exchange of knowledge on physical education and sport as well as sport pedagogy.

In many countries, national physical education teacher associations publish journals devoted to informing teachers and coaches. Therefore, they contain a mixture of practical and theoretical articles (including ones with a sport pedagogical nature). A few examples include:

the *Journal of Physical Education, Recreation and Sport and Strategies: A Journal for Physical and Sport Educators* are published by AAHPERD in the United States; the *British Journal of Physical Education; Sportunterricht and Sportpädagogik in Germany; Bewegungserziehung* in Austria; *Education Physique et Sport and Science* and *Hyper* in France; the *Japanese Journal of Physical Education;* the *Canadian Journal of Health, Physical Education and Recreation; Healthy Lifestyles Journal* in Australia; and in Portugal, *Boletim da Sociedade Portuguesa de Educação Fisica,* the Journal of SPEF (Portugese Society of PE).

On occasions, sport pedagogical articles may also be found within many of the sub-discipline's journals, as well as the *Research Quarterly for Exercise and Sport,* as previously noted, the German *Sportwissenschaft,* as well as in general educational research journals (e.g., *Educational Researcher, Journal of Curriculum Studies, Journal of Research and Development in Education* and *Journal of Teacher Education*).

2.2. Reference Books, Encyclopaedias, etc.

An Encyclopaedia or Handbook on sport pedagogy has yet to be published. However, there are some encyclopaedic-type texts in sports science with overviews of sport and pedagogy and related research methods which provide information about physical education, selected themes in sport pedagogy and related research methods. These include:

Bain, L. (1990). Physical Education Teacher Education. In W.R. Houston (Ed.), *Handbook of Research on Teacher Education*. New York: Macmillan.

Beashel, P. and Taylor, J. (Eds.). (1996). *Advanced Studies in Physical Education and Sport*. London: Nelson.

Beyer, E. (1992). *Wörterbuch der Sportwissenschaft. Dictionary of Sport Science. Dictionnaire des Sciences du Sport* (2nd ed.). Schorndorf: Hofmann.

Hardy, C. and Mawer (Eds.). (1999). *Learning and Teaching in Physical Education*. London: Falmer Press.

Houston, W.R. (Ed.). (1990). *Handbook of Research on Teacher Education*. NY: Macmillan.

Piéron, M., and Gonzalez Valeiro (Eds.). (2000). *Ten Years of José Maria Cagigal Scholar Lectures*. La Coruña, Spain: AIESEP.

Röthig, P. et al. (Hrsg.). (2003). *Sportwissenschaftliches Lexikon* (7. Aufl.). Schorndorf: Hofmann.

Silverman, S. and Ennis, C. (1996). *Student Learning in Physical Education*. Champaign, IL: Human Kinetics.

Steinhardt, M. (1992). Physical Education. In P. Jackson (Ed.), *Handbook of Research on Curriculum*. New York: Macmillan.

Another valuable reference source which provides an overview of special areas of research in Sport Pedagogy is the special issue of Sport Pedagogy by ICSSPE:

Haag, H. (1994). Sport Pedagogy. *Sport Science Review*, 3(1).

In addition, many national text books have been published which provide sources and documentation on sections of the body of knowledge for sport pedagogy. These have largely been reviewed from a national-cultural perspective and with different approaches:

Balz, E. and Kuhlmann, D. (2003). *Sportpädagogik. Ein Lehrbuch in 14 Lektionen*. Aachen: Meyer and Meyer.

Bräutigam, M. (2006). *Sportdidaktik. Ein Lehrbuch in 12 Lektionen* (2. Aufl.).

Aachen: Meyer and Meyer.

Dietrich, K. and Landau, G. (1990). *Sportpädagogik*. Reinbek: Rowohlt.

Grupe, O. and Krüger, M. (1996). Sport Pedagogy: Anthropological Perspectives and Traditions. In P. Schempp (Ed.), *Scientific development of Sport Pedagogy* (pp. 103-122). Münster: Waxmann.

Haag, H. and Hummel, A. (Hrsg.). (2001). *Handbuch Sportpädagogik*. Schorndorf: Hofmann.

Haag, H. and Strauß, B. (Hrsg.). (2006). Themenfelder der Sportwissenschaft: Schorndorf: Hofmann.

Naul, R. and Hardman, K. (Eds.). (2002). *Sport and Physical Education in Germany*. London, New York: Routledge.

Piéron, M. (1991). *Pedagogie des Activités Physiques et du Sport*. Paris: Editions Revue Education Physique et Sport.

Riordan, J. and Jones, R. (Eds.). (1999). *Sport and Physical Education in China*. London, New York: Spon.

Prohl, R. (2006). *Grundriss der Sportpädagogik* (2. Aufl.).Wiebelsheim: Limpert.

Siedentop, D. (Ed.). (1994). *Sport Education: Quality PE through Positive Sport Experiences*. Champaign, IL: Human Kinetics.

2.3. Book Series

Important book series have been published by some national professional associations or sport science associations like the American Association of Health, Physical Education, Recreation and Dance (AAHPERD) in the USA, the Physical Education Association of the United Kingdom (PEAUK), the Association Francophone pour la Recherche en Activités Physiques Sportives (AFRAPS, France), and the Deutsche Vereinigung für Sportwissenschaft (dvs) in Germany, as well as by some international publishing houses in the fields of physical education and sport pedagogy (e.g. Human Kinetics, McGraw Hill, Hofmann, Meyer and Meyer, Spon, Routledge). Numerous series within and outside these publishing houses are produced to introduce, promote, develop and enhance specific physical activities and sports at different performance levels.

2.4. Conference/Workshop Proceedings

The five international associations which constitute the International Committee of Sport Pedagogy (ICSP) of the International Council for Sports Science and Physical Education (ICSSPE) are: the Association Internationale des Ecoles Supérieures d'Education Physique (AIESEP); Fédération Internationale d'Education Physique (FIEP); International Association of Physical Education and Sport for Girls and Women (IAPESGW); International Federation of Adapted Physical Activity (IFAPA); and the International Society of Comparative Physical Education and Sport (IS-CPES). ICSP was established in1984 and has organised annual and/or biennial conferences and congresses at regional, continental and global levels. Since 1963, AIESEP, for example, has published 35 volumes of conference proceedings edited by the local conference organisers and published with many companies around the world (e.g. Proceedings from 2006 AISEP World Congress in Finland). Since 1978, ISCPES has published volumes of conference proceedings regularly with Human Kinetics, Meyer and Meyer and FIT (Fitness Information Technology). Some proceedings of Pre-Olympic Scientific Congresses since 1973 have also been published and contain documentary sections or chapters on symposia and workshops with a sport pedagogy focus. Occasionally, separate texts based on sport pedagogically orientated symposia as part of Pre-Olympic Scientific Congresses have also been published. Examples include:

Doll-Tepper, G. and Brettschneider, W.-D. (Eds.). (1996). *Physical Education and Sport. Changes and Challenges. World Congress Physical education and Sport '94.* Aachen: Meyer & Meyer.

Heikinaro-Johansson, P., Telama, R. and McEvoy, E. (Eds.). (2007). The Role of Physical Education and Sport in Promoting Physical Activity and Health. Proceedings of the 2006 AIESEP World Congress.

Kang, S. (Ed.). (1990). *New Horizons in Human Movement. Seoul Olympic Scientific Congress Proceedings.* Dankook University, Soul: Sport Science Institute.

Naul, R., Hardman, K., Piéron, M. and Skirstad, B. (Eds.). (1998). *Physical Activity and Active Lifestyle of Children and Youth.* Schorndorf: Hofmann.

Noguera, M. (Ed.). (1995). *Actas Congresso Cientifco Olimpicio 1992 Pedagogia y Educacion Fisica Comparada.* Vol III. Malaga: Instituto Andulaz del Deporte.

Piéron, M. and Graham, G. (Eds.). (1986). *Sport Pedagogy. 1984 Olympic Scientific Congress Proceedings,* Vol 6. Champaign, IL: Human Kinetics.

Tolleneer, J. and Renson, R. (Eds.). (2000). *Old Borders, New Borders: Sport and Physical Education in a Period of Change*. Aachen: Meyer and Meyer.

2.5. Data Banks

No specific data banks exist for sport pedagogy. Valuable sources are available in the SIRC-SPORTSearch data bank in Canada and at the German Federal Sport Science Institute in Bonn with "Spolit". Many descriptors in the field of physical education for international literature searches exist, however, the lack of some special descriptors render the body of knowledge in sport pedagogy incomplete. Unfortunately, no international data bank exists to provide statistical resources on research topics like youth sports and physical education teacher training etc.

2.6. Internet Sources

Multilingual

SportQuest
www.sirc.ca - SPORTQuest is an important 'first stop' for sport, sport science and physical education information on the web. This site contains high quality links in many languages. It is produced and updated by the Sport Information Resource Centre (SIRC) in Canada. Physical education is included in the "List of Topics" under "Sport Science".

ICSSPE Conference Calendar
www.icsspe.org - A current list of forthcoming international congresses, conferences, symposia and seminars dealing with sport science and physical education. Update monthly.

In English

Physical Education - The Role of Physical Education and Sport in Education (SPINED)
http://spined.cant.ac.uk – SpinEd is an international research project commissioned by the International Council for Sports Science and Physical Education (ICSSPE) and funded by the International Olympic Committee (IOC). The project

aims to gather and present evidence regarding the benefits of quality Physical Education and Sport to schools. In addition, the website provides international academic references and relevant links.

PE Central
www.pecentral.org/ - A comprehensive resource with a wide variety of links to other internet sites. The site is divided into: News; Teaching; Fitness; Advocacy; Technology; Miscellaneous and a PE Forum where users can post messages to discussion groups.

Physical Education Digest
www.pedigest.com/ - A 36-page quarterly magazine that provides the latest ideas, tips, coaching cues and research on sports, fitness and physical education topics from around the world - condensed into brief, easy-to understand articles. Each issue represents the best information selected from almost 5000 pages of marginal material.

Listserves in Physical Education/Physical Activity
Physical Activity and Public Health On-Line Network. Please contact:
PHYS-ACT@VM.SC.EDU
Australian Physical Education Discussion Listserve. Please contact: Austpe-l@ hms.uq.edu.au

In French:

Les listes de diffusion francophones.
www.fcomte.iufm.fr - Rubrique "Recherche" : e-journal de la Recherche sur l"Intervention en Education Physique et Sports (eJRIEPS)" is now available on the web.

In German:

Physical Education - Sportpaedagogik Online
www.sportpaedagogik-online.de - Das Internet als sportpädagogisches Nachschlagwerk und Diskussionsforum.

3. Organisational Network

3.1. International Level

As mentioned in section 2.4 Conference/ Workshop Proceedings, five member associations of ICSSPE represent the field of sport pedagogy at the international level and constitute the International Committee of Sport Pedagogy (ICSP). They are: AIESEP, FIEP, IAPESGW, IFAPA and ISCPES. AIESEP is a network organisation of higher education institutes in physical education, but also offers individual memberships to scholars in sport pedagogy; FIEP supports sport pedagogical research but is more practically oriented in promoting physical education at schools around the world; IAPESGW deals with sport pedagogical issues concerning women and girls both from a scientific and practical point of view; IFAPA is a special organisation for adapted physical activities and deals with sport pedagogical problems among people with disabilities; ISCPES is an international society for research and teaching comparative physical education and sport pedagogy but also includes cross-cultural aspects of the history, sociology and psychology of sports.

In the network of the International Council of Health, Physical Education, Recreation, Sport and Dance (ICHPER SD), there are continental sub-divisions in Europe, Africa, Asia and North America in which special interest groups (SIGs) represent physical education and sport pedagogy as well.

3.2. Regional Level

At the regional level, there are well known sub-divisions of the International Council for Health, Physical Education, Recreation, Sport and Dance (ICHPERSD), which is affiliated with the American Alliance for Health, Physical Education, Recreation and Dance (AAHPERD) in the United States. A new division has recently emerged in Africa, the African Association for Health, Physical Education, Recreation, Sport and Dance (AFAHPER.SD). Focusing on the comparative domain, a new body has been instituted in Asia, the Asian Society of Comparative Physical Education and Sport (ASCPES). In Europe, a College of Sport Science has been founded (ECSS), which includes scholars from the field of sport pedagogy. Furthermore, several national physical education teacher associations have combined to establish the European Union of Physical Education Associations (EUPEA).

3.3. Specialised Centres

There are several international university research units around the world (e.g. USA, UK, Australia, Finland, Belgium, Portugal, Germany, Japan and other countries) which have established special laboratories, institutes, chairs and working groups focusing on selected topics of research e.g. in teaching physical education and physical education curriculum studies.

Some international university research units around the world include: Ohio State University, USA; University of South Carolina, USA; Technical University of Lisbon, Portugal; University of Liége, Belgium; University of Tübingen, University of Kiel and the University of Bielefeld, Germany; University of Queensland, Australia.

Centres particularly involved in youth sport activities and health studies include: University of Exeter and University of Loughborough, Great Britain; University of Jyväskylä, Finland; Free University of Brussels, Belgium; and University of Paderborn in Germany). Many other universities investigate these and other areas within the field of sport pedagogy.

Specialised centres also exist that are affiliated to a university but function at the governmental level, such as: LIKES at the University of Jyväskylä; Center for Disease Control, Atlanta, Georgia (www.cdc.gov/), or the Tucker Centre for Research on Girls and Women in Sport (www.education.umn.edu/tuckercenter/). Others may operate similarly but at non-governmental levels.

More initiatives are needed, however, to further develop specialised centres to improve research in physical education and sport pedagogy.

3.4. Specialised International Degree Programmes

Not applicable.

4. Appendix Material

4.1. Terminology

Piéron, M., Cheffers, J. and Barrette, G. (1990). An Introduction to the Terminology of Sport Pedagogy. Liège: ICSP.

This collection and description of terms has an American-centric focus as well as an orientation towards the empirical-analytical paradigm of the body of knowledge. A broader cross-cultural impact is provided in the dictionary by Beyer, which lacks, however, an extended collection of special terms related to sport pedagogy:

Beyer, E. (1992). *Wörterbuch der Sportwissenschaft. Dictionary of Sport Science. Dictionnaire des Sciences du Sport* (2nd ed.). Schorndorf: Hofmann.

Another dictionary on sport, physical education and sport science, provides full text in English and ten languages in the index:
Haag, H. and Haag, G. (2003). *Dictionary: Sport – Physical Education – Sport Science.* Kiel: ICS.

4.2. Position Statement

There are several position statements, ranging from the UNESCOC International Charter of Physical Education and Sport (1978), to the World Summits on Physical Education (Berlin, November, 1999 and Magglingen, December 2005: see www.icsspe.org) on which occasion the ICSP prepared declarations of support for school physical education. In addition, FIEP, a member association of the ICSP published "A new concept of Physical Education" in 2000/01.

4.3. Varia

Not applicable.

4.4. Free Statements

Not applicable.

Athletic Training and Therapy

Catherine Ortega and Michael S. Ferrara

Contact

Dr. Catherine Ortega
World Federation of Athletic Training and Therapy
University of Texas Health Science Center at San Antonio
School of Allied Health Sciences
7703 Floyd Curl Drive, MSC 6247
USA
Phone: +1 210 567 8750
Fax: +1 210 567 8774
Email: ortegac2@uthscsa.ed

1. General Information

1.1. Historical Development

In early Greece, the establishment of the Panhellenic Games - the forerunner to the modern-day Olympics - led to the introduction of the terms coach and trainer to describe an athlete's primary health care providers. These early health care practitioners possessed a remedial knowledge of diet, rest and exercise, as well as the effect of each on physical development and performance. They utilised basic tools such as hot baths, massage, anodynes and other measures to condition and treat the athletic competitors. The education, function and role of these early coaches and trainers/therapists evolved through the years, and today we refer to these health care professionals as athletic trainers, athletic therapists, sport rehabilitators, biokineticists and sport physiotherapists, to name a few. The professional titles of the allied health practitioner differ from one country to another. Still, the populations these allied health professionals work with and the goals of health care outcomes are closely related, and in some cases are virtually the same.

1.2. Function

Athletic trainers and therapists are qualified health care professionals educated in the management of problems and conditions related to physical activity. Working closely with physicians and other health care personnel, the athletic trainer or therapist functions as an integral member of the health care team in various settings. Athletic trainers work with medical personnel, athletic personnel, individuals involved in physical activity and parents of young athletes in the development and coordination of efficient and responsive health care delivery systems.

1.3. Body of Knowledge

The discipline of athletic training/therapy brings together theoretical and applied perspectives from several inter-related bodies of knowledge. The practitioner's professional preparation is directed toward the development of specified competencies in the following areas:

- Risk management and injury prevention,
- Pathology of injuries and illnesses,
- Assessment and evaluation,
- Acute care of injury and illness,
- Pharmacology,
- Therapeutic modalities,
- Therapeutic exercise,
- General medical conditions and disabilities,
- Nutritional aspects of injury and illness,
- Psychosocial intervention and referral,
- Health care administration; and
- Professional development and responsibilities.

1.4. Methodology

Various methodological approaches and tools are utilised in athletic training, therapy and related research. Research can be categorised as follows:

Basic Science – includes controlled laboratory studies in the sub-disciplines of exercise physiology, biomechanics, motor behavior and others that relate to athletic training and sports medicine.

Clinical Studies – includes assessments of the validity, reliability and efficiency of clinical procedures, rehabilitation protocols, injury prevention programmes, surgical techniques and related practices.

Educational Research –a broad category ranging from basic surveys to detailed athletic training/sports medicine curriculum development. Studies in this category generally include assessments of student learning, teaching effectiveness (didactic or clinical), educational materials and curriculum development.

Sports Injury Epidemiology – includes studies of injury patterns among athletes. These studies generally encompass large-scale data collection and analysis. Surveys and questionnaires may be classified in this category, but are more likely to come under the Observational/Informational Studies category.

Observation/Informational Studies – includes studies involving surveys, questionnaires and descriptive programmes that relate to athletic training and sports medicine.

1.5. Relationship to Practice

Athletic training and athletic therapy have numerous applications in the related field of allied health care. In cooperation with physicians and other allied health personnel, the athletic trainer and therapist function as integral members of the athletic/sport- related/physical activity health care team in their respective country. They are employed in secondary schools, colleges and universities, sports medicine clinics, professional sports programmes, biomechanics laboratories, academic settings and other athletic health care settings. They also participate in extensive clinical affiliations with athletic teams or physically active individuals under alternative settings. At the secondary school, college and university levels, the athletic trainer or therapist is responsible for the safety and quality of health care services for student athletes. Professional sports require the practitioners to work with elite athletes year round in rehabilitation, conditioning and player development. Within sports medicine clinics and hospitals, athletic trainers and therapists work with a diverse patient population and with a variety of other health care professionals. In an industrial setting, they help reduce time lost due to injury and help maintain cost-effectiveness.

The following professional organisations are affiliated with the World Federation of Athletic Training and Therapy:

- Association of Chartered Physiotherapists in Sports Medicine (UK)
- Association of Chartered Physiotherapists in Sports Medicine (UK)
- Biokinetics Association of South Africa
- Canadian Athletic Therapists' Organization
- Federazione Italiana Fisioterapisti
- Japan Amateur Sports Association
- Japan Athletic Trainers' Association
- Korean Association of Certified Exercise Professionals
- National Athletic Trainers' Association (USA)
- Ontario Athletic Therapists' Association
- Society of Tennis and Medicine
- Spanish Association of Sports Nurses
- Taiwan (Republic of China) Athletic Trainers' Society

The relationship among the professional organisations of the Federation is not possible without an ongoing and intensive exchange of information between researchers and practitioners. Affiliations with numerous organisations enhance the Federation's ability to provide this vital exchange of information.

1.6. Future Perspectives

For the profession, the future holds further interactions with worldwide leaders in athletic training, athletic therapy, biokinetics, sport physiotherapy and sports rehabilitation.

The disciplines hope to improve collaboration among health care professionals throughout the world in the development of new and innovative strategies to improve the health care of individuals participating in physical activity.

2. Information Sources

2.1. Journals

American Journal of Sports Medicine
Athletic Therapy Today
British Journal of Sport and Medicine
Journal of Athletic Training
Journal of Sports Rehabilitation

2.2. Reference Books/Encyclopedia etc.

ACSM (1997). *Exercise Management for Persons with Chronic Diseases and Disabilities*. Champaign, IL: Human Kinetics.

ACSM (2000). *Guidelines for Exercise Testing & Prescription, 6th ed.* Philadelphia, PA: Williams & Wilkins.

Agostini, R. (1994). *Medical and Orthopedic Issues of Active and Athletic Women.* St. Louis, MO: Mosby-Year Book, Inc.

American Academy of Pediatrics. (1997). *Preparticipation Physical Examination, 2nd ed.* Dubuque, IA: McGraw-Hill.

American Red Cross. (1993). *CPR for the Professional Rescuer.* StayWell.

Anderson, M., & Hall, S. (1997). *Fundamentals of Sports Injury Management*. Philadelphia, PA: Lippincott, Williams & Wilkins.

Arnheim, D., & Prentice, W. (2000). *Principles of Athletic Training*, 10th ed. Dubuque, IA: McGraw-Hill.

Arnheim, D., & Prentice, W. (1999). *Essentials of Athletic Training*, 4th ed. Dubuque, IA: WCB McGraw-Hill.

Bates, B., Bickley, L., & Hoekelman, R. (1995). *Guide to Physical Examination and History Taking, 6th ed*. Philadelphia, PA: Lippincott, Williams & Wilkins.

Baumgartner, T. A., & Strong, C. H. (1994). *Conducting and Reading Research in Health and Human Performance*. Dubuque, IA: WCB Publishers.

Booher, J., & Thibodeau, G. (2000). *Athletic Injury Assessment, 4th ed*. Dubuque, IA: McGraw-Hill.

Ciccone, C. (1996). *Pharmacology in Rehabilitation*. Philadelphia, PA: FA Davis.

Clark, N. (1997). *Nancy Clark's sports nutrition guidebook*. Champaign, IL: Human Kinetics.

Clarkson, H., & Gilewich, G. (2000). *Musculoskeletal Assessment*: Joint Range of Motion and Manual Muscle Strength, 2nd ed. Philadelphia, PA: Lippincott, Williams & Wilkins.

Crosby, L., & Lewallen, D. (1995). *Emergency Care & Transportation of the Sick & Injured, 6th ed*. Boston, MA: Jones & Bartlett Publishers.

D'Orazio, B. (2001). *Back Pain Rehabilitation, 2nd ed*. Boston, MA: Andover Medical Publishers.

Fritz, S. (1995). *Mosby's Fundamentals of Therapeutic Massage*. St. Louis, MO: Mosby Lifeline.

Gallaspy, J., & May, J. (1996). *Signs & Symptoms of Athletic Injuries*. St. Louis, MO: Mosby.

Gallup, E. (1995). *Law and the Team Physician*. Champaign, IL: Human Kinetics.

Greenberger, N. (1993). *History Taking and Physical Examination*. St. Louis: Mosby-Year Book.

Hall, C., & Brody, L. (1999). *Therapeutic Exercise: Moving Toward Function*. Philadelphia, PA: Lippincott, Williams, & Wilkins.

Hall, S. (1995). *Basic Biomechanics*. St. Louis, MO: Mosby Publishing.

Hamill, J., & Knutzen, K. (1995). *Biomechanical Basis of Human Movement*. Baltimore, MD: Williams, & Wilkins.

Heil, J. (1993). *Psychology of Sport Injury*. Champaign, IL: Human Kinetics.

Kendall, F., McCreary, E., & Provance, P. (1993). *Muscles: Testing and Function:*

With Posture and Pain, 4th ed. Philadelphia, PA: Lippincott Williams & Wilkins.

Kettenbach, G. (1995). *Writing SOAP Notes, 2nd ed.* Philadelphia, PA: FA Davis Co.

Kisner, C., & Colby, L. (1996). *Therapeutic Exercise: Foundations & Techniques, 3rd ed.* Philadelphia, PA: FA Davis Co.

Knight, K. (1995). *Cryotherapy in Sport Injury Management.* Champaign, IL: Human Kinetics.

Konin, J., & Wikstein, D. (1997). *Special Tests for Orthopedic Examination.* Thorofare, NJ: SLACK Inc. Publishers.

Konin, J. (1997). *Clinical Athletic Training.* Thorofare, NJ: SLACK Inc.

McArdle, W., Katch, F., & Katch, V. (2001). *Exercise Physiology: Energy, Nutrition and Human Performance, 5th ed.* Philadelphia: Lippincott Williams & Wilkins

Mellion, M., Walsh, W., Madden, C., & Putukian, M. (2001). *The Team Physician's Handbook, 3rd edition.* Philadelphia, PA: Lippincott, Williams, & Wilkins.

Moore, K., & Dalley, A. (1999). *Clinically Oriented Anatomy.* Philadelphia, PA: Lippincott, Williams & Wilkins.

NATABOC. (1997). *Role Delineation Study: Athletic Training Profession, 4th ed.* Research Triangle Park, NC: Columbia Assessment Services, Inc.

O'Keefe, M., & Limmer, D. (1998). *Emergency Care, 8th ed.* Upper Saddle River, NJ: Brady Prentice Hall.

Payton, O. (1995). Research: *The Validation of Clinical Practice, 3rd ed.* Philadelphia, PA: FA Davis.

Perrin, D. (1995). *Athletic Taping & Bracing.* Champaign, IL: Human Kinetics.

Perrin, D. (1999). *Isokinetic Exercise and Assessment.* Champaign, IL: Human Kinetics.

Perrin, D. (1999). *The Injured Athlete, 3rd ed.* Philadelphia, PA: Lippincott, Williams & Wilkins.

Peterson, M. (1996). *Eat to Compete, 2nd ed.* St. Louis, MO: Mosby.

Pfeifer, R., & Mangus, B. (1998). *Concepts of Athletic Training, 2nd ed.* Boston, MA: Jones & Bartlett.

Powers, S., & Howley. E. (1996). *Exercise Physiology, 3rd ed.* Dubuque, IA: Brown & Benchmark.

Prentice, W. (1999). *Rehabilitation Techniques in Sports Medicine, 3rd ed.* Dubuque, IA: McGraw-Hill.

Prentice, W. (1999). *Therapeutic Modalities in Sports Medicine, 4th ed.* New York, NY: McGraw-Hill.

Rachlin, E. (1994). *Myofascial Pain and Fibromyalgia: Trigger Point Management.* St. Louis, MO: Mosby-Year Book, Inc.

Rankin, J., & Ingersoll, C. (2001). *Athletic Training Management: Concepts & Application, 2nd ed.* Dubuque, IA: McGraw-Hill.

Ray, R. (1997). *Management Strategies in Athletic Training.* Champaign, IL: Human Kinetics.

Ray, R., & Wiese-Bjornstal, D. (1999). *Counseling in Sports Medicine.* Champaign, IL: Human Kinetics.

Ray, R. (1995). *Case Studies in Athletic Training Administration.* Champaign, IL: Human Kinetics.

Snider, R. (1997). *Essentials of Musculoskeletal Care.* Rosemont, IL: American Academy of Orthopaedic Surgeons.

Starkey, C., & Ryan, J. (2002). *Evaluation of Orthopaedic & Athletic Injuries, 2nd ed.* Philadelphia, PA: FA Davis Co. Publishers.

Starkey, C. (1999). *Therapeutic Modalities, 2nd ed.* Philadelphia, PA: FA Davis and Co.

Street, S., & Runkle, D. (2000). *Athletic Protective Equipment: Care, Selection and Fitting.* Dubuque, IA: McGraw-Hill.

Thomas, C. (1993). *Tabers Cyclopedic Medical Dictionary, 19th ed.* Philadelphia, PA: FA Davis.

Thomas, J., & Nelson, J. (2001). *Research Methods in Physical Activity, 4th ed.* Champaign, IL: Human Kinetics.

Thomas, J. (1998). *Drug, Athletes & Physical Performance.* New York, NY: Plenum Publishing.

Thompson, F. (1994). *Manual of Structural Kinesiology, 12th ed.* St. Louis, MO: Mosby-Year Book, Inc.

Tippett, S., & Voight, M. (1995). *Functional Progression for Sports Rehabilitation.* Champaign, IL::Human Kinetics.

Torg, J. (1991). *Athletic Injuries to the Head,* Neck & Face, 2nd ed. St. Louis: C. V. Mosby.

Torg, J., & Shepard, R. (1995). *Current Therapy in Sports Medicine.* St. Louis, MO: Mosby.

Tritschler, K. (2000). *Barrow & McGee's Practical Measurement and Assessment, 5th ed.* Philadelphia, PA: Lippincott, Williams & Wilkins.

Turner, L. (1993). *Life Choices: Health Concepts & Strategies, 2nd ed.* West Publishing Co.

Valmass, R. (1996). *Clinical Biomechanics of the Lower Extremity.* St. Louis, MO: Mosby.

Williams, M. (1995). *Introduction to Nutrition for Fitness & Sport, 4th ed.* Dubuque, IA: Brown & Benchmark.

Ziegler, T. (1997). *Management of Bloodborne Infections in Sport: A Practical Guide for Sports Healthcare Providers and Coaches.* Champaign, IL: Human Kinetics.

2.3. Book Series

Not applicable.

2.4. Conference/Workshop Proceedings

The World Congress of the World Federation of Athletic Training & Therapy is held every other year with an international rotation. Health care professionals from around the globe meet to share information and knowledge related to the prevention, treatment and management of sports injuries. The event includes presentations of scientific papers and case studies, and the opportunity to participate in clinical hands-on workshops. The 3rd World Congress of the World Federation of Athletic Training and Therapy was a combined meeting with the British Association of Sport and Exercise Medicine and the Association of Chartered Physiotherapists in Sports Medicine and a celebration of the 500th Anniversary of the Royal College of Surgeon of Edinburgh. In 2007 the World Congress was held in Tokyo, Japan by the Japan Sports Association.

Congress proceedings are printed and distributed to the membership.

2.5. Data Banks

Several athletic training and therapy organisations maintain their own databases of injury statistics and on the related health care materials for athletes and those who are physically active.

One such resource, the National High School Sports Injury Registry, is designed to be an ongoing source of information regarding the frequency, type and severity of injuries that occur in sports at the high school level in the United States. The registry tracks injuries in american football, wrestling, baseball, softball, field hockey, girls volleyball, boys and girls basketball and football, offering athletic trainers immediate access to a wealth of data that can be utilised in the areas of research, education and public relations.

2.6. Internet Sources

The World Federation of Athletic Training & Therapy
    www.wfatt.org/
Canadian Athletic Therapists Organization
    www.athletictherapy.org
The National Athletic Trainers' Association
    www.nata.org
Association of Chartered Physiotherapists in Sports Medicine (UK)
    www.acpsm.org
Biokinetics Association of South Africa
    www.biokinetics.org.za
Japan Athletic Trainers' Organization
    www.jato-trainer.org
Japan Sports Association
    Tanaka-n@japan-sports.or.jp

3. Organisational Network

3.1. International Level

The World Federation of Athletic Training & Therapy (WFATT) is an international coalition of national organisations of health care professionals in the fields of sport, exercise, injury/illness prevention and treatment. The Federation strives to promote the highest quality of health care and functional activity through the collaborative efforts of its members.

The first official World Congress of the World Federation of Athletic Training & Therapy was held in Los Angeles, California, in 2001 and will continue to meet every two years at various international locations.

WFATT membership is open to professional associations and organisations whose scope of practice includes the prevention, care and rehabilitation of athletic and sports-related injuries and conditions. More than one association/organisation from a country can be approved for membership in the WFATT.

3.2. Regional Level

Not applicable.

3.3. National Level

At the national level, the charter associations of the WFATT assist athletic trainers and therapists in advancing, encouraging and improving the profession through educational and practical programmes. These associations are:

Association of Chartered Physiotherapists in Sports Medicine (ACPSM) (UK)
Biokinetics Association of South Africa (BASA)
Canadian Athletic Therapists Organization (CATA)
Japan Amateur Sports Association (JASA)
Japan Athletic Trainers' Association (JATO)
National Athletic Trainers' Association (NATA) (US)
Taiwan (Republic of China) Athletic Trainers' Society
Additional national member associations are:Italian Federation of Physiotherapists (FIF)
Japan Athletic Trainers' Association for Certification
Korean Association of Certified Exercise Professionals
Ontario Athletic Therapists' Association
Spanish Association of Sport Nurses (AED)
Society of Tennis Medicine and Science (STMS)

Educational institutions and associate member organisations partner with WFATT to advance mutual goals. These organisations include:

Board of Certification, Inc, United States
Murdoch University, Australia
University of Bedfordshire, United Kingdom

3.4. Specialised Centres

Not applicable.

3.5. Specialised International Degree Programmes

Not applicable.

4. Appendix Material

4.1. Terminology

Not applicable.

4.2. Position Statements

The National Athletic Trainers' Association (NATA) has released the following position statements, available online at www.nata.org:

- Blood Borne Pathogens Guidelines
- Fluid Replacement for Athletes
- Lightning Safety for Athletics and Recreation
- Physically Active Definition
- Management of Sport Related Concussion
- Exertional Heat-Related Illnesses
- Emergency Planning in Athletics
- Management of Asthma in Athletes
- The Canadian Athletic Therapists Association (CATA), another member organisation of WFATT, also compiles consensus statements. These can be obtained via correspondence at www.athletictherapy.org.

4.3. Varia & 4.4. Free Statement

Not applicable.

Comparative Physical Education and Sport

Ken Hardman

Contact

Prof. Dr. Ken Hardman
School of Sport & Exercise Science
University of Worcester
Henwick Grove
Worcester, WR2 6AJ
UNITED KINGDOM
Email: k.hardman@worc.ac.uk / ken.hardman@tiscali.co.uk

ISCPES General Enquiries:
Dr. Darwin Semotiuk
ISCPES President
The University of Western Ontario
London Ontario N6A 3K7
CANADA
Email: semotiuk@uwo.ca

ISCPES Membership:
Dr. Richard Baka
School of Human Movement
School of Kinesiology
Victoria University
P.O. Box 14428
Melbourne, Victoria 8001
AUSTRALIA
Email: Richard.Baka@vu.edu.au

1. General Information

1.1. Historical Development

Comparative studies in Society and Education have their origins in explorers' and travellers' accounts of customs and practices, usually stemming from journeys based on simple curiosity in the strange and exotic and later, commercial enterprise; they preceded the nineteenth century quest for knowledge and emulation of foreign schools' practices through purposeful observation. Whilst Comparative Physical Education and Sport is a relatively young area of study in a formal sense, the quest for knowledge about practices and systems has been in evidence since the Prussian Count Leopold Berchtold included physical education and sport in a 400-page questionnaire for travellers in 1789. Some 20 years later, Frenchman, Auguste Basset wrote on *The Utility of Making Observations in Foreign Countries Concerning Their Different Modes of Education and Instruction.* The pioneering work of Berchtold and Basset was developed by another Frenchman, Marc Antoine Jullien, when in 1817, he published a series of questions on public education, including physical education in European countries. Jullien's work, like that of others in the early nineteenth century, was characterised by the pragmatism of cultural borrowing.

The pragmatism associated with detailed observation for potential 'cultural borrowing', however, was concerned more with the 'What' and less with the 'Why' and 'How'. From these seeds grew a comparative methodology movement, which called for a more comprehensive analytical and explanatory approach. This was an approach, largely evolved by Sir Michael Sadler, which was grounded in the establishment of general principles, and when set in a comparative framework, led to an analysis of similarities and differences. Historical inquiry and explanation were at the core of such comparative study, one outcome of which was that 'national character' is a major determinant in shaping a system of education.

Comparative study was progressed beyond the historical explanatory approach to a semi-scientific orientation by Moehlmann (1963), who suggested a theoretical model to ensure systematic analysis of contemporary trends and long-range cultural factors. Moehlmann argued that long-range interactive factors including population demographics, spatial (geographical) and temporal (historical) concepts, language (communication systems), art, philosophy (value choices), religion, social structure, governmental systems, economics, science and technology, health and educational process determined the profile of an education system. Systematic exploration and analysis of national practices in education were progressed through the structured approach devised by Bereday (1964) comprising four stages: **description; interpretation** (or **explanation); juxtaposition;** and **comparison**. Thus, by the mid-1960s, comparative studies in education had progressed from individual intuitive and descriptive 'raw data' and historical techniques to more sophisticated systematic methods of analysis, drawing largely from social science methods of investigation and involving interdisciplinary 'team' approaches.

In 1970, American sports historian, John E. Nixon, reported an increasing interest in international aspects of physical education, testimony to which was the plethora of descriptive articles contributed to professional journals by American physical educators. In the main, these articles represented information derived from observational educational or 'touristic' visits to be shared with colleagues. They did not qualify, in Nixon's (1970) view, as comparative research reports and reflected the broader situation of comparative studies in physical education and sport trailing behind reported research in the 'parent' area, 'comparative education'. Indeed, texts concerned with comparative and international issues and dimensions were at that time a rarity. Morton (1953) and Louis and Louis (1964), had produced descriptions of sport in the Soviet Union and Nixon himself had co-edited (1968) a text with C.L. Vendien containing information on health, physical education and recreation in a number of countries around the world. As Nixon (1970) conceded, these publications were "...At best... examples of the early phases of the descriptive stage of comparative research" with "physical education in an early stage of growth in research in comparative, international and developmental studies in the United States" (p.8). The general tone of Nixon's commentary was rather American-centric but it might well have been less so had he been aware of the work of then active researchers in other countries. For elsewhere in the academic physical education and sport arena, more scholarly activity had been, and was being, carried out and reported. In the United Kingdom, for example, pioneering work had been undertaken by Molyneux (1962), Sullivan (1964) and Anthony (1966), who completed a doctoral study in the genre in 1971.

After 1970, comparative and international studies in physical education and sport were subject to a relatively significant development of interest and scholarly activity, the latter being especially marked by two seminal texts in the field: Bennett, Howell and Simri (1975) and Riordan (1978). The present existence of a range of publications, an international and several national societies, seminars and conferences with a comparative focus suggest that the field of study has markedly progressed as an area of academic and professional activity. A major initiative in its international development was the formation of the International Society for Comparative Physical Education and Sport (ISCPES) in 1978/9, which symbolically marked the coming of age of as a genre domain of study. Subsequent to its formation, various ISCPES publications collectively demonstrate substantial progress in nature, scope and methodological procedures and applications in the comparative genre. Significantly, they also reveal enrichment through contributions from a wider academic and scholarly community beyond physical and sport educators to embrace historians, psychologists, social psychologists and sociologists and others with vested interests in socio-cultural and pedagogical domains. The maturity of ISCPES was symbolically represented in invitations in 1988 (Japan) and 1995 (Kuwait) to conduct Workshops aimed to bring together experts in various aspects respectively of *Physical Education and Sport for All*. Of special interest to the national partner agencies were comparative methodological procedures and sports policies and practices in other countries.

Over the last decade or so in the Anglo-Saxon speaking world, there has been a steep decline in the number of higher education institutions offering courses or modular units under the specific nomenclature of *Comparative Physical Education and Sport*. Hence, whilst it is far from being extinct, evidence of which lies particularly in 'first order' comparisons, the body of knowledge thesis is becoming increasingly tenuous. In some countries, what used to appear under the title has generally been subsumed within themes or under topics such as *International Dimensions* or features in cross-disciplinary courses variously covering *Issues, Politics, Economics, Policies, Sociological and/or Cultural Perspectives* and so on. More usually than not, "Sport" prevails over "Physical Education" in such course units. In North America, a similar trend can be discerned particularly in re-orientation to courses in *International Studies and International Development* with a Sport and Physical Education focus. For all of these thematic/topic approaches, interpretation and understanding of content can be more open-ended in outcomes: comparisons can be directly and explicitly made or can be indirect, implicit or in the mind of the learner/reader.

Somewhat counter to this general trend are developments elsewhere in the world. In central and eastern Europe, where 40 years of socialism inhibited exchange of knowledge, and in some Asian and Middle Eastern countries, particularly in those countries not especially subjected to western, particularly American influences, there has been increasing interest in international and comparative studies in the last decade or so. Testimony to this are an Asian Conference on Comparative Physical Education and Sport Symposium organised by the Sport Social Science Branch of the Chinese Sports Science Society in Shanghai with published proceedings in 1995, and the more recent biennial International Conferences held in the Islamic Republic of Iran since the late 1990s. Journal publications, which carry international and comparative articles (continental regional e.g. *Asian Journal of Physical Education*; national and institutional e.g. *Kinesiology; International Scientific Journal of Kinesiology and Sport* of the Faculty of Physical Education, University of Zagreb, Croatia; *Acta Universitatis Carolinae; "Gymnica"* of Charles University, Prague, Czech Republic; and *Man-Movement* of the Academy of Physical Education, Wroclaw and Polish Scientific Society of Physical Culture), and translations into Chinese, French, Japanese, Polish and Spanish languages of the Proceedings of the World Summit on Physical Education, Berlin, 1999 also attest to international development. This more positive trend of published articles, which have a comparative, international and cross-cultural focus, can also be observed in other well established and respected international, single, multi- and cross-discipline journals, a selected list of which was included in *International Journal of Physical Education*, volume XXXVIII, Issue 3, 3rd Quarter, 2001 (Hardman, 2001, p. 99).

A positive development in the comparative field has been the academic, professional, but most significantly, political interest generated in publication of data derived from a range of international, national and regional surveys and longitudinal literature reviews. Examples are found in: the 1998-1999 ICSSPE supported and International Olympic Committee (IOC) sponsored *World-wide Survey on the State and Status of Physical Education in Schools* (Hardman and Marshall, 2000); the Council of Europe Survey of Physical Education in Member States (Hardman, 2002); the North Western Counties PE Association follow-up world-wide survey on the situation of school physical education endorsed by ICSSPE and UNESCO, undertaken as part of the United Nations 2005 Year of Sport and Physical Education (refer Hardman & Marshall, 2006); and the European Parliament Project (2006-2007) concerned with the situation and future sustainability of School Physical Education in European Union countries (Hardman, 2007). Such international and comparative-focused studies have helped to place physical education in schools on the world political agenda. The inclusion of school physical education on inter-governmental and non-governmental agencies' agendas demonstrates the values of engagement in comparative studies: provision of information; increase in knowledge of one's own and others' 'worlds'; and potential facilitation of amelioration.

With the concept and contexts of globalisation, renewed interest has arisen in international dimensions of physical education and sport with varying engagement of inter-governmental agencies, national and regional governments and international, national and regional non-governmental organisations as well as a range of social and educational institutions and individuals involved in either overarching or specific development initiatives. The designated 2004 European Year of Education through Sport and United Nations 2005 Year of Sport and Physical Education demonstrate the significance of physical education and sport to international communities. The protagonist push for harmonisation in physical education in Europe is another indicator of international interest and its process has been, and is being, aided and abetted by European Union programs such as Erasmus and Socrates. The trend for harmonisation was clearly articulated in the 1999 Bologna Agreement to create a common model for Higher Education in Europe with institutions subsequently encouraged to develop a framework of comparable and compatible qualifications for their programs. Hence, a four year (2003-2007) Erasmus funded Thematic Network Project, *Aligning European Higher Education Structure in Sport Science* (AEHESIS), emerged. Drawing on the pilot project *Tuning Educational Structures in Europe* (the so-called *Tuning Project*) methodologies, the AEHESIS Project established innovative guidelines specifically for the broadly defined sport sector (Health and Fitness, Physical Education, Sport Coaching and Sport Management) for the development of curricula and quality assurance systems for study programs. It is anticipated that the results of the Project might influence national policies in the sector and provide guideline models or frameworks and benchmarks for non-participating organisations (refer AEHESIS Project Annual Reports for further details).

Notably, for comparative genre scholars, this renewed interest in international issues marks a 'back to basics' approach within the area of physical education and sport. Whether this is indicative of 'turning back the clock' or the 'wheel coming round full circle' is debatable. What is clear is that the renewed interest in international issues is conducive to the generation of rich data, which can be compared and utilised for ameliorative development, a fundamental purpose of the comparative study domain. In essence, the situation mirrors the position on research in comparative, international and development studies identified by Nixon almost 40 years ago.

1.2. Function

Comparative physical education and sport as an area of study draws from a number of disciplines and hence, is seen to be multi- and inter-disciplinary in nature and scope. Specifically, as an area of scholarly activity, the genre seeks through the establishment of reliable data to: (a) provide information on the 'worlds' of others; (b) foster knowledge about one's own 'world' through confrontation with alternatives; and (c) amelioration through learning about and from others. Of crucial importance to these processes is the necessity of discovering and revealing shaping influences, which through cross-analysis, provide causal connections, and hence, explanations. It is in this way that deeper insight into, and understanding of, the processes and products of delivery are acquired.

1.3. Body of Knowledge

A persistent problem is to obtain a common understanding of what constitutes comparative studies. Arguably, comparative physical education and sport might be more closely identified with method rather than with a distinct body of knowledge. The critical term is *comparative*. The International Society for Physical Education and Sport (ISCPES) defines comparative study as *investigation into and comparison of two or more units (countries, cultures, ideologies, regions, states, systems, institutions, populations)* mostly occurring in different geographical settings. Examples of phenomena to be compared include: school systems (or elements) of physical education and sport models in a macro or micro context. Usually the phenomena associated with such units are universal, but cross-culturally and cross-nationally, they may differ in focus and substance. Comparativists study how and why they differ. Comparative analyses involve those directing and initiating research, which explore the suitability of new elements from other cultures for inclusion in their programs. Besides the comparative dimension, the domain encompasses issues related to studies of countries (so-called mono-national, first order comparative studies), education for internationalism and development assistance in its research and teaching dimensions.

1.4. Methodology

The field of comparative physical education and sport has travelled a similar route to that of comparative education, from which it has adopted the various methodological approaches to comparative study. Thus, there have been initially facile, then expert, travellers' accounts, area and national studies, followed by cultural borrowing or infusion (as in former British colonies) of ideas and systems, and the international exchange of information and utilisation of cross-cultural or transnational 'Problem' approaches, which focus on specific issues e.g. 'Excellence Systems', 'Sport for All Programs', 'Women in Sport', 'Coach Education' etc. After the early historico-cultural explanatory traditions and social scientific approaches to analysis, a number of classification frameworks or schema were developed and used to examine physical education and sports systems. They range from the simple, first hand reports (Vendien and Nixon, 1968; Johnson, 1981), to detailed delineation of shaping or influential determinant factors (Sturzebecker, 1967; Bennett, 1970) and elaborate conceptual frameworks based on established schema in a range of related fields and disciplines (Morrison, 1979). All, however, emphasise that physical education and sport should be seen as part of the societal setting in which they exist. This overview of the progress of methodologies serves to illustrate that comparative study has moved on from early descriptive narratives of the 'what', through the formative historico-explanatory tradition, to comprehensive and systematic methods of data collection in the tradition of the social sciences to reveal the 'why' and 'how' of developed and developing systems.

At present, comparative physical education and sport studies' methodology is deemed to embrace a range of analytical tools to be applied to comparative data. Comparative study no longer attempts to define a single methodology and no one single method is developed as canon. In recent years, comparative education scholars have adopted a range of methodological approaches to develop ways of dealing with complex issues. These eclectic and pluralistic approaches provide means of dealing with a broad range of issues. Empirical quantitative approaches establishing correlations have been enriched by the qualitative paradigm seeking to achieve understanding and interpretation of processes and reveal causality.

1.5. Relationship to Practice

Whilst it is clear that physical education and sporting activity do have an ubiquitous global presence, they are at the same time subject to culturally specific 'local' (national) interpretations, policies and practices. Inevitably, similarities and differences are encountered at these levels. This is a feature that demonstrates both diversity and complexity in process and product as well as in the influential factors, which have acted collectively and inter-dependently to 'shape' a delivery system. Ideological variants, for example, reinforce the argument of similarities and differences and the diversities evident at local, regional and national levels. Such diversity supports the thesis that 'localisation' within 'globalisation' can and does exist. Even in regions where there have been common ideologies, such as the former 'socialist bloc' of central and eastern European states with their centralised systems, relevant research points to substantial variations in aspects of the delivery services. Typical of this were the variations evident in the development of young, talented athletes to levels of excellence. The 'localisation/globalisation' debate is also manifest in the national settings of European Union countries, where efforts to bring about congruence and harmony in programs have to recognise the existence of deep-seated diversities of trans-national contexts. The embedded traditions in physical education and sport in European countries are inextricably bound up with historical antecedents and are inevitably culture bound. These are features that are fundamental to understanding, when curriculum planners strive for uniformity and standardisation.

A potential pitfall within the domain of comparative and international studies lies within the sphere of the 'truth' of 'fact', often witnessed in discrepancies between principles and practices or, for example, in government policy rhetoric and its actual implementation. Illustrations of gaps between policy promises and actual practice are seen in the recent international surveys of the situation of physical education in schools (refer Hardman & Marshall 2000; Hardman 2002; Hardman & Marshall, 2005; and Hardman, 2007). Despite such constraints, comparative study can and does facilitate an awareness of possibilities for amelioration of existing structures and mechanisms through processes of adoption or adaptation to suit local (national), socio-cultural, economic and environmental circumstances. The international dimension within comparative research can, and does, inform policy at inter- and national governmental agency levels as testified by UNESCO, World Health Organisation, Council of Europe and European Parliament and a plethora of national governments' responses to findings from international comparative surveys.

1.6. Future Perspectives

For future developments in the comparative physical education and sport domain, mixed messages are evident. Generally, there is a shift away from the mono- and multi-national 'area' approaches of the latter years of the 20th century to thematic or topic approaches. This shift is seen in two developments: (i) the disappearance of Comparative Physical Education and Sport modules or course units from university level programs and replacement by interdisciplinary issues-based units, which focus on international themes such as the situation of physical education in schools, gender, disability, or topics such as politics or youth sport, with a comparative dimension or from a comparative perspective; and (ii) in titles and contents of published texts including books, journal articles and reports. However, in some regions of the world, and mainly those that are economically emerging or developing or have recently politically and ideological re-aligned, there is interest in features and systems in national entity contexts. Central to future initiatives will be the continuing globalisation versus localisation debate, the role of ethernet communication, and increasingly sophisticated methodological procedures to enable validity and reliability of data when crossing cultural and other divides.

2. Information Sources

2.1. Journals

The International Society for Physical Education and Sport has published an international journal since 1999, entitled *International Sports Studies* (formerly *Journal of Comparative Physical Education and Sport*). As a multi-disciplinary and cross-cultural perspectives' publication, it is essentially concerned with research and scholarship in the social sciences that focuses upon international studies of physical education and sport.

Examples of other journals, which carry comparative, international and cross-cultural articles include:

International Journal of Physical Education (ICSSPE);
FIEP Bulletin (FIEP);
Journal of the International Council for Health, Physical Education, Recreation, Sport and Dance (ICHPER.SD) and its autonomous regional derivatives (African, American, Asian and Australian Council for Health, Physical Education and Recreation);
European Physical Education Review (NWCPEA/Sage);
Journal of Sports History;
International Journal of Sports Sociology;
Sportwissenschaft (Verlag Karl Hofmann);
Kinesiology. International Journal of Fundamental and Applied Kinesiology (a Croatian based journal);
Human Movement (a Polish based journal).

Journals in Spanish language include:

Lecturas: Educación Física y Deportes, Revista Digital
Apunts: Educación Física y Deportes
Revista de Educación Física
PERSPECTIVAS de la actividad física y el deporte
Quaderns Didáctics: De las Ciencies aplicadas a L'Esport
Gaceta GYMNOS
RED: Revista de Entrenamiento Deportivo
Revista Española de Educación Física y Deporte
Pedagogía en movimiento.

A host of national journals also contain international related articles or items; Physical Education Matters, a UK based journal, for example, regularly features an international news section.

Affinity journals include: Compare, Comparative Education and Comparative Education Review.

2.2. Reference Books, Encyclopaedias, etc.
(a selection of recent texts)

Bartlett, R., Gratton, C. and Rolf, C. (2003). *Encyclopaedia of International Sports Studies*. London, Taylor and Francis.

Haag, H. (Ed.). (2004). *Research Methodology for Sport and Exercise Science*. Schorndorf, Verlag Karl Hofmann.

Haag, H. and Haag, G. (Eds.). (2003). *Dictionary. Sport, Physical Education, Sport Science*. Kiel, Institut für Sportwissenschaften.

Klein, G. and Hardman, K. (Eds.). (2007). *L'éducation physique et l'éducation sportive dans l'Union européenne*. Dossier EP.S No.71, vol.1. Les Éditions Revue EP.S., Paris.

Klein, G. and Hardman, K. (Eds.). (2008). *L'éducation physique et l'éducation sportive dans l'Union européenne*. Dossier EP.S No.72, vol.2. Les Éditions Revue EP.S., Paris.

Pühse, U. and Gerber, M. (Eds.). (2005). *International Comparison of Physical Education. Concepts, Problems, Prospects*. Oxford, Meyer and Meyer Sport (UK) Ltd.

Riordan, J. and Krüger, A. (Eds.). (2003). *European Cultures in Sport. Examining the Nations and Regions*. Bristol, Intellect Books.

2.3. Book Series

The ISCPES Book Series (terminated in 2004) was published by Routledge (Spon, UK). Structurally, the Series was divided into two types of text: volumes, which essentially have an 'area', i.e. mono-national focus, and alternately volumes which address 'problems', i.e. topics or themes in cross-cultural and/or international settings. Two 'area' study volumes were published: *Sport and Physical Education in China* (Jones & Riordan, 1999); and *Sport and Physical Education in Germany* (Naul & Hardman, 2002). A third title specifically addressed a gender-related topical theme (a 'problem' study approach): *Social Issues in Women and Sport – International and Comparative Perspectives* (Hartmann-Tews & Pfister, 2003).

ISCPES Monographs include:

Hardman, K. (Ed.). (1988). *Physical Education and Sport in Africa. ISCPES Monograph*. Manchester, University of Manchester. (This comprises an edited collection of papers concerned with themes and issues in several African countries: Botswana, Kenya, Nigeria, South Africa and Tanzania).

Hardman, K. (Ed.). (1989). *Physical Education and Sport under Communism. IS-CPES Monograph.* Manchester, University of Manchester. (This Monograph is composed of papers presented at the inaugural Seminar meeting of the British Interest Network for Comparative Physical Education and Sport held at the University of Manchester in December, 1988).

Hardman, K. (1996). *Foundations in Comparative Physical Education and Sport.* Manchester, Centre for Physical Education & Leisure Studies, University of Manchester.

Maeda, M., Ichimura, S. and Hardman, K. (Eds.). (1996). *Physical Education and Sport in Japan.* Manchester, University of Manchester.

Hardman, K. (Ed.). (1996). *Sport for All: Issues and Perspectives in International Context. ISCPES Monograph.* Manchester, University of Manchester.

Spanish language series include:

López de D'Amico, R. y Bolívar, R. (Ed.). (2007). *La actividad física en el desarrollo humano.* Caracas: Serie de libros arbitrado del Vicerrectorado de Investigación y Postgrado UPEL.

Girginov, V., Parry, J. y López de D'Amico, R. (2007). *Los juegos olímpicos explicados. Una guía de estudio a la evolución de los juegos olímpicos modernos.* Caracas: Ministerio del Poder Popular para el Deporte.

López de D'Amico, R y Murillo, J. (2005). *Ballet aplicado al entrenamiento deportivo.* Maracay: UPEL Maracay.

Prado, J y Gonzalez, V. (Ed.) (2007). *La Educación Física y el Deporte en la República Bolivariana de Venezuela.* Mérida Universidad de los Andes.

Bolívar, G. (2007). *Educación mediante la Recreación. Modelo teórico para el logro de los objetivos educacionales.* Caracas: Serie de libros arbitrado del Vicerrectorado de Investigación y Postgrado UPEL

2.4. Conferences/Proceedings/Workshops

Between 1978 and 2000, ISCPES published the proceedings of its eleven Biennial Conferences:

- Wingate Institute, Israel, 1978;
- University of Dalhousie, Halifax, Nova Scotia, Canada, 1980;
- University of Minnesota, Minneapolis, USA, 1982;
- Malente, Kiel, Federal Republic of Germany, 1984;
- University of British Columbia, Vancouver, Canada, 1986;
- Chinese University of Hong Kong, 1988;
- Bisham Abbey, Marlow, England, 1990;
- University of Houston, Texas, USA, 1992;
- Charles University, Prague, Czech Republic, 1994;
- Hachi-ohji, Tokyo, Japan, 1996;
- Catholic University, Leuven, Belgium, 1998.

Since the Leuven Conference, papers have featured variously as a double 'special' issue of the *International Sports Studies* journal (12th Biennial Conference, Maroochydore, Queensland, 2000) and in CD Rom format (13th Biennial Conference, Windsor, Canada, 2001 and 14th Biennial Conference, Melbourne, 2006).

At the regional level, examples of conference and workshop proceedings include:

Oropeza, R. & López de D'Amico, R. (2005). Jornada internacional en investigación de la actividad física.
Proceedings of the Regional Latin America ISCPES Congress. Maracay: UPEL Maracay; González, V., Abreu, J. & Guerrero, G. (2007). II, Congreso internacional de recreación y turismo. Book of proceedings. Maracay: EDUFISADRED; López de D'Amico, Gonzalez, V. & Murillo, J. (2006). Congreso Latinoamericano de la ICHPER-SD. Book of proceedings. Cojedes: Universidad Iberoamericana del Deporte; and INDER. (2007). Retos actuales y perspectivas para la educación física y el deporte contemporáneo. Cumbre Regional ISCPES. CD Rom proceedings. Varadero-Cuba: INDER

2.5. Data Banks

Wilcox, R. (Ed.). (1986). Comparative physical education and sport directory. New York, Adelphi University. The Directory is divided into three sections:

1. Biographical profiles
2. Agencies offering information and funding potential; Journals identified as potential sources for publication; Key to Institutions of Higher Education; Key to academic, professional and research bodies
3. Index by nation of residence

Educational Resources Information Centre (ERIC) has over 1 million periodical articles (CIJE) and educational reports (RIE).

Heracles is a French language sport and physical education database produced by the SPORTDOC national network and the Institut National du Sport et de l'Education Physique in Paris.

SISA Sportlit Database has over 20,000 records (the majority is South African). It is produced by the South African (SA) Department of Sport and Recreation with support from the SA National Sports Council and the SA National Olympic Committee.

SPOLIT database has many records, nearly half of which are in English. It is produced by the Federal Institute of Sport Science in Cologne, Germany.

SPORTDISCUS has numerous records in a wide range of languages. It is produced by the Sport Information Centre in Gloucester, Ontario, Canada.

Sport and Leisure Research on Disc covers UK dissertations and theses at doctoral and masters levels in all areas of sport, recreation and leisure from 16 contributing universities.

In Latin America, the following data banks are available:

Latindex (www.latindex.unam.mx/); Scielo (www.scielo.org/index.php?lang=en); and Fonacit (www.fonacit.gob.ve/publicaciones/indice.asp).

2.6. Internet sources (Web sites, list serves, etc.)

AEHESIS Project: www.aehesis.com
EUPEA: www.eupea.com
ICSSPE: www.icsspe.org
ISCPES: www.iscpes.org

www.thenapa2.org
www.inclusivesports.org
www.erasmusmundus.be

3. Organisation Network

3.1. International Level

A major initiative in the international development of the comparative physical education and sport domain was the formation of the International Society for Comparative Physical Education and Sport (ISCPES). ISCPES was founded initially as the International Committee on Comparative Physical Education and Sport in December 1978 on the occasion of an international seminar on comparative physical education and sport at the Wingate Institute, Israel. The name change to ISCPES occurred at the second international seminar on Comparative Physical Education and Sport in Halifax, Nova Scotia in 1980.

Over three decades, ISCPES has become acknowledged as the leading body in comparative physical education and sport studies. The Society is a research and educational organisation, which has the expressed purpose of supporting, encouraging and providing assistance to those seeking to initiate and strengthen research and teaching programs in comparative physical education and sport throughout the world. Specifically, ISCPES holds biennial conferences, distributes research findings and information, supports and cooperates with local, national and international organizations with similar goals and latterly, regional conferences/workshops. In addition to its range of publications, it has sponsored (in the form of patronage) and spawned, comparative/cross-cultural/ international research projects. The scope of the Society's academic mandate is affected by members' interests and research needs. Members are drawn primarily from all sub-disciplines of sport science with an interest in cross-cultural and international issues.

The ISCPES Executive Board members serve for a period of four years on a rolling basis to preserve an element of continuity. The Board currently comprises: Darwin Semotiuk (President, Canada), Rosa Lopez D'Amico (Venezuela), Scott Martyn (Canada), Anthony Church (Canada), Richard Baka (Australia), Herbert Haag

(Germany), Ken Hardman (UK), Lateef Amusa (South Africa), Hai Ren (Peoples's Republic of China), Walter Ho (Macao) and Jan Tolleneer (Belgium).

In 1986, ISCPES gained membership of ICSSPE and is represented on the Associations' Board and is a constituent member of the International Committee of Sport Pedagogy (ICSP). ISCPES also works closely with other international associations such as, for example: the International Society for History of Physical Education and Sport (ISHPES) and the International Associations for the Philosophy of Sport (IAPS) and Sociology of Sport (ISSA). All of these associations contribute to comparative literature, research and study in comparative physical education and sport.

Other international bodies contributing to comparative literature and with international policies in the physical education and sport domain are: the United Nations Educational, Scientific and Cultural Organisation (UNESCO); the World Health Organisation (WHO); the International Olympic Committee (IOC); the International Federation for Physical Education (FIEP); and the International Council for Health, Physical Education, Recreation, Sport and Dance (ICHPER.SD). ICHPER.SD has a number of Special Interest Group Commissions, of which Comparative Physical Education and Sport is one.

3.2. Regional Level

American Alliance for Health, Physical Education, Recreation and Dance has an International Relations Council.
European Physical Education Association (a forum, which comprises representatives of national Physical Education Associations).
FIEP Europe.
Latin American Association of Socio-cultural Studies in Sport.

3.3. National Level

National societies known to have existed but which now appear to be inactive comprise: the British Network Interest Group for Comparative Physical Education and Sport (1988-96), the Japan Society for Comparative PE and Sport and a Comparative PE and Sport Society in the Federal Republic of Germany. There is an established German-Japanese PE and Sport group, which meets every two years for comparative exchanges.

3.4. Specialised Centres

Beyond Higher Education Institutions' programs, such as the Estudios en Educación Física, Salud, Deporte, Recreación y Danza (EDUFISADRED), UPEL - Pedagógico de Maracay, Maracay, Venezuela, there are no known, dedicated Comparative PE and Sport specialised centres.

3.5. Specialised International Degree Programmes

A number of university level institutions offer programs that have an international orientation or dimension. At the regional, continental level, international Master's programs have been developed in Physical Education and Adapted Physical Activity. In September 1999, an innovative European Masters in Physical Education degree program was introduced. The program involves a set of core European dimension modules, 'home' university elective modules and a comparative (European dimension) research project. To date related intensive course programs have been held in Jarandilla, Spain, Extramadura University, Badajoz, Spain and Technical University, Lisbon, Portugal.

There is a well-established European one-year post-graduate program co-ordinated by the Catholic University, Leuven, Belgium, that provides research and teaching methodology in Adapted Physical Activity (APA), as well as the social, pedagogical and technical aspects of physical activity adapted to the needs of disabled persons. This European Masters in Adapted Physical Activity exists alongside a recently introduced Erasmus Mundus Master in Adapted Physical Activity. Both programs consist in total of 60 European Credits (ECTS) with English as the official language of instruction. Another disability model inspired program is THENAPA II: "Ageing and disability - a new crossing between physical activity, social inclusion and life-long well-being". The associated thematic network aims to: a) define the current situation at European Higher education institutions in relation to the extent the subject of Adapted Physical Activity (APA) is included in the curricula of the future service providers; and b) create a basic profile and implement the subject of adapted physical activity for the elderly in European higher education curricula in order to compensate for the current lack of information and resources in the specific domain.

4. Appendix Material

4.1. Terminology

Inherent with studying phenomena from different geo-political and socio-cultural locations are challenges in interpretation of linguistic terms and conceptual variations over time and space. These are areas for concern, for language and terminology, together with concepts, present particular problems in studies with a comparative or international focus. In translation, some words lose their original meaning because they are culture-bound. Terms may and do differ from country to country. The European region, with its diversity of languages and cultures, serves to illustrate the point. In France, physical education appears as *l'Éducation Physique et Sport (physical education and sport)* in schools; in Germany, the term *Sport/Sportunterricht (sport/sport instruction or teaching)* was generally adopted in the 1970s onwards with the *physical educator* termed the *sports teacher*; in the former divided Germany (1949-1990), the generic term for physical education in the two decades after the Second World War was *Leibeserziehung* in the then Federal Republic (West Germany) and *Körpererziehung* in the Democratic Republic (East Germany), the latter influenced by a post-World War II sovietisation process, in which *physical culture* and cultivation of the *socialist personality* had an important role throughout central and eastern Europe; since, the year 2000, several Länder in reunified Germany have introduced a re-conceptualised form of physical education, *Bewegungserziehung (movement education)*; in Sweden the term in general use is *idrott i hälsa (sport and health)*; whereas in the United Kingdom, the term *physical education* is used. Taking these terminology divergences into account, it is not wholly surprising that different and various forms of the subject exist in terms of the curriculum. These differences may be in strict or liberal regulatory implementation of the physical education curriculum, in general or precise prescriptions for content, in traditional and/or new aims and objectives, in central governmental and/or local school-based concepts, in teacher- or student-centred teaching concepts, in sport or movement-based skill concepts, in process and/or product approaches, in diverse and sometimes even contradicting concepts of physical education teacher training. Similar evolutionary developments can be seen elsewhere in Europe, where imported and indigenous ideas have merged in a host of 'melting pots' as the various national systems demonstrate.

These examples illustrate the difficulties not only between countries with different languages but also between countries which share a common language, but which have distinctively different ideological settings determining cultural norms and values. Thus, terminology and terminological issues are pervasive areas of debate within comparative physical education and sport studies, especially in the context of research validity when collecting and interpreting data across cultural divides. 'Back translation' of research instruments such as questionnaire and interview schedules is an imperative in cross-cultural studies. Increasingly sophisticated methodologies drawn from other disciplinary areas are being employed to assist in terminological validation of research data.

4.2. Position Statement(s)

Refer Section 1.3. Body of Knowledge.

4.3. Varia

For the lead comparative physical education and sport domain organisation, the recent resurgence of interest in international issues has coincided with developmental initiatives within and by ISCPES:

- regional conferences in Maracay, Maturin and Rubio,Venezuela, (October 2005) and Varadero, Cuba, (April 2006)
- a new promotional brochure available in 6 languages
- establishment of regional and country representatives
- a graduated annual membership fee system (individual and institutional) based on national economic status
- creation of a web-site (www.ISCPES.org) and review of its constitution.

References

Anthony, D.W.J. (1966). Comparative Physical Education. *Physical Education*, 58 (175), 70-73.

Bennett, B.L. (1970). A Historian Looks At Comparative Physical Education. *Gymnasion*, VII, Spring, 11-12.

Bennett, B.L., Howell, M.L. and Simri, U. (1975). *Comparative Physical Education and Sport*. Philadelphia, Lea & Febiger.

Bereday, G.Z.F. (1964). *Comparative Method in Education*. New York, Holt, Reinhart & Winston.

Hardman, K. (2001). *International Journal of Physical Education*, volume XXXVIII, Issue 3, 3rd Quarter, 96-103.

Hardman, K. (2002). *Council of Europe: Committee for the Development of Sport (CDDS) European Physical Education/Sport Survey*. MSL-IM 16 (2002) 9. Strasbourg, Council of Europe.

Hardman, K. (2007). *Current Situation and Prospects for Physical Education in the European Union*. European Parliament, Directorate General Internal Policies of the Union, Policy Department Structural and Cohesion Policies, Culture and Education. IP/B/CULT/IC/2006_100. PE 369.032. 12/02.

Hardman, K. and Marshall, J.J. (2000). *World-wide Survey of the State and Status of School Physical Education. Final Report*. Manchester, University of Manchester.

Hardman, K., and Marshall, J.J. (2005). *Follow up Survey on the State and Status of Physical Education Worldwide*. 2nd World Summit on Physical Education, Magglingen, Switzerland, 2-3 December.

Hardman, K. and Marshall, J.J. (2006). Update on Current Situation of Physical Education in Schools. *ICSSPE Bulletin*, 47, May.

Johnson, W. (1980). *Sport and Physical Education Around the World*. Champaign, IL, Stipes Publishing Co.

Louis, V. and Louis, L. (1964). *Sport in the Soviet Union*. London, Pergamon Press.

Moehlmann, A.H. (1963). *Comparative Educational Systems*. New York, Center for Applied Research in Education, Inc.

Molyneux, D.D. (1962). *Central Government Aid to Sport and Physical Recreation in Countries of Western Europe*. Department of Physical Education, University of Birmingham.

Morrison, D.H. (1979). Towards a Conceptual Framework for Comparative Physical Education. In M.L.

Morton, H.W. (1953). *Soviet Sport*. New York, Collier Books.

Nixon, J.E. (1970). Comparative, International and Developmental Studies in Physical Education. *ICHPER Journal*, VIII (1), Spring, 4-9.

Petry, K., Froberg, K. and Madella, A. (Eds.). (2004). *Report of the First Year*. AEHESIS Thematic Network Project. Cologne, the Department of Leisure Studies, German Sport University.

Petry, K., Froberg, K. and Madella, A. (Eds.). (2005). *Report of the Second Year*. AEHESIS Thematic Network Project. Cologne, the Department of Leisure Studies, German Sport University.

Petry, K., Froberg, K. and Madella, A. (Eds.). (2006). *Report of the Third Year*. AEHESIS Thematic Network Project. Cologne, the Department of Leisure Studies, German Sport University.

Riordan, J. (1978). *Sport in Soviet Society*. Cambridge, Cambridge University Press.

Sturzebecker, R.L. (1967). Comparative Physical Education. *Gymnasion*, IV, Spring-Autumn, 48-49.

Sullivan, D.T. (1964). *A Comparative Study of Physical Education in the USSR and in England*. Unpublished M.A. Thesis, University of London.

Vendien, C.L. and Nixon, J.E. (1968). *The World Today in Health, Physical Education and Recreation*. Englewood Cliffs, New Jersey, Prentice-Hall Inc.

Doping in Sport

Lauri Tarasti

Contact

Lauri Tarasti
Supreme Administrative Court
Vanhaväylä 33C
00830 Helsinki
Finland
Phone: +358 9 1853315
Email: lauri.tarasti@kolumbus.fi

1. General Information

1.1. Historical Development

Doping is an old phenomenon. Efforts to enhance performance with artificial sub-stances have been known for nearly as long as competitive sport itself. However, doping, as a modern term, appeared first at the end of the 1950s and was defined at the beginning of the 1960s by the International Olympic Committee (IOC). At that time, the control testing technology was able to detect only a few doping substances, mainly stimulants.

Doping's public break-through occurred in the1960s. Although The Council of Europe adopted its first resolution against doping in 1967 (*Resolution on the Doping of Athletes 67/12*), no real doping control existed before sports medicine developed a reliable test for detection of anabolic steroids in the early 1970s. The radioimmunoassay test was first officially used at the Montreal Olympic Games in 1976 but it was a more sensitive detection method based on gas chromatography/mass spectrometry that made testing effective by the early 1980s. The list of prohibited substances had anabolic steroids added to it in 1976, testosterone in the early 1980s, beta-blockers and blood-doping in 1985, diuretics in 1987 and EPO in 1990.

For many years, testing only occurred during competitions. Out-of-competition testing began in the 1980s and the first International Association of Athletics Federations (IAAF) "flying squad" of testing officers started its world-wide testing in 1990. Today, anti-doping organisations conduct approximately 200,000 tests annually, of which the majority are out-of-competition tests.

EPO (erythropoietin) and growth hormone are still difficult substances to be detected and genetic technology will become possible in the near future. The race between doping control and doping substances can be expected to continue.

Doping has attracted a lot of publicity and attention from the media. Sport organisations have taken a strong position against doping and governments have paid much attention to the battle against it. Doping has resulted in many structural consequences in sport. Due to heavy sanctions being imposed for doping, the protection of athletes´ rights have had to be organised. The first internal tribunals in sport were established in 1982 (the IAAF Arbitration Panel) and, at the initiative of the International Olympic Committee, the Court of Arbitration for Sport (CAS) in 1983 mainly in order to deal with doping cases. The CAS is currently the first tribunal for international disputes related to doping.

After many doping scandals in sport, especially in the Tour de France cycling race of 1998, the IOC convened a large anti-doping congress in 1999 in Lausanne, where the World Anti Doping Agency (WADA) was established. This unique international foundation is based on the cooperation between sport organisations and governments. The WADA is also financed and equally lead by sport organisations and governments.

The biggest achievement of the WADA has so far been the world-wide *World Anti Doping Code* (WADC), which was first version accepted in 2003 and updated with a revised version in 2007.

In addition, governments have accepted the UNESCO's *International Convention Against Doping in Sport*, which came into force in 2007. In the Convention, governments are bound to follow the principles of the WADC.

1.2. Function

The idea of the World Anti Doping Code is to harmonize anti-doping rules and measures in the battle against doping in all sports and in all countries. Nearly all international sport federations have accepted the WADC. Therefore WADA and the WADC are in a central position in most matters, both scientific and practical, concerning doping.

This must be kept in mind when pointing out that doping in sport, or the battle against doping, touches many areas of sport science. The starting point is sports medicine where substances, which are prohibited as doping substances, are detected. This happens within the WADA after consultations with all stakeholders. The determination is final and shall not be subject to challenge (Article 4.3.3 of the WADC).

Anti-doping rules are juridical norms and belong to the area of sports law. In fact, these rules have been in an important position ever since sports law developed into its own area of a sport science.

In coaching science, the differences between allowed substances and prohibited doping substances must be understood and, while "chemical training" is an important part of training today, coaches and athletes must have sufficient knowledge to be able to differentiate between the various substances.

Ethics of doping is examined in the philosophy of sport. It is a question of education and sport pedagogy: which values of sport are taught in physical education?

Doping as a phenomenon of modern sport is studied in the sociology of sport and also in sport history and the political science of sport.

Doping also has connections with criminality. Many doping substances are also illegal drugs. The illegal trade of doping substances is wide. The largest group using these substances are body builders, not elite athletes.

1.3. Body of Knowledge

The *revised World Anti Doping Code* was accepted in Madrid in November 2007. The new Code shall apply in full after January 1, 2009. It consists of 25 articles and definitions. Most of them are mandatory for the signatories. However, in a given doping case, the solution must be based on the athlete's own sport federation's anti-doping rules, because sport takes place under the jurisdiction of sport federations (and the IOC).

The WADC binds its signatories, mostly sport federations. Governments are not signatories, although most of them have accepted the principles of the WADC as it has been pointed out in the above-mentioned UNESCO's Convention.

In addition, WADA has accepted mandatory international standards for different technical and operational areas as the international standard for testing, the international standard for laboratories, the prohibited list (of doping substances and prohibited methods) and the international standard for therapeutic use exemptions. WADA can also accept non-mandatory models of best practice and guidelines as recommendations. Such models have been made available to different groups of signatories and for the investigation of "non-analytical" offences.

The juridical side of doping control is finally solved by the CAS acting as the court of last instance. The decisions under the WADC in cases arising from competition in an international event or in cases involving international-level athletes may be appealed exclusively to the CAS. In cases involving national-level athletes the decisions may be appealed to an independent and impartial body in accordance with the rules established by the anti-doping organization concerned.

Governments have accepted to follow the principles of the WADC as said before. But at the same time many governments especially in Europe have included doping crimes in their criminal code or in a separate act. The supervision of doping offences by the authorities has consequently expanded.

Today we can have two sanction systems in doping cases depending on the country concerned:

- sport organizations and their tribunals applying the above-mentioned anti-doping rules mainly based on the WADC with disciplinary sanctions and
- state criminal courts applying state legislation concerning doping punishments.

The difference between private and public law has separated these systems from each other.

1.4. Methodology

Methodology when studying doping depends on the relevant area of sport science.

1.5. Relationship to Practice

See under Sections 1.2. Function and 1.3. Body of Knowledge above.

2. Information Sources

2.1. Journals

WADA publishes a periodical journal "Play True" three times per year. It is their official magazine and is available in magazine format in English and French and in text format in Spanish. See www.wada-ama.org/en/dynamic.ch2?pageCategory.id=274

Some national anti-doping organizations have their own newsletters. Otherwise, doping is dealt with in the journals of various sport sciences especially in Sports Medicine and Sports Law. See these sections in this Directory.

2.2. Reference Books, Encyclopaedias, etc.

WADA has a Digital Library, which is a global Clearing House of available informational and educational anti-doping material created by WADA's stakeholders. See www.wada-ama.org/en/dynamic.ch2?pageCategory.id=540

For some key books on doping, see:

Houlihan, Barrie (1999). *Dying to Win. Doping in Sport and the Development of Anti-Doping Policy*, Strasbourg, Council of Europe.

Siekmann, Robert C.R. and Soek, Janwillem (Eds). (2007). *The Council of Europe and Sport. Basic documents*, The Hague, T.M.S. Asser Press.

Soek, Janwillem (2006). *The Strict Liability Principle and the Human Rights of Athletes in Doping Cases*, The Hague, T.M.S. Asser Press. This includes a list of CAS decisions on doping 1992-2005.

Blackshaw, Ian S., Siekmann, Robert C.R and Soek, Janwillem (Eds). (2006). *The Court of Arbitration for Sport 1984-2004*, The Hague, T.M.S. Asser Press. Includes 33 articles, many of them on doping

For a wide list of reference books, journals etc. see Janwille Soek´s above-mentioned book, The Strict Liability Principle and the Human Rights of Athletes in Doping Cases (2006).

2.3. and 2.4. Book Series and Conference/Workshop Proceedings

See Section 2.2. Reference Books, Encyclopaedias, etc. above.

2.5. & 2.6. Data Banks and Internet Sources

The internet pages of WADA www.wada-ama.org contain some presentations on legal articles, case law and national laws.

The internet pages of the CAS www.tas-cas.org has the most recent decisions rendered by them.

3. Organisational Network

3.1. International Level

WADA

The WADA is the world-wide agency in the battle against doping. The WADA is a private foundation in accordance with Swiss legislation but with its headquarters located in Montreal, Canada. It has four regional offices (Lausanne, Tokyo, Cape Town and Montevideo).

The WADA is financed and lead equally by sport organizations and governments. The WADA's annual budget is approximately US$25 million. The WADA's first president was Mr Richard Pound (Canada), who came from the sport organizations´ side and the second president from 2008 is Mr John Fahey (Australia), from the governments´ side.

In accordance with Article 6.1 of the WADC, all samples taken in doping testing shall be analysed only in WADA-accredited laboratories or as otherwise approved by the WADA. The choice of the WADA-accredited laboratory (or other laboratory or method approved by the WADA) used for the sample analysis shall be determined exclusively by the anti-doping organization responsible for results management. There are now 33 such international laboratories in the world. They have examined around 200 000 tests annually, of which approximately 2% have produced positive findings (i.e. doping detected).

The WADA also maintains a special computer file called ADAMS *(the Anti-Doping Administration and Management System)*, which is a web-based database management tool for data entry, storage, sharing and reporting, designed to assist stakeholders and the WADA in their anti-doping operations in conjunction with data protection legislation.

The Court of Arbitration for Sport (CAS)

The IOC ratified the statutes of the CAS in 1983. The CAS has a very important position as the last juridical appealing instance in international doping cases in accordance with the WADC. The CAS gives the final interpretation on the WADC.

The *Code of Sports-related Arbitration* includes four procedures:

* the ordinary arbitration procedure
* the appeals arbitration procedure (consisting of doping cases)
* the advisory procedure
* the mediation procedure.

The CAS has established ad hoc divisions for large events such as the Olympic Games with the task of settling, within a 24-hour time-limit, any disputes arising during the Games.

The seat of the CAS is in Lausanne and the CAS has regional offices in Sydney and New York.

The CAS has nearly 300 arbitrators of which three usually sit on the Panel for a single case. The CAS has rendered approximately 75 decisions a year.

3.2. Regional Level

Not applicable. See under Section 3.1 International Level above.

3.3. National Level

Nearly all countries have their own national anti-doping organizations and often, more than one. This is because national sport federations are usually also anti-doping organizations, not only in results management and sanctioning, but sometimes also in testing. If a country has a national anti-doping agency, usually financed at least partially by the government, doping control can be centralized and be one of its duties.

Information on doping is often available from the national organizations of sports medicine of over 100 countries. The International Federation of Sports Medicine (FIMS) is the central organization of these national bodies (see Section Sports Medicine in this Directory). Some information on doping may also be available from sports law associations in over 20 countries, and also from the two international sports law organizations (see Section Sports Law in this Directory):

1. International Association of Sports Law (IASL)
   Prof. James Nafziger, President
   Web: www.iasl.org
   Email: info@iasl.org

2. International Sport Lawyers Association (ISLA)
   Dr. Jochen Fritzweiler, President
   Web: www.isla-int.com
   Email: dr.fritzweiler@t-online.de

3.4. Specialised Centres

See under section 3.3. National Level above.

3.5. Specialised International Degree Programmes

There are some such programs available in:

Australia (e.g. the University of Melbourne graduate.law.unimelb.edu.au/index.cfm?objectId=CFC49992-1422-207C-BA8CF019103AB750
United States (e.g. National Sports Law Institute of the Marquette University Law School www.law.marquette.edu, email: matt.mitten@marquette.edu)
and in Europe.

4. Appendix Material

4.1. Terminology

The terminology follows the definitions in the WADC.

4.2. Position Statement(s)

See the internet pages of the WADA (www.wada-ama.org), the IOC (www.olympic.org) and the international sport federations

4.3. Varia

Not applicable.

4.4. Free Statement

Not applicable.

Health-Enhancing Physical Activity

Pekka Oja

Contact

Dr. Pekka Oja
UKK Institute
P.O. Box 30
33501 Tampere
Finland
Phone: +358 3 282911
Email: pekka.oja@uta.fi

1. General Information

1.1. Historical Development

During the past several decades, there has been a progressive decline of physical activity in people's daily living in industrialised countries. For a majority of people, little physical effort is involved any more in their work, domestic chores, transportation and leisure. Driven by the facts that physical inactivity is a major risk factor for the most common non-communicable diseases and that physical activity can counteract many of the ill effects of inactivity, the study of interrelationships between physical activity and health has emerged as a new area of research, closely related to sport sciences.

Although research interest on physical activity and health dates back to the 1950s, the breakthrough in scientific evidence on health benefits of physical activity largely took place in the 1980s and 1990s. "Health-enhancing physical activity" (HEPA) has emerged as a research field, drawing its substance from diverse areas of physical activity and health sciences with strong elements of both basic and applied research. The accumulating evidence-base of the health benefits of physical activity is increasingly being adopted in major health policies of the World Health Organisation, regional organisations, such as the European Commission, and national governments. The current HEPA movement is an open, multi-disciplinary network of scientists, policy makers and practitioners aiming at the realisation of the health potential of physical activity for public health.

1.2. Function

The research field of health-enhancing physical activity has developed in order to provide a knowledge base for understanding of the significance and role of sport, exercise and physical activity for the health, function and well-being of all people. It is a multi-disciplinary research field, encompassing a wide spectrum of sport and health science disciplines. These include established sport science disciplines such as exercise physiology, sport psychology, sociology of sport and adapted physical activity. Pertinent medical and health science disciplines are epidemiology, clinical medicine, rehabilitation, sports medicine, preventive medicine, behavioural medicine, health education and health promotion. In addition, knowledge from related research and policy areas, such as environmental and urban planning, transport and geography, is being increasingly applied in the study and promotion of physical activity and health.

The primary goal of health-enhancing physical activity research is to establish a reliable evidence-base for the promotion of sport , exercise and physical activity for the health and well-being of individuals, communities and populations. The research focuses on establishing the links between physical activity, fitness and health, identifying the determinants of physical activity, developing methods for measuring health-related physical activity and fitness, and developing and evaluating ways to promote physical activity for health.

1.3. Body of Knowledge

Systematic collection, review and analysis of the scientific evidence of the health benefits of physical activity have taken place largely during the past two decades.

Two consecutive consensus conferences in Canada reviewed and evaluated existing evidence on the interrelationships between physical activity, fitness and health (Bouchard et al., 1990 and 1994). A further critical evaluation was conducted by the US Surgeon General (US Department of Health and Human Services, 1996). This report concluded that "Promotion of physical activity is important in the whole population, because it benefits growth and development in children and youth, prevents many diseases in adults, helps maintaining functional capacity in elderly, and supports independent life-style in ageing people".

Subsequent statements by authoritative bodies such as the US Centre for Disease Control and Prevention together with the American College of Sports Medicine (Pate et al., 1995) and the World Health Organisation, together with the Fédération Internationale de Médecine du Sport (FIMS) (1995) put forward recommendations for the promotion of physical activity for public health. More recently, the World Health Organisation (WHO) issued the Global Strategy on Diet and Physical Activity (2004) and guidelines for implementation (WHO, 2006). WHO Europe published the European Charter on Counteracting Obesity in 2006 and a follow-up on physical activity in 2007, and the European Commission has placed physical activity firmly in its public health (EC, 2007a) and sport (EC, 2007b) policies.

Simultaneously, with accumulating evidence of the health benefits of physical activity, understanding of the specific characteristics of health-enhancing physical activity has evolved. The broadly adopted HEPA recommendation by the US Centre for Disease Control and Prevention and the American College of Sports Medicine (Pate et al., 1995) states that "Every US adult should accumulate 30 minutes or more of moderate-intensity physical activity on most, preferably all, days of the week." According to the WHO Global Strategy on Diet, Physical Activity and Health (2004), "Different types and amounts of physical activity are required for different health outcomes: at least 30 minutes of regular, moderate-intensity physical activity on most days reduces the risk of cardiovascular disease and diabetes, colon cancer and breast cancer. Muscle strengthening and balance training can reduce falls and increase functional status among older adults. More activity may be required for weight control." Many countries, such as England (Department of Health, 2004), Switzerland (Swiss Federal Office of Sports, 2004) and Finland (Fogelholm et al., 2006), have recently issued their own recommendations for health-enhancing physical activity, taking into account not only the adult population but also children, adolescents and elderly people. The latest recommendations have been issued jointly by the American College of Sports Medicine and American Heart Association for adults (Haskell et al., 2007) and for elderly people (Nelson et al., 2007).

Estimates of current levels of physical activity in European Union (EU) countries suggest that around two thirds of the adult population do not reach recommended levels of physical activity for health (Sjöström et al., 2006). Activity levels vary widely across the countries ranging, from 45% to 23% of sufficiently active people. Information available from North America, Australia and some other countries suggest a similar situation in many other parts of the world. Thus, a majority of the world's industrialised populations, and increasingly those in the developing world, is insufficiently active for their health and could greatly benefit public health through physical activity.

While the evidence on health risks of inactivity and health benefits of increased activity, as well as the characteristics of HEPA, are becoming well established, the ways in which to effectively increase physical activity in individual, group, community and population levels remains a challenge for the research community. Yet, theoretically based models and practices of physical activity promotion are developing based on the principles of health promotion (Green and Kreuter, 1991), health behaviour change (Glanz et al., 1997) and ecological models of health behaviour (Sallis and Owen, 1999). The latter suggests that in addition to individual factors, the social and physical environments should also be targeted for interventions on population physical activity.

1.4. Methodology

As a multi-disciplinary area, HEPA employs a wide variety of research methods, applicable to the relevant scientific disciplines. These include physiological, psychological, sociological and assessment methodology as used in sport sciences, while epidemiological, clinical and basic medical methodology is utilised in studying the health effects of physical activity. Methods from health education and health promotion research are applied when studying how to change physical activity behaviour of individuals, communities and populations. Increasingly, methods used in transport, environmental and geography research are also being applied.

1.5. Relationship to Practice

Health-enhancing physical activity research is primarily practice-oriented in that its eventual goal is a positive impact on public health. The key HEPA message from the population perspective – a modest increase in daily physical activity – has significant promotional implications on individual, community, environmental and policy level. Physical activity needs to be promoted not only as sports and exercise but also as lifestyle activity that can be incorporated into everyday activities. Promotional measures need to foster cultural, social and environmental support for people to engage in physical activity as part of day-to-day living.

The HEPA research on individual physical activity behavioural change is anchored on a number of behavioural modification theories and models. An integrated construct, the Trans-theoretical Model (Prochaska and Marcus, 1994), provides theoretical bases for individual physical activity counselling and guidance in different settings. Ecological models of health behaviour describe multiple levels of social, cultural and physical environment factors relevant to health behaviour change (Sallis and Owen, 1999) and provide directions for environmental and policy interventions. Brassington and King (2004) recently reviewed the literature on the theoretical bases for HEPA promotion. The current evidence on individual and small group interventions has been reviewed by Biddle (2004), on community interventions by Cavill and Foster (2004), and on environmental interventions by Owen and Salmon (2004).

Examples of high level policy initiatives include the WHO's Global Strategy on Diet, Physical Activity and Health (2004), the European Charter on Counteracting Obesity (WHO Europe, 2006), and the European Union's White papers on nutrition (EC, 2007a) and sport (EC, 2007b).

1.6. Future Perspectives

Although a solid scientific evidence base on the relationships between physical activity, fitness and health has been established, considerable challenges continue to face HEPA research. Of particular importance is to further pursue understanding of the dose-response relationships of physical activity and different health outcomes. This kind of knowledge is needed to effectively apply physical activity as part of health promotion. As much of the current knowledge on the health benefits of physical activity comes from epidemiological studies, more experimental research remains to be done to better understand the dose-response relationships of physical activity and specific health outcomes and the biological mechanisms underlying the changes.

Scientific bases of how to make individuals, communities and populations more active still remains relatively unexplored. In order to fully appreciate the public health potential of physical activity, more research-based knowledge is needed. This requires multi-disciplinary efforts including not only sport and health sciences but also environmental, transportation, urban and community planning, architectural and economical research approaches.

Research concerning health-enhancing physical activity needs to also be practice-driven in the future. It has to serve the needs of decision-makers and professionals who design and implement health policies and practices.

2. Information Sources

Due to the multidisciplinary nature of HEPA, relevant scientific information is published in and accumulates through a variety of sources covering varying areas of sport, medical, health and behavioural sciences.

2.1. Journals

American Journal of Health Promotion
American Journal of Public Health
Annual Review of Public Health
British Medical Journal
Health Education & Behaviour
Health Promotion International
Journal of Ageing and Physical activity
Journal of the American Medical Association (JAMA)
Journal of Epidemiology and Community Health
Journal of Physical Activity and Health
Medicine and Science in Sports and Exercise
New England Journal of Medicine
Patient Education and Counselling
Preventive Medicine
Scandinavian Journal of Medicine and Science in Sports
Sports Medicine

2.2. Reference Books, Encyclopaedias, etc.

Biddle, J.H., Fox, K.R. and Boutcher, H. (Eds.). (2000). *Physical Activity and Psychological Well-Being*. London: Routledge.

Bouchard, C., Blair, S. and Haskell, W. (Eds.). (2007). *Physical Activity and Health*, Champaign, IL. Human Kinetics Publishers, Inc.

Bouchard, C., Shephard, R.J., Stephens, T., Sutton, J.R. and McPherson, B.M. (Eds.). (1990). *Exercise, Fitness and Health. A Consensus of Current Knowledge*. Champaign, IL: Human Kinetics Publishers Inc.

Bouchard, C., Shephard, R.J. and Stephens, T. (Eds.). (1994). *Physical Activity, Fitness and Health. International Proceedings and Consensus Statement*. Champaign, IL: Human Kinetics.

Glanz, K., Lewis, F.M. and Rimer, B.K. (Eds.). (1997). *Health Behavior and Health Education*. 2nd ed. San Fransisco: Jossey-Bass Inc.

Green, L. and Kreuter, MV. (1991). Health Promotion Planning. An Educational and *Environmental Approach*. Palo Alto: Mayfield Publishing Company.

Hardman, A. and Stensel, D. (2003). *Physical Activity and Health: the Evidence explained*. London, Roudledge.

Leon, A.S. (Ed.). (1997). *Physical Activity and Cardiovascular Health: A National Consensus*. Champaign, Ill: Human Kinetics Publishers Inc.

Oja, P. and Borms, J. (Eds.). (2004). *Health Enhancing Physical Activity*. ICSSPE, Perspectives: The Multidisciplinary Series of Physical Education and Sport Science, Vol. 6. Oxford, Meyer & Meyer Sport.

Sallis, J.F. and Owen, N. (1999). *Physical Activity and Behavioral Medicine*. Thousand Oaks, California: Sage Publications.

US Department of Health and Human Services (1999). *Promoting Physical Activity: A guide for Community Action*. Atlanta, GA: U.S Department of Health and Human Services, Centers for Disease Control and Prevention, National Center for Chronic Disease Prevention and Health Promotion.

2.3. Book Series

American College of Sports Medicine. *Exercise and Sport Sciences Reviews*. Williams and Wilkins.

Borms, J., Hebbelink, M. and Hills, A. (Series Editors). *Medicine and Sport Science*. S. Karger.

Dose-response issues concerning physical activity and health: An evidence-based symposium. Medicine and Science in Sports and Exercise (2001). Supplement to 33(6), S345-S641.

Pan EU Survey of Consumer Attitudes to Physical Activity, Body Weight and Health. Public Health Nutrition (1999). 2(1A), 77-160.

Physical Activity Interventions (theme issue). American Journal of Preventive Medicine (1998). 4(15), 255-437.

Promotion of Health-Related Physical Activity among Adults. Patient Education and Counselling (1998). 33 (Suppl. 1), S1-S120.

Physical Activity Guidelines for Adolescents (Special Issue). Sallis, J.F. (Ed.). (1994). Pediatric Exercise Science, 6, 299-463.

2.5. Databases and Electronic Journals

Databases

Cochrane   Systematic reviews in medicine.
                                  More information about the Cochrane collabo-
                                  ration: www.cochranelibrary.com/enter/
                                  Link to the database:
                                  www3.interscience.wiley.com/cgi-bin/
                                  mrwhome/106568753/HOME

Eric   Pedagogy. Available in many different data-
                                  bank services, eg. Ebsco, DataStar, CSA

Medline   Medicine. National Library of Medicine, USA.
                                  Available in many different databank or library
                                  network services, eg. Ovid,
                                  Free of charge:
                                  PubMed www.ncbi.nlm.nih.gov/entrez/query.
                                  fcgi

Cinahl   Nursing and Allied Health. Available in many
                                  different databank and library network
                                  services, eg. Ovid

Pedro   Physical therapy evidence database, free of
                                  charge
                                  www.pedro.fhs.usyd.edu.au/

Physical Education Index/CSA   www.csa.com/

PsycLit/ PsycInfo/ PsycArticles Psychology. Available in many different data-
                                  bank services, eg. DataStar, Ebsco

Sport	Sports sciences, international, maintained in Canada. Available in many different databank services, eg. Ebsco, DataStar
Spolit	Sports sciences, Germany. Available free of charge: www.bisp-datenbanken.de
Sociology.	Available in different databank and library network services, eg. CSA

Databases and full text journals offered by the publishers

Ebsco:
   www.ebscohost.com/
ScienceDirect:
   www.sciencedirect.com/
SpringerLink
   www.springerlink.com/home/main.mpx
Web of Science:
   http://scientific.thomson.com/products/wos/

Open access journals

Free medical journals
   www.freemedicaljournals.com/
Directory of open access journals (DOAJ)
   www.doaj.org/ljbs?cpid=20/
PubMedCentral
   www.pubmedcentral.nih.gov/
PLOS Medicine
   http://medicine.plosjournals.org/perlserv/?request=index-html&issn=1549-1676

2.6. Internet Resources

American College of Sports Medicine
  www.acsm.org/
US Centres for Disease Control and Prevention, Health Topic: Physical Activity and
  Health
  www.cdc.gov/nccdphp/dnpa/physical/index.htm
EU: Nutrition and physical activity (NP) Network
  http://ec.europa.eu/health/ph_determinants/life_style/nutrition/nutrition_en.htm
EU platform for action on Diet, Physical Activity and Health
  http://ec.europa.eu/health/ph_determinants/life_style/nutrition/platform/
  platform_en.htm
European Network for Action on Ageing and Physical Activity (EUNAAPA)
  www.eunaapa.org/
International inventory of documents on physical activity promotion (HEPA Europe,
  WHO)
  data.euro.who.int/PhysicalActivity/?TabID=107125
PELINKS4U: Health, fitness & nutrition
  www.pelinks4u.org/sections/health/health.htm
Sport Information Resource Centre (SIRC). Sportquest, Canada
  www.sirc.ca/online_resources/sportquest.cfm
WHO: Global Strategy on Diet, Physical Activity and Health
  www.who.int/dietphysicalactivity/en/
WHO: Move for Health Day
  www.who.int/moveforhealth/en/
WHO Europe, European Network for the Promotion of Health-enhancing Physical
  Activity:
  www.euro.who.int/hepa

3. Organisational Network

The international health-enhancing physical activity research and promotion community is loosely organised and operates mainly as part of relevant scientific
organisations' activities or as informal and unstructured networks. Only recently
have HEPA focused organisations been established.

3.1. International Level

HEPA research is integrated into the congresses of several international scientific organisations as one actively pursued research topic. These organisations/congresses include:

Fédération Internationale de Médecine du Sport (FIMS)
    www.fims.org
International Society of Behavioural Medicine (ISBM)
    www.isbm.info/
International Society of Behavioural Nutrition and Physical Activity (ISBNPA)
    www.isbnpa.org
World Sport for All Congress organised by IOC
    www.olympic.org/uk/organisation/commissions/sportforall/index_uk.asp

International HEPA networks include:

Global Alliance on Physical Activity
    www.globalpa.org.uk/cdc-who.html
Agita Mundo
    www.agitamundo.org/
International Physical Activity and Environment Network (IPEN)
    www.ipenproject.org

3.2. Regional Level

Research organisations:

American College of Sports Medicine (ACSM)
    www.acsm.org
European College of Sport Sciences (ECSS)
    www.ecss.de/

Networks:

European Network for the Promotion of Health-Enhancing Physical Activity (HEPA Europe)
    www.euro.who.int/hepa
Physical Activity Networks of the Americans (RAFA/PANA)
    www.rafapana.org

3.3. National Level

At the national level, HEPA research is appearing as a new topic on the working agenda of many national scientific organisations in the areas of sport and health sciences. Increasing activities are seen particularly in Europe, North America and Australia.

3.4. Specialised Centres

While HEPA is being introduced increasingly to programs of sport and health science universities and research institutions throughout the world, only a few institutions focus primarily on HEPA research. Three WHO collaborating centres have HEPA as their special focus area:

UKK Institute, Centre for Health Promotion Research, Tampere, Finland
    www.ukkinstituutti.fi
Physical Activity and Health Branch, US Centers for Disease Control & Prevetion, Atlanta, USA
    www.cdc.gov/nccdphp/dnpa
Department of Preventive Medicine and Public Health, Tokyo Medical College, To-kyo, Japan
    www.tokyo-med.ac.jp

4. Appendix material

4.1. Terminology

Not applicable.

4.2. Position Statements/Recommendations

American College of Sports Medicine Position Stand (1998). The recommended Quantity and Quality of Exercise for Developing and Maintaining Cardiorespiratory and Muscular Fitness, and Flexibility in Healthy Adults. *Medicine and Science in Sports and Exercise*, 30:975-991.

American College of Sports Medicine Position Stand (1998). Exercise and Physical Activity for Older Adults. *Medicine and Science in Sports and Exercise*, 30(6), 992.1008.

American Heart Association (1992). Statement on Exercise. *Circulation*, 86(1), 2726-2730.

American Heart Association (2001). Exercise Standards for Testing and Training. A Statement for Healthcare Professionals from the American Heart Association. *Circulation*, 104(14), 1694-1780.

Fletcher, G.F., Balady, G., Froelicher, V.F. et al. (1995). Exercise Standards: A Statement for Healthcare Professionals from the American Heart Association Writing Group. *Circulation*, 91:580-615.

Franklin, B.A., Whaley, M.H. and Howley, E.T. (Eds.). (2000). *ACSM's Guidelines for Exercise Testing and Prescription*. Philadelphia, Pa: Lippincott Williams & Wilkins.

Haskell, W., Lee, I-M., Pate, R., Powell, K., Blair, S., Franklin, B., Macera, C., Heath, G., Thompson, P. and Bauman, A. (2007). Physical Activity and Public Health: Updated Recommendations for Adults from the American College of Sports Medicine and the American Heart Association. *Medicine and Science in Sports and Exercise*. 39(8), 1423-1434.

Nelson, M., Rejeski, W., Blair, S., Duncan, P., Judge, J., King, A., Macera, C. and Castaneda-Sceppa, C. (2007). Physical Activity and Public Health in Older Adults: Recommendation from the American College of Sports Medicine and the American Hearth Association. Medicine and Science in Sports and Exercise. 39(8), 1435-1445.

Pate, R., Pratt, M., Blair, S., Haskell, W.L., Macera, C.A., Bouchard, C. et al. (1995). Physical Activity and Public Health. A Recommendation from the Centres for the Disease Control and Prevention and the American College of Sports Medicine. *Journal of American Medical Association*, 273(5), 402-407.

Sallis, J.F. and Patrick, K. (1994). Physical Activity Guidelines for Adolescents: Consensus Statement. *Pediatric Exercise Science*, 6(4), 302-314.

Saris W.H.M., Blair S.N., van Baak M.A., Eaton S.B., Davies P.S.W., Di Pietro L.,

Fogelholm M., Rissanen A., Schoeller D., Swinburn B., Tremblay A., Westerterp K.R. and Wyatt H. (2003). How much physical activity is enough to prevent unhealthy weight gain? Outcome of the IASO 1st Stock Conference and consensus statement. *Obesity Reviews* 4, 101-114.

Swiss Federal Office of Sports (2004). Health Enhancing Physical Activity Recommendations. www.hepa.ch/gf/gf_baspo/HEPA_recommendations_e.pdf

US Department of Health and Human Services, Centres for Disease Control and Prevention, National Centre for Chronic Disease Prevention and Health Promotion (1996). *Physical Activity and Health: A Report of the Surgeon General.* Pittsburg, Pa: President's Council on Physical Fitness and Sports.

World Health Organisation (WHO) and Federation of Sports Medicine (FIMS) Committee on Physical Activity for Health (1995). Exercise for Health. *Bulletin of the World Health Organisation*, 73(2), 135-136.

4.3. Varia & 4.4. Free Statement

Not applicable.

References

Biddle, S. (2004). Individual and small group interventions. In: Oja, P. and Borms, J. (Eds.) *Health Enhancing Physical Activity.* ICSSPE, Perspectives: The Multidisciplinary Series of Physical Education and Sport Science, Vol. 6. Oxford, Meyer & Meyer Sport.

Brassington, G. and King, A. (2004). Theoretical considerations. In: Oja, P. and Borms, J. (Eds.) *Health Enhancing Physical Activity.* ICSSPE, Perspectives: The Multidisciplinary Series of Physical Education and Sport Science, Vol. 6. Oxford, Meyer & Meyer Sport.

Cavill, N. and Foster, C. (2004). Community interventions. In: Oja, P. and Borms, J. (Eds.) *Health Enhancing Physical Activity.* ICSSPE, Perspectives: The Multidisciplinary Series of Physical Education and Sport Science, Vol. 6. Oxford, Meyer & Meyer Sport.

Cavill, N., Kahlmeier, S. and Racioppi, F. (2006). *Physical activity and health in Europe: evidence for action.* WHO Press, Geneva, Switzerland.

Department of Health, Physical Activity, Health Improvement and Prevention (2004). *At least five a week. A report from the Chief Medical Officer.* www.

dh.gov.uk/PublicationsAndStatistics/Publications

European Commission (2006). *European Union Platform on Diet, Physical Activity and Health.* www.ec.europa.eu/health/ph_determinants/life_style/nutrition/platform/platform_en.htm European Commission (2007 a). *White Paper on Nutrition* www.ec.europa.eu/health/ph_determinants/life_style/nutrition/key-docs_nutrition_en.htm

European Commission (2007 b). *White Paper on Sport* www.ec.europa.eu/sport/whitepaper/wp_on_sport_en.pdf

Green, L.W. and Kreuter, M.W. (1999). Health promotion planning: an educational and ecological approach. 3rd ed., Mountain View, Ca: Mayfield.

Haskell, W., Lee, I-M., Pate, R., Powell, K., Blair, S., Franklin, B., Macera, C., Heath, G., Thompson, P.and Bauman, A. (2007). Physical Activity and Public Health: Updated Recommendations for Adults from the American College of Sports Medicine and the American Heart Association. *Medicine and Science in Sports and Exercise.* 39(8), 1423-1434.

King, A.C. (1994). Community and public health approaches to the promotion of physical activity. *Medicine and Science in Sports & Exercise.* 26:1405-1412.

Nelson, M., Rejeski, W., Blair, S., Duncan, P., Judge, J., King, A., Macera, C. and Castaneda-Sceppa, C. (2007). Physical Activity and Public Health in Older Adults: Recommendation from the American College of Sports Medicine and the American Hearth Association. *Med. Sci. Sports Exrc.* 39(8), 1435-1445.

Oja, P. and Borms, J. (Eds.). (2004). *Health Enhancing Physical Activity.* ICSSPE, Perspectives: The Multidisciplinary Series of Physical Education and Sport Science, Vol. 6. Oxford, Meyer & Meyer Sport.

Owen, N. and Salmon, J. (2004). Environmental interventions. In: Oja, P. and Borms, J. (Eds.) *Health Enhancing Physical Activity.* ICSSPE, Perspectives: The Multidisciplinary Series of Physical Education and Sport Science, Vol. 6. Oxford, Meyer & Meyer Sport.

Prochaska, J.O. and Marcus B.H. (1994). The transtheoretical model: applications to exercise. In: Dishman, R.K. (Ed.). *Advances in Exercise Adherence.* Champaign, IL: Human Kinetics.

Sallis, J.F. and Owen, N. (1997). Ecological models. In: Glans, K., Lewis, F.M. and Rimer, B.K. (Eds.). *Health Behavior and Health Education.* 2nd Edition. San Fransisco: Josey Bass Inc.

World Health Organisation (2004). *Global Strategy on Diet, Physical Activity and Health.* www.who.int/dietphysicalactivity/goals/en/index.html

World Health Organisation, Regional Office for Europe (2006). *European Charter on Counteracting Obesity*. WHO Regional Office for Europe, Copenhagen, Denmark. www.euro.who.int/Document/E89567.pdf

World Health Organisation, Regional Office for Europe (2007). *Steps to Health. A European Framework to Promote Physical Activity for Health*. WHO Regional Office for Europe, Copenhagen, Denmark.

Acknowledgement

Birgitta Järvinen, the librarian of the UKK Institute in Tampere Finland, has kindly provided the information for sections 2.5. Databases and Electronic Journals and 2.6. Internet Resources.

Physical Education

Richard Bailey

Contact

Prof. Dr. Richard Bailey
Physical Education and Sport Group
School of Education
University of Birmingham
Birmingham, B15 2TT
United Kingdom
Email: baileyrichard1@mac.com

1. General Information

This chapter focuses on international developments in Physical Education. In order to avoid duplication with other sections of this Directory, it reports on recent concerns for the state and status of Physical Education. Readers are advised also to consult the chapter on Sport Pedagogy in this Directory, as the two concepts are sometimes considered either synonymous or closely related (for example, Special Interest Groups in both Australia and the United Kingdom, as well as an international research journal, are entitled 'Physical Education and Sport Pedagogy'). More commonly, however, sport pedagogy assumes reference to Physical Education as well as related contexts such as sports coaching. The term Physical Education, initially predominant among Anglophone countries, is usually taken to refer to a narrower range of concerns, that area of the school curriculum concerned with structured and supervised physical activities during the school day (Bailey, 2006).

1.1. Historical Development

Recent years have seen a flurry of policy and advocacy activity directed at claiming and securing the place of Physical Education in the school curriculum. To some extent, however, concerns for the place of the subject can be traced back many years to the origin of formal schooling (Kirk, 1992). This seems to be a legacy of ancient and enlightenment thinking that taught that the body is, in Plato's words, the 'grave of the soul', and that the body and soul are not only separate but opposed and unequal. Descartes' view that 'the human body may be considered as a machine' neatly captures the dominant understanding of the role of the body in education and development: it needs to be maintained, but such maintenance surely does not constitute the real business of schooling any more than it is of life (Best, 1978). Thus Physical Education has traditionally been interpreted by curriculum designers and central governments in terms of body maintenance, rather than learning. Barrow (1982) captures the spirit of this relegation of Physical Education to a marginal position when he writes that it surely qualifies as a part of schooling, since it contributes to children's health and fitness, but it does not qualify as a proper educational activity. It is not surprising then, that when educational institutions are forced to prioritise or economise, those subjects at the margins are most likely to be sacrificed.

It was during the 1970s and 1980s that academics and practitioners from around the world began reporting a decline in the position and security of Physical Education. The year 1978 saw the creation of the International Charter of Physical Education and Sport, which was created under the auspices of the United Nations Educational, Scientific and Cultural Organisation (UNESCO). The Charter stated a set of Articles including the following:

Article 1.   The practice of physical education and sport is a fundamental right for all
Article 2.   Physical education and sport for an essential element of lifelong education in the overall education system
Article 3.   Physical education and sport programmes must meet individual and social needs
Article 11. International co-operation is a prerequisite for the universal and well-balanced promotion of physical education and sport.

Yet, despite the UNESCO Charter, there was a continuing decline and in some cases disappearance of the subject in the 1990s which resulted in widespread anxiety and prompted a series of conferences, advocacy statements and lobbying activities (Hardman, 2006). Perhaps the most significant of these was the World Summit on Physical Education, which was initiated by the International Council for Sport Science and Physical Education (ICSSPE). The World Summit took place in Berlin in November, 1999, with the support of a host of international agencies including the International Olympic Committee (IOC), UNESCO, and the World Health Organisation (WHO). The Summit was a coming together of individuals and groups from many countries seeking to assert the value of Physical Education, and to reverse what was perceived as an unprecedented diminishment of its security within school curricula. A central element of the Summit was a report of the first worldwide survey of the "state and status of Physical Education" (Hardman and Marshall, 1999; 2000). This survey reported a precarious predicament of Physical Education within schools, and suggested that, in some countries, the elements necessary for even basic provision - protected time allocation, accessible curriculum content, suitable resources and facilities and trained teachers - were simply not present. In other countries, whilst Physical Education enjoyed a relatively secure curriculum position, questions were raised about the quality of children's experiences, and about the equity of the opportunities presented to children from different groups. The authors of the survey concluded that 'physical education has been pushed into a defensive position. It is suffering from decreasing curriculum time allocation, budgetary controls with inadequate financial, material and person-nel resources, has low subject status and esteem, is being ever more marginal-ised and undervalued by the authorities' (Hardman & Marshall, 1999).

The Worldwide Survey was accompanied by a set of position papers that argued the 'case for Physical Education (see for example, Talbot, 1999). Taken together, these documents prompted the Summit to conclude with the formulation of Action Agendas. In the same month as the World Summit, an Appeal was made to the UNESCO General Conference and the third Meeting of Ministers and those respon-sible for Physical Education and Sport (MINEPS III) in Punta del Este, Uruguay. The final report of MINEPS III articulates a situation that would continue to occur over the course of the following decade, namely a significant difference between policy and practice. The report stated that "twenty-one years after the proclamation of the International Charter, physical education and sport at school did not seem to be national priorities and were often the target of budget cuts" (Savolainen, 1999, pp. 3-4).

Since then there have been a number of international, regional and national de-velopments which, on the one hand, have reasserted the dangers of the reduced status of Physical Education within school curricula, and on the other, moved the discussion forward articulating a programme of international and national develop-ment. Examples of this progression include the Regional Seminar in Africa and for Latin America and the Caribbean (Bamako and Havana in 2003), MINEPS IV (Ath-ens, Greece, 2004) and the Porto Novo Draft Quality Reference Framework (Benin 2005). To some extent multi-national agencies recognised the relevance of these initiatives. It may be claimed that this is reflected in the fact that 2004 was named the European Year of Education through Sport, closely followed by the United Na-tion's declaration of 2005 as the International Year for Sport and Physical Educa-tion. A supplementary approach was taken by ICSSPE when it commissioned the Sport in Education (SpinEd) project, which was an international research study ulti-mately aiming "to gather and present evidence to policymakers regarding the ben-efits to schools of good quality physical education and sport" (Bailey and Dismore, 2004). The findings of the SpinEd project were presented at the 2004 Pre-Olympic Congress in Thessaloniki, Greece, and the European Year Closing Manifestation, and furthermore, the findings provided a framework for the Physical Education and Sport element of MINEPS IV. The headline findings were that Physical Education had the potential to make distinctive contributions to a range of educational and societal values, including physical health, the development of social skills, the improvement of emotional and affective wellbeing, and could relate to improved academic performance.

Despite these initiatives, there continues to be cause for concern. A Second World Summit took place in Magglingen, Switzerland, in 2005 and included presentations on the international situation of Physical Education from two groups of researchers: Hardman and Marshall (2006) provided an update of their Worldwide Survey, and Pühse and Gerber (2006) introduced a new study. Despite the differences of methodology and sources of evidence, both of these studies reported that, whilst some countries had made good progress, many continued to demonstrate poor quality or no Physical Education provision.

1.3. Function

There is no internationally agreed definition or statement of function of Physical Education. However, Bailey and Dismore (2004) surveyed more than fifty countries to generate a 'functional definition' (that is a description of what happens rather than an analytical account), as follows: "Physical Education refers to those structured, supervised physical activities that take place at school and during the school day".

1.4. Body of Knowledge

Physical Education research draws on an extremely wide range of bodies of knowledge. To provide a concise overview, given the focus of this Chapter, the main areas of research include the following:

Comparative Physical Education and Sport (e.g., Hardman and Marshall, 2000; Pühse and Gerber, 2006)
Philosophical Justifications for Physical Education (e.g., McNamee, 2006; Siedentop, 2002)
Didactics and Pedagogy (e.g., Amade-Escot, 2000; Kirk and Macdonald, 1998; Rink, 2001)

Gender (e.g., Oliver and Lalik, 2001; Gieß-Stüber, 2006)

Inclusive / Adapted Physical Education (e.g., Fitzgerald, 2005; DePauw and Doll-Tepper, 1989)

Policy Analysis (e.g., Houlihan, 2002; Hummel and Schierz, 2006)

Empirical Reviews of Outcomes of Physical Education (e.g., Bailey, 2006; Bailey, Armour, Kirk, Jess, Pickup & Sandford, 2006).

1.5. Methodology

Kirk, Macdonald and O'Sullivan (2006) provide a summary of the dominant methodological perspectives in international Physical Education research. Of course, a list like this cannot be sensibly considered comprehensive, but it does provide a useful starting point. The methodological perspectives include:

* Behaviour analysis, which is concerned with the explanation and description to enable the prediction and control of behaviour (Ward, 2006);
* Interpretive perspectives, such as symbolic interactionism, hermeneutics and phenomenology (Pope, 2006);
* Socially critical perspectives, which aim for the emancipation of those involved, and leads to radical challenges in their conclusions and practices (Devís- Devís, 2006);
* Postmodern, poststructural and postcolonial perspectives (Wright, 2006); and
* Feminist perspectives (Nilges, 2006).

1.6. Relationship to Practice

Physical Education might be said to have greatest value and justification in a balance between theory and practice: it problematises and theorises the real world of teachers and students; and draws out practical implications for theoretical advances. If this is accepted, then a perusal of the academic literature suggests that the balance has not been struck in recent years, since the great majority of articles bear only a marginal relationship to actual Physical Education lessons. In English language research in particular, sociological and social policy theorising has dominated the literature; pedagogy competes neither in terms of quantity of papers nor influence on other researchers.

1.7. Future Perspectives

Expressions of concern for the marginal state and status of Physical Education are likely to continue for some years, as there is little evidence of a large-scale acceptance of the need for a compulsory, core Physical Education in most countries. The most plausible mechanisms for such a change to take place include a greater understanding of the political dimension of educational change, and the continued accumulation of evidence for the benefits of Physical Education for young people and for schooling systems.

References

Bailey, R.P. (2006). Physical Education and Sport in Schools: A review of benefits and outcomes. *Journal of School Health*, 76(8), pp. 397-401.

Bailey, R.P. and Dismore, H.C. (2004). Sport in Education (SpinEd): Initial report of findings. Invited presentation at the 2004 Pre-Olympic Congress, Thessaloniki, Greece.

Bailey, R.P., Armour, K., Kirk, D., Jess, M., Pickup, I. and Sandford, R. (2006). *The Educational Benefits Claimed for Physical Education and School Sport: An academic review*. Macclesfield: British Educational Research Association.

Barrow, R. (1982). *The Philosophy of Schooling*. Brighton: Wheatsheaf.

Best, D. (1978). *Philosophy and Human Movement*. London: Allen and Unwin.

Devís- Devís, J. (2006). Socially Critical Perspectives in Physical Education (pp. 37-58), in: D. Kirk, D. Macdonald and M. O'Sullivan (Eds) (2006). *The Handbook of Physical Education*, London, Sage.

Fitzgerald, H. (2005). Still feeling like a spare piece of luggage? Embodied experiences of (dis)ability in physical education and school sport. *Physical Education and Sport Pedagogy*, 10(1), pp. 41-59.

Hardman, K. (2005). Foreword. In U. Pühse and M. Gerber (eds), *International Comparison of Physical Education*. Oxford: Meyer and Meyer.

Hardman, K. (2001). World-wide Physical Survey on the State and Status of Physical Education in Schools (pp. 15-36). In G. Doll-Tepper and D. Scoretz (eds) *World Summit on Physical Education*. Berlin: International Council for Sport Science and Physical Education.

Hardman, K. and Marshall, J. (2000). *World-wide Survey of the State and Status of School Physical Education: Final report to the International Olympic Commit-*

tee. Manchester, UK: University of Manchester.

Houlihan, B. (2002). Sporting Excellence, School and Sports Development: The politics of crowded policy spaces. *European Physical Education Review* 6 (2), 171-194.

Kirk, D. (1992). Defining Physical Education: *The social construction of a school subject in postwar* Britain. London: Falmer.

Kirk, D. and Macdonald, D. (1998). Situated Learning in Physical Education. *Journal of Teaching in Physical Education*, 17 (3), 376-387.

Kirk, D., Macdonald, D. and O'Sullivan, M. (2006). *The Handbook of Physical Education*, London, Sage.

McNamee, M. (2005). The Nature and Values of Physical Education. In Green, K. and Hardman, K. (Eds), *Physical Education: Essential issues*, London Sage Publications.

Nilges, L. (2006). Feminist Strands, Perspectives, and Methodology for Research in Physical Education, in: D. Kirk, D. Macdonald and M. O'Sullivan (Eds) (2006). *The Handbook of Physical Education*, London, Sage.

Oliver, K.L. & Lalik, R. (2001). The Body as Curriculum: Learning with adolescent girls. *Journal of Curriculum Studies* 33 (3), 303-333.

Pope, C. (2006). Interpretive Perspectives in Physical Education Research (pp. 21-36), in: D. Kirk, D. Macdonald and M. O'Sullivan (Eds) (2006). *The Handbook of Physical Education*, London, Sage.

Rink, J. E. (2001). Investigating the Assumptions of Pedagogy. *Journal of Teaching in Physical Education*, 20, 112-128.

Savolainen, K. (1999). Third International Conference of Ministers and Senior Officials Responsible for Physical Education and Sport (MINEPS III): Final Report. Paris: UNESCO.

Wright, J. (2006). Physical Education Research from Postmodern, Poststructural and Postcolonial Perspectives (pp. 59-75), in: D. Kirk, D. Macdonald and M. O'Sullivan (Eds) (2006). *The Handbook of Physical Education*, London, Sage.

Siedentop, D. (2002). Junior Sport and the Evolution of Sport Cultures. *Journal of Teaching in Physical Education*, 21 (4), 392-401.

Talbot, M. (2001). The Case for Physical Education (pp. 39-50). In G. Doll-Tepper & D. Scoretz (eds), *World Summit on Physical Education*. Berlin, International Council for Sport Science and Physical Education.

Ward, P. (2006). The Philosophy, Science and Application of Behavior Analysis in Physical Education (pp. 3-20), in: D. Kirk, D. Macdonald and M. O'Sullivan (Eds) (2006). *The Handbook of Physical Education*, London, Sage.

2. Information Sources

Readers are encouraged to consult the Sport Pedagogy chapter in this Directory for a more general coverage of relevant information sources for Physical Education. The sections that follow seek to highlight sources that are especially relevant to the focus of this chapter.

2.1. Journals

Most specialist journals report findings related to the state and status of Physical Education in schools, as do those from related areas such as sport science, education and health. Those that have included relevant findings in recent years include the following:

Journal of Teaching Physical Education (US);
QUEST (US);
Research Quarterly for Exercise and Sport (US)
Journal of School Health (US);
AVANTE (Canada);
Sport, Education and Society (UK);
Physical Education and Sport Pedagogy (UK);
International Journal of Physical Education (Germany);
African Journal of Physical Education, Recreation and Dance (South Africa);
Asian Journal of Physical Education (Taiwan);
International Journal of Sport and Health Sciences (Japan);
Science et Motricité (France);
Sportwissenschaft (Germany);
Healthy Lifestyles Journal (Australia).

2.2. Reference Books, Encyclopedias, etc.

The outstanding reference resource for research in Physical Education is:

Kirk, D., Macdonald, D. and O'Sullivan, M. (2006). *The Handbook of Physical Education*. London: Sage.

Other worthwhile sources of information are 'Readers', or collections of important articles. Three readers have been published in recent years in English:

Green, K. and Hardman, K. (Eds) (1998). *Physical Education: A reader*. Aachen: Meyer and Meyer.

Green, K. and Hardman, K. (Eds) (2005). *Physical Education: Essential issues*, London: Sage Publications.

Bailey, R.P. and Kirk, D. (2008). *The Routledge Reader in Physical Education*. London: Routledge.

2.3. Book Series

Most of the major publishers of Physical Education literature (e.g., Routledge, Hofmann, Meyer and Mayer, Human Kinetics) produce series related to Physical Education. A recent international series, which focuses on multi-national perspectives on Physical Education is International Studies in Physical Education and Youth Sport (Routledge).

2.4. Conferences/Workshops Proceedings

A landmark conference report was the proceedings of the first World Summit in Berlin:

Doll-Tepper, G. & Scoretz, D. (eds) (2001). World Summit on Physical Education. Berlin, International Council for Sport Science and Physical Education.
This contains the key findings of the Worldwide Survey, as well a number of advocacy statements from leading international thinkers.

Other valuable information can be found in the proceedings of conferences of organisations that constitute the International Council for Sport Pedagogy (ICSP). These organisations are listed in section 3.1.

2.5. Data Banks

Apart from the generic sport science databases described elsewhere, there are no specific databases for Physical Education. However, two important sources of information on the state of Physical Education around the world are:

Pühse, U. and Gerber, M. (Eds) (2006). International Comparison of Physical Education. Oxford: Meyer and Meyer.
Hardman, K. and Marshall, J. (2000). World-wide Survey of the State and Status of School Physical Education: Final report to the International Olympic Committee. Manchester, UK: University of Manchester.
Both of these contain reports from informants regarding the position of Physical Education at the time of writing.

2.6. Internet Sources

SportQuest
    www.sirc.ca
ICSSPE Conference Calendar
    www.icsspe.org
Physical Education - The Role of Physical Education and Sport in Education (SPINED) -
    spined.cant.ac.uk
PE Central -
    www.pecentral.org
Physical Education Digest -
    www.pedigest.com
Physical Activity and Public Health On-Line Network. Please contact:
    PHYS-ACT@VM.SC.EDU
Australian Physical Education Discussion Listserve. Please contact:
    Austpe-l@hms.uq.edu.au
Intervention en Education Physique et Sports (eJRIEPS)
    www.fcomte.iufm.fr
Sportpaedagogik Online
    www.sportpaedagogik-online.de

3. Organisational Network

3.1. International Level

The main umbrella organisation for Physical Education organisations is the International Committee for Sport Pedagogy (ICSP), which operates under the auspices of ICSSPE. Its membership includes the five subject-specific associations:

- the Association Internationale des Ecoles Supérieures d'Education Physique (AIESEP);
- the Fédération Internationale d'Education Physique (FIEP);
- the International Association of Physical Education and Sport for Girls and Women (IAPESGW);
- the International Federation of Adapted Physical Activity (IFAPA); and
- the International Society for Comparative Physical Education and Sport (ISCPES).

The International Council for Health, Physical Education, Recreation, Sport and Dance (ICHPER-SD) is fairly large organisation with an international membership. Its structure and scope is somewhat different to that of ICSP and, indeed, ICSSPE, as it is not an umbrella organisation for other groups.

3.2. Regional Level

Both ICSSPE and ICHPER-SD utilise regional networks to promote their work and support the role of Physical Education. There are also region-specific groups, such as the European Physical Education Association (EPEA), the African Association for Health, Physical Education, Recreation, Sport and Dance (AFAHPERD) and the American Alliance for Health, Physical Education, Recreation and Dance (AAH-PERD).

Some regional or national groups of Physical Education researchers form Special Interest Groups (SIG). Among the most active are the British Educational Research Association's Physical Education and Sport Pedagogy SIG, the American Educational Research Association's Research on Learning and Instruction SIG and the Australian Association for Research in Education's Health and Physical Education SIG.

3.3. Specialised Centres

Please consult the Sport Pedagogy chapter.

3.4. Specialised International Degree Programmes

A European Physical Education Masters has been in place since 2006 and is of-fered by four partner universities in Italy, Austria, Denmark and Norway with asso-ciate partners located in Germany and the UK. The two-year programme includes coursework and an internship.

4. Appendix Material

4.1. Terminology

Definitions of Physical Education vary around the world and are often debated in academic forums. Definitions and interpretations of Physical Education are, at times, assumed and not clearly specified. It is typically understood that physical education takes place in a structured and supervised manner in the school set-ting. The 'functional definition' from Bailey & Dismore (2004) best represents this approach (see section 1.3). Definitions and terminology related to Physical Educa-tion are also discussed and challenged in the related fields of Sport Pedagogy and Comparative Physical Education.

4.2. Position Statement

There are several position statements, ranging for example from the UNESCO International Charter of Physical Education and Sport (1978), to the World Sum-mits on Physical Education (Berlin, November, 1999 and Magglingen, December 2005: see www.icsspe.org). In addition, FIEP, a member association of the ICSP published "A new concept of Physical Education" in 2000/01.

4.3. Varia & 4.4. Free Statements

Not applicable.

Sport and Development

Jackie Lauff, Bert Meulders and Joseph Maguire

Contact

Jackie Lauff
Freie Universität Berlin
Unit for Inclusive Education, Physical Activity and Sport
Fabeckstrasse 69
Berlin, Germany
Email: jakwak@hotmail.com

Bert Meulders
Department Humane Kinesiologie
K.U.Leuven
Tervuursevest 101
B-3001 Leuven, Belgium
Phone: +32 16 329056
Fax: +32 16 329196
Email: Bert.Meulders@faber.kuleuven.be

Prof. Dr. Joseph Maguire
School of Sport & Exercise Sciences
Loughborough University
LE11 3TU Loughborough, United Kingdom
Phone: +44 1509 223328
Fax: +44 1509 223971
Email: J.A.Maguire@lboro.ac.uk

1. General Information

1.1. Historical Development

Sport and Development is an emerging discipline within sport science and physical education. As an academic field this area is very new and is attracting much interest around the world. In the 1990's, early pioneers in this field included organisations such as SCORE, MYSA and Right To Play that began to use sport in a developmental context.

Over the last decade, sport and development has become an established policy area with international recognition. World leaders began to recognise the power of sport and its values at the 2000 United Nations Millennium Summit and at the 2002 Special Session on Children (UNICEF, 2008). In 2002, international attention intensified with the creation of a UN Inter-Agency Task Force of ten organisations that convened to review activities involving sport within the UN system. The Task Force developed a comprehensive report titled, "Sport for Development and Peace: Towards Achieving the Millennium Development Goals," which concluded that sport is a powerful and cost-effective way to advance the Millennium Development Goals. (United Nations, 2003).

A number of high-profile conferences emerged and these served to further reinforce the role of sport as a means to promote health, education, development and peace. In February 2003, the first International Conference on Sport and Development took place in Magglingen, Switzerland which highlighted that development organisations and sports organisations not only share common goals, but also common objectives and could largely benefit from world-wide cooperation. Another international conference, the 'Next Step,' took place in Amsterdam, the Netherlands with the aim of gathering practitioners and planners together to focus on the development benefits of investing in sport.

At the end of 2003, the UN General Assembly adopted a resolution (58/5) entitled, "Sport as a means to promote Education, Health, Development and Peace," calling on governments, UN funds and programmes and sport-related institutions to promote the role of sport and physical education for all when furthering their development programmes and to include sport and physical education as a tool to contribute to broader aims of development and peace (United Nations, 2003).

In 2004, Right to Play co-hosted a Roundtable Forum entitled, "Harnessing the Power of Sport for Development and Peace" during the (XXVIII) Olympic Games in Athens, Greece. The Roundtable Forum brought together political leaders and experts in development to showcase the potential of sport in achieving social, economic, health and development goals and thus initiating the first steps towards the creation of a policy framework for sport and development around the world. The main overall outcome of the Athens Roundtable Forum was the establishment of the Sport for Development and Peace International Working Group as a four-year policy initiative to articulate and promote the adoption of sport and development policies by governments (SDPIWG, 2008). Also in 2004, the UN adopted a second Resolution on Sport and Development (59/10) which expanded on these goals to further include aspects of youth, gender equality and social inclusion (United Nations, 2004).

The United Nations declared 2005 the International Year of Sport and Physical Education which provided a unique opportunity to focus the world's attention on the importance of sport in society and on how sport and physical education programmes can be used as tools to help combat challenges such as extreme poverty, conflict and HIV/AIDS and help achieve the Millennium Development Goals.

Building on previous international policy developments in sport science and physical education and a wealth of sport development experience, these latest international milestones have connected the policy areas of sport and development cooperation in recognition of their similar goals and purpose. Moreover, these developments have stimulated a rapid growth in the implementation of sport and development projects world-wide and engaged a broad spectrum of organisations, institutions and agencies.

Function Terms

The function of sport and development is more than developing sporting systems, infrastructure and opportunities, although this was central to the activities of the first so-called sport and development NGOs. The aim of Sport and Development is to work in partnership with people, organisations and agencies across the world to help assist countries with the creation of their own sporting systems, promote the power of sport as a tool for human and social development and form and maintain strategic partnerships with international partners. Sport and development terminology is used to cover many different areas and most generally it is used to indicate the use of sport to promote and address specific societal goals. These goals include areas such as health, economic development, gender, peace, disability, trauma and child development.

Development is a process of enlarging people's choices and increasing the opportunities available to all members of society. Based on the principles of inclusion, equity and sustainability, emphasis is on the importance of increasing opportunities for the current generation as well as generations to come. Sport can impact on development outcomes and help to build these capabilities in individuals and communities.

According to the Final Report of the Sport for Development and Peace International Working Group:

Strong programs combine sport and play with other non-sport components to enhance their effectiveness and are delivered on an integrated basis with other local, regional and national development and peace initiatives so that they are mutually reinforcing. Programs seek to empower participants and communities by engaging them in the design and delivery of activities, building local capacity, and pursuing sustainability through collaboration, partnerships and coordinated action.

(SDPIWG, 2008)

Sport and physical education play a vital role at all levels of society. For the individual, sport enhances one's personal abilities, general health and self-knowledge. On the national level, sport and physical education contribute to economic and social growth, improve public health, and bring different communities together. On local and global levels, if used consistently, sport and physical education can have a long-lasting positive impact on development, public health, peace and the environment. Sport is a social and cultural phenomenon which can have positive and negative effects in different contexts. The potentially negative impacts of sport need to be safeguarded against to ensure that sport is in fact meeting identified objectives. Negative aspects of sport include doping, corruption, child labour amongst others and if used in the wrong way, sport can have detrimental effects on development outcomes.

Access to and participation in sport and physical education provide an opportunity to enjoy social and moral inclusion for populations otherwise marginalised by social, cultural or religious barriers. Cooperative sport programmes can play an important role in peace-building, conflict resolution and social inclusion. Through sport and physical education, individuals can experience equality, freedom and a dignifying means for empowerment, which can be particularly valuable for girls and women, people with disabilities, people living in conflict areas and those recovering from trauma (NCDO, 2004).

1.2. Body of Knowledge

The academic field of Sport and Development is relatively young and can be considered in its 'embryonal' stages. A body of evidence has been developed through research, project evaluations and case studies. In 2006, the Sport for Development and Peace International Working Group collated the existing literature and research surrounding Sport and Development in the following categories:

Individual Development
Health Promotion and Disease Prevention
Promoting Gender Equality
Social Integration and the Development of Social Capital
Peace Building and Conflict Prevention or Resolution
Post-Disaster Trauma Relief and Normalisation of Life
Economic Development
Communication and Social Mobilisation

Expanding on this collection of evidence, a series of literature reviews were conducted in 2007 by the University of Toronto, commissioned by the Sport for Development and Peace International Working Group (SDPIWG, 2007).

1.3. Methodology

There is no uniform methodology that is used in the implementation or evaluation of Sport and Development projects. Depending on the specific goals, a project may include methodologies from various fields. For example, a project that uses sport to achieve health outcomes might use methods from health promotion. Similarly, a project using sport to educate and raise awareness of HIV/AIDS prevention might utilise methods from epidemiological studies.

Strategies and methods of monitoring and evaluation of Sport and Development programmes are appearing as governments, institutions and organisations strive to build on the evidence base of Sport and Development. Empowerment, participatory approaches and sustainability are very central themes in the discourse of best practices and whilst these are not evaluation instruments, they are key considerations in programme development. Influences of development theory are also found in applications of logical framework approaches and innovative evaluation methods such as participatory video techniques (Biddle, 2006).

As Sport and Development continues to develop as an academic discipline, influences from other fields such as sociology, psychology and sport management will continue to influence data collection, analysis and the interpretation of Sport and Development research.

1.4. Relationship to Practice

While the reach of global sport is perhaps not in question, the actual impact of modern achievement sport has been and continues to be across the planet and whether it is conducive to the United Nations Millennium Goals needs to be assessed (Maguire, 2005). After a period of rapid growth, there have been claims that sport and development projects need to introduce quality measures to ensure that they do, in fact, meet intended development aims (van Eekeren, 2006).

The absence of a strong body of compelling evidence in support of Sport and Development is repeatedly identified as a barrier to convincing policy makers and private sector donors to increase support for the field. (SDPIWG, 2006). Moreover, sport has received virtually no academic attention in mainstream development texts (Desai & Potter, 2002; Potter, 2004).

Monitoring and evaluation is gaining increasing attention and new tools are being developed to enable more effective data collection and analysis of programme effectiveness (Coalter, 2006). Given the large number of existing sport and development projects across the world, there is much potential for academic research to collaborate with existing stakeholders. A realistic picture of global sport and its impact on people, nations and civilisations across the globe is needed.

1.5. Future Perspectives

With the international focus and attention on sport and development, there needs to be increased research output to evaluate and guide policy development in this growing area of development. Closer working relationships between the sport sector and the development sector are essential in creating cooperative research and development. Stronger links are also emerging between the sport and business sectors (May & Phelan, 2007).

Furthermore, the Sport for Development and Peace International Working Group identify a number of future needs for the Sport and Development arena to strengthen network infrastructure including:

Articulation of a common vision and tangible aims
A framework for action for stakeholders from all sectors
Opportunities for sharing evidence, success stories and failures
Development in progress monitoring and reporting
Coordinated research and scientific investigation
Greater emphasis on the power of sport and its impact on women and people with disabilities

Recent years have seen great expansion in networking, online collaboration and sharing of information, knowledge and project examples. Tertiary training programmes are needed to further enhance the quality of programme delivery, the capacity of project staff and the volume of evidence-based research.

2. Information Sources

2.1. Journals

There are no journals dedicated to this specific topic. Research on this topic does, however, appear in a range of social sciences, sport science and physical education journals.

2.2. Reference Books

Coalter, F. (2007). A wider social role for sport: Who's keeping the score? Routledge.

Coalter, F. (2006). Sport-in-Development: A Monitoring and Evaluation Manual, London, UK Sport.

Keim, M. (2003). Nation-building at play. Meyer and Meyer, Germany.

Chappell, R. (2007). Sport in Developing Countries, Roehampton University, UK.

2.3. Book Series

Not applicable.

2.4. Conference/Workshop Proceedings

The outcomes of various Sport and Development Conferences are contained in proceedings, reports and declarations.

UK Sport. (2007). Report on the Next Step Conference in Windhoek, Namibia 2007.

Vanden Auweele, Y., Malcolm, C. and Meulders, B. (Eds.) (2006). Sport and Development. Cape Town: Iannoo Campus.

Second Magglingen Conference Sport and Development, November, 2005.

Sport and Development Conference (2005). Economy, Culture and Ethics, Bad Boll, Germany.

UK Sport. (2005). Report on the Next Step Conference in Zambia.

Magglingen Declaration, 2005.

Magglingen Declaration and Recommendations, 2003.

2.5. Data Banks

An International Platform on Sport and Development is maintained by the Swiss Academy for Development and serves as an information resource centre dedicated entirely to Sport & Development. It is also a communication tool for those with an interest in Sport & Development to share ideas, information and experience.

International Platform on Sport and Development
www.sportanddev.org

2.6. Internet Sources

International Olympic Committee
    www.olympic.org/uk/organisation/missions/humanitarian/index_uk.asp
Literature Reviews on Sport for Development and Peace
    http://iwg.sportanddev.org/data/htmleditor/file/Lit.Reviews/literaturere-viewSDP.pdf
Sport for Development and Peace International Working Group
    http://iwg.sportanddev.org
Sport for Development and Peace: Preliminary Report
    http://iwg.sportanddev.org
Sport and Development
    www.sportdevelopment.org (Netherlands site)
Sport and Development Toolkit
    www.toolkitsportdevelopment.org
UN Report on Sport for Development and Peace
    www.worldvolunteerweb.org
    USA_SportDevelopment.pdf
UN Office of Sport for Development and Peace
    www.un.org/themes/sport/
Value of Sport Monitor
    www.uksport.gov.uk/pages/international_development/

3. Organisational Networks

3.1. International Level

Numerous United Nations Funds and Programmes, international government and non-governmental organisations are actively involved in Sport and Development. These include:

ICSSPE
ICSA
International Labour Office (ILO)
International Olympic Committee (IOC)
International Paralympic Committee (IPC)
FIFA
Kicking Aids Out
NCDO
Netherlands Ministry of Health, Welfare and Sport (VWS)
Netherlands Olympic Committee and Confederation of Sport (NOC*NSF)
Platform on Sport and Development Cooperation
Right to Play
Sports sans frontiers
Streetfootballworld
SCORE
Swiss Agency for Development Cooperation (SDC) / Swiss Academy for Development (SAD)
UEFA
UN Programmes and Funds
UN Office of Sport for Development and Peace
UNHCR
UNICEF
UNDP
UNEP

3.2. Regional Level

At the regional level, some developing countries are leading the way with creating regional networks for Sport and Development. Additionally, further resources are being developed by individual regions and these details can be found in the Sport for Development and Peace International Working Group's Interim Report.

3.3. National Level

Australian Sports Commission
Commonwealth Games Canada
FK Norway
KNVB
MYSA
National Sports Council Zambia
Norwegian Olympic Committee and Confederation of Sports (NIF)
South African Sports Commission
Sport Canada
UK Sport

3.4. Specialised Centres

University of Toronto
Katholieke Universiteit Leuven
University of Stirling

3.5. Specialised International Degree Programmes

Not applicable.

4. Appendix Material

4.1. Terminology

As with other academic disciplines, terminology and definitions are often challenged and critiqued and there are many different interpretations of the concept of Sport and

Development. The terms 'Sport and Development', 'Sport for Development', 'Sport for Development and Peace' are often used interchangeably in practice and there is ongoing academic debate surrounding these definitions. The terms 'Sport Plus' and 'Plus Sport' offer accepted differentiations regarding the role of sport in development.

Sport Plus' focuses on sport-related outcomes with the benefits of learning new sports skills and/or improved health and social integration through direct involvement in sports. While health and development outcomes can accompany these sports activities, they are not the primary objectives.

The 'Plus Sport' approach emphasises sport as a means to an end and uses sport-based initiatives or sport networks paired with public health, conflict resolution or other methodologies to achieve development gains.

4.2. Position Statement(s)

Resolution European Parliament on sport and development
    www.europarl.eu/int
South African White Paper "Getting the Nation to Play"
    www.srsa.gov.za/whitepaper.asp
Towards Achieving the Millennium Development Goals through Sport
    www.un.org/themes/sport/reportE.pdf
United Nations Resolutions
    www.un.org/sport2005/resources/resolution.html

4.3. Varia

Not applicable.

4.4. Free Statement

As this policy area continues to grow, the relationships between sport and development and other thematic disciplines in sport science and physical education should be strengthened. The scope of Sport and Development has great potential for cross-disciplinary partnerships to further the research agenda and improve theoretical and practical applications in this diverse field.

References

Biddle, S. (2006). Defining and Measuring Indicators of Psycho-social Wellbeing in Youth Sport and Physical Activity, in Auweele, Y., Malcolm, C. and Meulders, B. (Eds.) (2006). Sport and Development. Cape Town: Iannoo Campus.

Coalter, F. (2006). Sport-in-Development: A monitoring and evaluation manual, University of Stirling. www.uksport.gov.uk/assets/File/News/monitoring_and_evaluation_140906.pdf, retrieved April 20, 2008.

Desai, V. and Potter, R.B. (2002). The Companion to Development Studies. London: Arnold.

International Business Leaders Forum. (2007). Shared Goals 2: Promoting private sector engagement in sport for development partnerships, Working Summary, International Business Leaders Forum.

Legat, P. (2007). Participatory Approaches in Sport for Development: Empowerment and democratization in a community based sports organisation in Zambia, Master thesis, Norwegian School of Sport Sciences, Oslo.

Maguire, J. (2005). Power, Cultures and Global Sport: Sport development or development through sport? In G. Doll-Tepper, V. Steinbrecher, D.Dumon & A. Chima (eds). International Forum on Sport and Development, Proceedings, [CD-ROM], Berlin.

May, G. and Phelan. J (2005). Shared Goals: Sport and Business in Partnerships for Development. London: IBLF.

NCDO, NOC*NSF. (2004). Toolkit Sport for Development [CD-ROM]. Amsterdam, The Netherlands.

Potter, R.B. (2004). Geographies of Development. Harlow: Prentice Hall.

SDPIWG. (2007). Literature Reviews on Sport for Development and Peace, Sport for Development and Peace International Working Group, http://iwg.sportanddev.org/data/htmleditor/file/Lit.%20Reviews/literature%20review%20SDP.pdf, retrieved, April 19, 2008.

SDPIWG. (2006). Sport for Development and Peace: From practice to policy. Preliminary report of the Sport for Development and Peace International Working Group, http://iwg.sportanddev.org/data/htmleditor/file/SDP%20IWG/Right%20to%20Play%20-%20From%20Practice%20to%20Policy%20book.pdf, retrieved April 19, 2008.

UNICEF, History of Sport and the UN, UNICEF, www.unicef.org/sports/index_40837.html, retrieved April 20, 2008.

United Nations. (2003). Towards Achieving the Millennium Development Goals Through Sport, United Nations, www.un.org/themes/sport/reportE.pdf, retrieved April 20, 2008.

United Nations, Resources: UN Resolutions, www.un.org/sport2005/resources/resolution.html, retrieved April 20, 2008.

van Eekeren, F. (2006). Sport and Development: Challenges in a new arena, in Auweele, Y., Malcolm, C. and Meulders, B. (Eds.) (2006). Sport and Development. Cape Town: Iannoo Campus.

Sport and Human Rights

Mary A. Hums, Eli A. Wolff and Meghan Mahoney

Contact

Dr. Mary A. Hums,
University of Louisville
Dept. of Health & Sport Science, Sport Administration
107 HPES/Studio Arts Building
Louisville, KY 40292 USA
Phone: +1-502-852-5908
Fax: 1-502-852-6683
Email: mhums@louisville.edu

Eli A. Wolff
Center for the Study of Sport in Society
Northeastern University
360 Huntington Avenue, Suite 350 RI
Boston, MA 02115
Phone: +1-617-373-8936
Fax: 1-617-373-4566
Email: e.wolff@neu.edu

Meghan O. Mahoney
Center for the Study of Sport in Society
Northeastern University
360 Huntington Avenue, Suite 350 RI
Boston, MA 02115
Phone: +1-617-373-3159
Fax: 1-617-373-4566
Email: me.mahoney@neu.edu

1. General Information

The field of sport and human rights examines the definition and understanding of sport as a human right, the right to participate in sport and human rights concerns within the realm of sport. The field of sport and human rights also explores the utilisation of sport as a vehicle to promote human rights. Developing and analysing the intersections of sport and human rights is essential to furthering a global awareness of social justice in and through sport.

1.1. Historical Development

Sport or physical activity has been included in a number of United Nations documents. Firstly, the articulation that sport is a human right is contained directly in Article 1 of the Charter on Physical Education and Sport put forth by UNESCO in 1978. Language regarding sport as a human right is also contained in the subsequent Convention on the Elimination of Discrimination against Women (1979), the International Convention Against Apartheid in Sports (1985), the Convention of the Rights of the Child (1989), and most recently, the Convention on the Rights of Persons with Disabilities (2006). In its 2003 document *Sport for Development and Peace: Towards Achieving the Millennium Development Goals*, the United Nations Inter-Agency Task Force on Sport for Development and Peace, sport and play are repeatedly acknowledged as a human right. In addition to United Nations documents, the Olympic Movement addresses human rights through its Olympic Charter. All of these documents place a responsibility on governments, non-governmental organisations, sport governing bodies and sport managers to make sure there is access for all to sport and physical activity.

In addition, numerous individuals and organisations have used sport to promote human rights. For example, the USA's John Carlos and Tommie Smith stood on the medal podium at the 1968 Mexico City Olympic Games with fists raised in the Black Power salute. Olympian Johann Koss founded Right to Play, which uses the power of sport to better people's lives. The Doves Olympic Movement initiative, based in Cyprus, created an interdisciplinary project that uses sports as a tool towards achieving the following educational objectives: (a) fundamental values and human rights; (b) global citizenship; (c) understanding and collaboration across cultures and diverse populations; (d) technological literacy; and (e) conflict management.

Peace Players International uses basketball and life skills programming to bring to-gether thousands of children from diverse cultural backgrounds. The Afghan Youth Sport Exchange seeks to create future sport leaders while working to address human rights issues through sport.

1.2. Function

The discussion of sport and human rights is intended to further knowledge and understanding of sport as a human right to examine the extent to which the prac-tice of sport is a human right. Work in this area examines human rights issues and needs within the sphere of sport. A second function of sport and human rights research explores how sport is utilised as a platform to promote human rights. This work looks at the way in which sporting activities and events serve as a vehicle for addressing broader human rights issues.

1.3. Body of Knowledge

The basic tenets of sport and human rights are outlined in the aforementioned United Nations documents. In addition, a number of sport scholars from differ-ent disciplines including sport sociology, sport philosophy, sport history, and more recently, sport management, have weighed in on the topic, as evidenced by the references listed below. It has also been a topic of interest to Olympic scholars and those who examine the concept and application of Olympism.

1.4. Methodology

The majority of the academic work in the area of sport and human rights has been qualitative in nature. Much of the work examines policy development in sport organisations. Scholars in the area of legal aspects of sport are also becoming interested in the topic.

1.5. Relationship to Practice

A number of sport and non-sport organisations work to further sport and human rights at different levels. The International Olympic Committee furthers the values of sport and human rights through programs such as Sport for All, the Women in Sport Commission, and the Olympic education programs of the International Olym-

pic Academy. The International Paralympic Committee released a Position Statement on Human Rights. Right to Play is committed to the basic ideals of promoting
sport as a human right. The International Football Federation (FIFA) and the World
Health Organization (WHO) partnered to promote a human rights approach to HIV/
AIDS. Non-sport organisations have become involved as well, as seen by the work
done by Handicap International and Landmine Survivors Network to support the
crafting and implementation of Article 30.5 of the Convention on the Rights of
Persons with Disabilities.

1.6. Future Perspectives

The study of sport and human rights is continuing to evolve. Work continues to
emerge examining the articulation of the practice of sport as a human right as
well as ongoing studies on a myriad of human rights issues and concerns within
sports. Further, there is ongoing analysis into the potential for sport activities
to serve as a vehicle for the promotion of human rights. Given our global society, with increased awareness of human rights violations, there will be increased
awareness of defining and implementing sporting rights, and utilising sport to
promote human rights in general. The sport and human rights field will continue to
be relevant and essential.

2.1. Journal Articles

Bhuvanendra, T.A. (1998-1999). Human rights in the realm of sport. *Olympic Re-
        view, 26*, 15-35.
Blauwet, C. (2005). Promoting the health and human rights of individuals with a
disability through the Paralympic Movement. International Paralympic Committee.
        Available at www.paralympic.org/release/Main_Sections_Menu/Development/
        Development_Programmes/Symposium/index.html .
Corbett, D. (2006). The politics of race and sport in the promotion of human
        rights. *ICSSPE Bulletin, 48*. Available at www.icsspe.org
DaCosta, L., Abreu, N. and Miragaya, A (2006). Multiculturalism and universal
proclaimed values in the Olympic Movement: An approach to human rights and
        sport development? *ICSSPE Bulletin, 48*. Available at www.icsspe.org
Donnelly, P. and Kidd, B. (2006). Achieving human rights in and through sports.
        ICSSPE, Bulletin, 48. Available at www.icsspe.org

Gilbert, K. (2006). The wrong way around. *ICSSPE Bulletin, 48.* Available at www. icsspe.org

Hums, M.A. and Wolff, E.A. (2006). Olympism and human rights: Ideals in action. *ICSSPE Bulletin, 48.* Available at www.icsspe.org

Kaufman, P. (2006) Activist athletes and human rights. *ICSSPE Bulletin, 48.* Available at www.icsspe.org

Kidd, B. and Donnelly, P. (2000). Human rights in sports. *International Review for the Sociology of Sport,* 35(2), 131–148.

Mbaye, K. (1998). Sport and human rights. *Olympic Review,* 24, 8-14

McArdle, D. (2006). Human rights and animal welfare: the implications of anti-hunting legislation in the United Kingdom. *ICSSPE Bulletin, 48.* Available at www.icsspe.org

Oglesby, C. (2006). Paths for women and men towards human rights in sport: The same, delayed or divergent? *ICSSPE Bulletin, 48.* Available at www.icsspe.org

Roy, E. (2006). The development of the human rights of individuals with disabilities in sport at the United Nations and beyond. *ICSSPE Bulletin, 48.* Available at www.icsspe.org

Roy, E. (2007). Aiming for inclusive sport. *Entertainment and Sport Law Journal,* 5(1). Retrieved 26 January 2007 from www2.warwick.ac.uk/fac/soc/law/elj/eslj/issues/volume5/number1/roy

Waldman, J. (2007). Best practice in human rights education: The SHR Sport and Human Rights Olympics. *Intercultural Education,* 18(3), 265-268.

Wolff, E.A. and Hums, M.A. (2006). Sport and human rights. *The Chronicle of Kinesiology & Physical Education in Higher Education,* 17(2), 3-4.

2.2. Reference Books, Encyclopedias, etc.

David, P. (2005). *Human rights in youth sport: A critical review of children's rights in competitive sports.* London: Routledge, Taylor & Francis.

Kidd, B. (2003). Athletes and human rights. Barcelona, Spain: Centre for Olympic Studies at the Autonomous University of Barcelona.

McArdle, D. and Giulianotti, R. (Eds.). (2006). *Sport, civil liberties, and human rights.* London: Routledge.

Sport for Development and Peace International Working Group. (2007).

Literature reviews on Sport for Development and Peace. Available at http://iwg.sportanddev.org

Wolff, E. and Hums, M.A. (Eds.). (2007). Sport and human rights. *International Council of Sport Science and Physical Education. Bulletin Special Edition on Sport and Human Rights, 48.*

Wolff, E.A., Hums, M.A. and Roy, E. (2007). *Sport in the UN Convention on the Rights of Persons with Disabilities.* Geneva, Switzerland: United Nations Office of Sport for Development and Peace.

2.6. Internet Sources

International Olympic Committee.
www.olympic.org
International Olympic Academy
www.ioa.org.gr
International Paralympic Committee
www.paralympic.org
Amnesty International
www.amnesty.org
International Working Group on Women and Sport
www.iwg-gti.org/
United Nations Sport for Development and Peace
www.un.org/themes/sport
International Year for Sport and Physical Education
www.un.org/sport2005/index.html
Office of the United Nations High Commissioner for Human Rights
www.ohchr.org/english/bodies/
UNESCO - Sport
http://portal.unesco.org/shs/en/ev.php-URL_ID=9534&URL_DO=DO_TOPIC&
URL_SECTION=201.html
World Health Organization
www.who.int
Council of Europe
www.coe.int/T/dg4/Sport/Default_en.asp
Convention on the Rights of the Child
www.unhchr.ch/html/menu3/b/k2crc.htm
Convention on the Elimination of All Forms of Discrimination against Women
www.hrweb.org/legal/cdw.html

Convention on the Rights of Persons with Disabilities
    www.un.org/disabilities/convention/conventionfull.shtml
Convention against Apartheid in Sports
    www.unhchr.ch/html/menu3/b/12.htm
Human Rights Yes
    www1.umn.edu/humanrts/edumat/hreduseries/TB6/
Compass - Sport
    www.eycb.coe.int/Compass/en/chapter_5/5_15.html

2.4. Conference Presentations

North American Society for the Sociology of Sport Conference Abstracts

2005 – Abstracts available at www.nasss.org/2005/2005abstracts.pdf

Gilbert, K. and Petri-Uy, M. (2005). Facing Reality: Resurrecting Disability Sport in Kosovo.

Hums, M.A. and Moorman, A.M. (2005). Sport as a Human Right: Role of the Olympic Movement.

2006 – Abstracts available at www.nasss.org/2006/abstracts2oct06.rtf

Corbett, D. (2006). The Politics of Race and Sport in the Promotion of Human Rights.

Hargreaves, J. (2006). Muslim Women in Sport: Islam, Agency, and Human Rights.

Roy, E. and Hums, M.A. (2006). Advancing the Human Rights of People with Disabilities in Sport: Developing the International Disability in Sport Working Group.

Wolff, E.A. and Fay, T. (2006). Rights from Wrongs: Applying Dershowitz in Sport.

2007 – Abstracts available at www.nasss.org/2007/2007finalProgram2.rtf

Hartmann, D. and Isett, C. (2007). Humanism Instead of Human Rights? The Challenge of Beijing 2008 to Olympic Idealism.

Kaufman, P., Hums, M.A. and Wolff, E.A. (2007). Introducing Sport and Human Rights in the Classroom.

Misener, L. and Mason, D. (2007). Rethinking Sporting Events through the Lens of Community Development.

North American Society for Sport Management Conference 2007

Hums, M.A. and Moorman, A. M. (2007). Sport and Human Rights: Developments in the Human Rights Movement and Disability Sport Initiatives. Available at www.nassm.com/files/conf_abstracts/2007_1572.pdf

Play the Game 2007

Hums, M.A. and Wolff, E.A. (2007). Promoting Human Rights through Sport: Can It Make a Difference?
Available at www.playthegame.org/upload/billeder%202007%20konference/speakers/speakerdocuments/humsmary.pdf

3. Organisational Networks

3.1. International Networks

International Disability in Sport Working Group
www.sportanddev.org/en/organisations/see-all-organisations/international-disability-in-sport-working-group-idiswg.htm
International Working Group on Women and Sport
www.iwg-gti.org/
Sport for Development and Peace International Working Group
http://iwg.sportanddev.org

3.4. Specialised Centres

Athletes for Human Rights Initiative, Center for the Study of Sport in Society at Northeastern University
www.sportinsociety.org

4. Appendix Materials

4.1. Terminology

Sport: In the context of human rights, sport is defined to include recreation, leisure, physical activity and play.

Human rights: Fundamental rights that belong to everyone in our global society.
Sport as a human right: The recognition that the practice of sport is a human right.
Sport as a platform for human rights: Sport activities can serve as a vehicle for
   the promotion of human rights, peace and development.

4.2. Position Statements

Brighton Declaration on Women and Sport
   www.iwg-gti.org/index.php?id=63
International Paralympic Committee Position Statement on Human Rights
   www.paralympic.org/release/Main_Sections_Menu/Development/Develop-
   ment_Programmes/Symposium/3.9_Position_Statement_on_Human_Rights_
   Attachment_revised.pdf
Magglingen Declaration
   www.magglingen2005.org/index.cfm?id=33
Accra Call for Action on Sport for Development and Peace
   http://iwg.sportanddev.org/floor/CMS?&server=iwg&lang=en&item_
   categoryID=13&item_ID=58

4.3. Varia

Brown University Royce Fellowship for Sport and Society
   http://swearercenter.brown.edu/whatwedo/FELLOWSHIP_SPORT.html

The Sport and Society Fellowship recognises Brown University undergraduates
who have a record of excellence in academics and sport. The program supports
innovative research or applied projects, exploring the intersection of sport and
human rights within a particular context.

4.4. Free Statement

Only by valuing sport as a universal human right and, equally, discovering how
sport's influence might be used to promote all human rights, can we move that
much closer to ultimate justice. Moving this important field forward, however, requires a universal acknowledgement and buy-in of the field as relevant and worth-

while. This entails: (a) the recognition of current practices as significant to human rights work both on a micro- and macro-level; (b) the translation of policy into support of practice, both directly and through response to human rights violations when they occur; (c) advanced qualitative and quantitative measurements of the effectiveness and awareness of human rights and sport policies; and (d) a more advanced framework for educating global citizens about the essential intersection of sport and human rights.

Sport Governance

Laurence Chalip, Mary A. Hums and Anastasios Kaburakis

Contact

Dr. Laurence Chalip
Sport Management Program
University of Texas
Bellmont Hall 222; D3700
Austin, TX 71712
U.S.A.
Phone: +1 512 471 1273
Fax:+1 512 471 8914
Email: LChalip@mail.utexas.edu

Dr. Mary A. Hums
University of Louisville
Dept. of Health & Sport Science, Sport Administration
107 HPES/Studio Arts Building
Louisville, KY 40292 USA
Phone: 1-502-852-5908
Fax: 1-502-852-6683
Email: mhums@louisville.edu

Dr. Anastasios Kaburakis
Lawyer
Assistant Professor of Sport Law and Sport Management
Director of Sport Management Graduate Program
Southern Illinois University Edwardsville
Department of Kinesiology
Vadalabene Center 1023, Campus Box 1126
Edwardsville, IL 62026-1126
Phone: +618 650-2033, (618) 650-3252
Fax: +618 650-3719
Email: tassos.kaburakis@gmail.com

1. General Information

The study of sport governance has emerged through hybridization of several disciplines, each of which boasts its own community of discourse. These include sport law, sport policy, sport sociology, sport economics and comparative studies of sport. Although each discipline contributes its particular insights, the growth of sport governance knowledge has been hampered by the challenges of obtaining appropriate information, often from governments and organisations that dislike scrutiny, and by the difficulties that arise when scholars from disparate fields endeavour to communicate across their respective paradigms.

1.1. Historical Development

Formal systems or sport governance can be traced to the earliest eras of recorded history, and seem to have emerged first as religious functions. The ancient Olympic Games, which were organised to honour the gods, are the best known and the most studied. Archaeological evidence suggests that formal sport competitions were also organised as religious functions by some pre-Columbian civilizations in the Americas. As the Olympic Games declined during the Roman era, other competitions, including chariot racing and gladiatorial combat emerged as popular but secular entertainments governed by systems of commerce, rather than the clergy. Sport remained secular during the middle ages, but governments became increasingly involved as sport was expected to serve as preparation for combat, rather than mere diversion. Royal families consequently saw sport as their jurisdiction and some monarchs went so far as to outlaw non-combat games. Secularization was fortified during the Protestant Reformation, as some Protestant sects discouraged sport participation. Nevertheless, competitions between clubs and villages required agreement on rules. Groups of aficionados came together to agree upon rules and ultimately to record and govern them. European games, rules and their systems for sport governance were spread to other continents by colonial administrations. As sport was also encouraged in some school systems, particularly in Britain and her colonies, systems for governing sport in schools became increasingly formal during the 19th century. By the late 19th century, a small group of European aristocrats formed what became the International Olympic Committee (IOC) to administer a quadrennial international sport festival. Since international competitions required national systems of governance to field teams, and international systems of governance to coordinate the rules of play and eligibility, additional governing bodies began to form, including National Sport Federations (sometimes called by other names, such as National Governing Bodies or National Sport Organisations), international sport federations (IFs) and National Olympic Committees (NOCs). To enable and enhance communication among these organisations, multi-sport associations were subsequently formed, including the Association of National Olympic Committees (ANOC), the General Association of International Sports Federations (AGFIS) and the International University Sports Federation (FISU). Additionally, the International Paralympic Committee (IPC) emerged as the international governing body for elite sport for athletes with disabilities. The increasingly salient presence of national and international sport governance organisations piqued government interest, particularly during the latter half of the 20th century, with the result that many national, state and city governments passed laws to regulate (and sometimes to fund) sport in their jurisdictions. Some national governments established ministerial-level portfolios to oversee national sport development. The United Nations, through UNESCO, grew its sports initiatives during the 1970s, and today maintains Sport Development for Peace under the United Nations Fund for International Partnerships. Because sport has an environmental impact, the United Nations (in conjunction with the IOC) also has incorporated sport into its Environment Programme. Although sport organisations have typically welcomed government funding, they have been less willing to embrace government authority. Consequently, the Court of Arbitration for Sport (CAS) and a number of associated national-level, sport-specific dispute resolution systems have been established since the 1980s as an alternative to public courts. At the start of this century, the emergence of sport-run international systems to police sport has been heralded by creation of the World Anti-Doping Agency (WADA) and a growing number of its national-level counterparts. Today, sport governance is characterized by a complex array of loosely networked national and international sport governing bodies, and emerging systems for sport-specific policing and arbitration.

1.2. Function

The study of sport governance endeavours to map and to understand the growing array of organisations and networks, as well as their internal systems of management and policymaking. There is not yet any commonly agreed set of research foci (or, for that matter, an agreed definition of sport governance). Researchers have been guided by their own intuitions and by the paradigms of their home disciplines.

1.3. Body of Knowledge

To date, there has been increasing work on legal issues, government policymaking, the challenges of developing sport and the economic rationale for sport policies. The mix of organisations and the complexity of networks have required a substantial amount of descriptive study simply to map the territory, but that work remains hampered by systems of academic prestige which award it scant status and few publication opportunities. The challenge has therefore been to theorize the work that is undertaken. To date, there has been little effort to develop sport-specific theories of governance, so theorization remains derivative from the home discipline of the researcher. Although the majority of studies focus on individual national cases or particular international organisations, an increasing number of comparative studies have trickled forth. Legal research is arguably the most adequately developed, insomuch as its volume of scholarship is greatest, and the requisite data are generally public. Work on specific government policies and policymaking has also grown in recent decades, although slowly. However, work on sport systems and networks remains problematic as the official public positions and descriptions of public and private sport organisations are often at odds with their actual practices of governance, and sport organisations (particularly the more powerful) are careful to promote their official face while concealing their inner workings. Indeed, one of the most significant contributions to the field has been work that identifies gaps between the official face and the actual practice of sport governance. Multi-disciplinary work has also contributed new synthetic insight, although multi-disciplinary studies remain rare.

1.4. Methodology

There is no established methodology or collection of methodologies for the study of sport governance. Researchers typically adopt methods that are familiar to their home disciplines. Studies have utilized surveys, interviews, participant and non-participant observation, and review of documents (particularly policy discourse and legal cases). Given that there can be a gap between a sport organization's official claims and its actual practices, there are grounds for being wary about the accuracy of survey and interview studies that are not bolstered by observation and/or review of documents. The best studies incorporate multiple methods.

1.5. Relationship to Practice

Although the study of sport governance holds significant promise for eventual contribution to the practice of sport governance, the field is not yet mature enough to boast a record of contributions to practice. The study of sport law, which is the most mature component of the field, has generated a substantial number of published cases and analyses that can and do have an impact on sport jurisprudence. The study of sport policy draws on the toolkits of policy analysis, which have demonstrated substantial utility for governance in other realms, but have not yet rendered an impact in sport (with the arguable exception of some work on sport economics, particularly work having to do with public subsidy of sport). In the early 1990s, there was an acrimonious debate among sport sociologists regarding the appropriateness of seeking to apply sport sociology to sport policy-making. Although some sociologists retain an interest in sport governance, those who inveigh against practical relevance have had the upper hand. Similarly, sport historians have remained wedded to narrative history, despite the demonstrated value of applied history to policymaking.

1.6. Future Directions

Sport governance is becoming ever more complex. Government and private organisations are increasingly intertwined and the international intricacies of sport governance are intensified by globalization. New organisations continue to emerge as new sports (e.g., paragliding, disc-golf, floorball) develop governance systems, as new sport events are created and as sport-specific policing and arbitration systems spread. The need to understand the dynamics of sport governance is consequently growing apace. To flourish, the field (as an area of study) requires a greater degree of cross-dialogue among contributing academic disciplines, and a more substantial commitment to multi-disciplinary and multi-method research.

2. Information Sources

2.1. Journals

Australian and New Zealand Sports Law Journal (Australia / New Zealand)
Causa Sport (Switzerland)
DePaul Journal of Sports Law & Contemporary Issues (USA)
Derecho Deportivo (Spain)
Desporto & Direito (Portugal)
European Sport Management Quarterly (EU)
Entertainment and Sports Law Journal (UK)
Entertainment and Sports Lawyer (USA)
International Journal of Sport Policy
International Review for the Sociology of Sport
International Sports Law Review
IUSPORT (Spain)
Journal of Legal Aspects of Sport (North America)
Journal of Sport Management (North America)
Journal of Sports Economics (North America)
Les Cahiers de Droit du Sport (France)
Marquette Sports Law Review (USA)
Nieuwsbrief Sport en Recht (Belgium)
Revista di Diritto ed Economia dello Sport (Italy)
Seton Hall Journal of Sport Law (Seton Hall University School of Law) (USA)
Sociology of Sport Journal (North America)
Sport Management Review (Australia / New Zealand)
The Journal of the Business Law Society (USA)
The Sports Lawyers Journal (USA)
Villanova Sports & Entertainment Law Journal (USA)
World Sports Law Report (UK)
Zeitschrift für Sport und Recht (Germany)

2.2. Reference Books

Chalip, L., Johnson, A. and Stachura, L. (Eds.). (1996). *National sport policies: An international handbook*. Westport, CT: Greenwood Press.

Green, M. and Houlihan, B. (2005). *Elite sport development: Policy learning and political priorities*. Abingdon, UK: Routledge.

Hoberman, J. and Møller, V. (Eds.). (2004). *Doping and public policy*. Odense: University Press of Southern Denmark.

Houlihan, B. and Green, M. (Eds.). (2008). *Comparative elite sport development: Systems, structures and public policy*. Oxford: Elsevier.

Hoye, R. and Cuskelly, G. (2007). *Sport governance*. Oxford: Elsevier.

Hums, M.A. and MacLean, J.C. (2004). *Governance and policy in sport organizations*. Phoenix, AZ: Holcomb-Hathaway.

Kluka, D., Stier, W. and Schilling, G. (2005). *Aspects of sport governance*. Berlin: ICSSPE.

Levermore, R. and Budd, A. (Eds.). (2004). *Sport and international relations: An emerging relationship*. London: Routledge.

Riordan, J. and Kruger, A. (Eds.). (1999). *International politics of sport in the 20th century*. London: E & FN Spon.

Thoma, J.E. and Chalip, L. (1996). *Sport governance in the global community*. Morgantown, WV: Fitness Information Technology.

Zintz, T. (2005). *Manager le changement dans les federations sportives en Europe*. Brussels: de Broeck.

2.3. Book Series

None available.

2.4. Conference Workshop Proceedings

Play the Game 2007 – Creating Coalitions for Good Governance in Sport. www.playthegame.org/Home/Conferences/Play_the_Game_2007/presentations.aspx

Play the Game 2005 – Governance in Sport – The Good, the Bad, and the Ugly www.playthegame.org/Home/Conferences/Play_the_Game_2005/Conference_presentations.aspx

2.5. Data Banks

None available.

2.6. Internet Sources

Association of National Olympic Committees
    www.acnolympic.org
Court of Arbitration for Sport
    www.tas-cas.org
Court of Justice of the European Communities
    http://curia.europa.eu/
EurActiv
    www.euractiv.com/en/sports
General Association of International Sports Federations
    www.agfisonline.com
International Olympic Committee
    www.olympic.org
International Paralympic Committee
    www.paralympic.org
International University Sports Federation
    www.fisu.net/site/medias/accueil.html
Sport and EU
    www.sportandeu.com/
Sport for Development and Peace (UN)
    www.un.org/themes/sport/
Sport Links Central
    www.sportslinkscentral.com
Sport and the European Union
    http://ec.europa.eu/sport/index_en.html
Sport for All
    www.tafisa.de/
WADA
    www.wada-ama.org

3. Organisational Networks

3.1. International Level

Association of National Olympic Committees
Association of Summer Olympic International Federations
Association of Winter Olympic Sports
General Association of International Sports Federations
International Sport Management Alliance
International Olympic Committee
International Paralympic Committee
International Sociology of Sport Association
International Sport Lawyers Association
International University Sports Federation
TAFISA

3.2. Regional Level

Association for the Study of Sport and the European Union
Asian Association for Sport Management
European Association for Sport Management
North American Association of Sports Economics
North American Society for Sport Management
North American Society for the Sociology of Sport
Sport Management Association of Australia and New Zealand
Sport and Recreation Law Association
Union of European Leagues of Basketball

3.3. National Level

National Collegiate Athletic Association (USA)
National Federations (each country; see relevant IF site for contacts)
National Olympic Committees (each country; see the IOC website for a list)

3.4. Specialised Centres

ASSER International Sport Law Centre, The Hague, The Netherlands
Centre d'Estudis Olímpics, Barcelona, Spain
Centre for Sport and Law Inc., Canada
Centre International D'Etude Du Sport, Neuchâtel, Switzerland.
Forschungsstelle für Sportrecht, Institut für Recht und Technik (IRUT), Germany
LA84 Foundation, Los Angeles, USA
National Sport Law Institute, Marquette University, USA

3.5. Specialised International Degree Programmes

Anglia Ruskin University: LLM International Sports Law, England
Master of Laws (LL.M.) in Sports Law for those with non-U.S. law degrees, Marquette University, USA
MSA specializing in managing sport governing organisations: International Academy of Sports Science and Technology, Switzerland

4. Appendix Material

4.1. Terminology

There is not yet an agreed set of definitions in the field. Hums and McLean (2004) define sport governance " the exercise of power and authority in sport organizations, including policy making, to determine organizational mission, membership, eligibility, and regulatory power, with the organization's appropriate local, national, or international scope."

4.2. Position Statement(s)

Not applicable.

4.3. Varia & 4.4. Free Statement

Not applicable.

Women and Sport

Darlene A. Kluka

Contact

Prof. Dr. Darlene A. Kluka
Barry University
School of Human Performance and Leisure Sciences
Department of Sport and Exercise Sciences
11300 NE 2nd Ave.
Miami Shores, Florida 33161-6695
USA
Phone: +1 305 899 3549
Email: DKluka@mail.barry.edu

1. General Information

The injection of equality between men and women in recent human history has challenged the foundation of social order in many cultures and societies throughout the world. In sport, women have been disadvantaged by being afforded relatively limited access to active participation. Women have also been visibly absent from decision making roles in sport, particularly in the areas of coaching, refereeing and administration/management. Increasing participation of women in and through sport, and particularly the empowerment that accompanies participation, come at a point in time when gender issues form part of a broader transformational movement, centering around inequality and the need for women's freedom and full participation in the whole of society. Sport has become more than a means to health and entertainment. It has become one of the fastest growing paths to perceived economic power and has become more politically attractive. There seems to be, then, a link between participation and roles of women in sport, their political attractiveness and women's freedom.

1.1. Historical Development

The past 60 years, in particular, have brought issues involving women and sport to the forefront of the global equality conversation. Arguably, the most significant declaration of the past half century in the Women and Sport Movement was the one developed in Brighton, England in 1994 at a conference entitled Women, Sport and the Challenge of Change. Known as the *Brighton Declaration*, the document has been adopted or endorsed by over 250 organisations to date. Its scope is "...addressed to governments, public authorities, organisations, businesses, educational and research establishments, women's organisations, and individuals who are responsible for, or who directly or indirectly influence the conduct, development or promotion of sport or who are in any way involved in the employment, education, management, training, development or care of women in sport." Ten principles were established that were aimed at the development of a sporting culture that "enables and values the full involvement of women in every aspect of sport." They include a framework for the following areas: (1) equity and equality in society and sport; (2) facilities; (3) school and junior sport; (4) developing participation; (5) high performance sport; (6) leadership in sport; (7) education, training and development; (8) sports information and research; (9) resources; and (10) domestic and international cooperation.

The women and sport conversation seemed to move quickly to a higher position on political and sporting agendas, and four years later, the 2nd World Conference on Women and Sport, *Call for Action*, was held in Windhoek, Namibia. Conference delegates produced a Call for Action to further the development of equal access and opportunities for females to participate in and through sport in its broadest sense. Conference participants had become increasingly convinced that closer cooperation with the Beijing *Platform for Action* (developed at the United Nations World Conference on Women in Beijing, 1996) and the United Nations Convention on the Elimination of All Forms of Discrimination Against Women would be beneficial to the women and sport movement. For the first time, the concept of women and physical activity was placed in the Beijing *Platform for Action*. In addition to reinforcing the principles of the Brighton Declaration, the attendees called for the following set of actions: (1) develop action plans with objectives and targets to implement the principles of the Brighton Declaration, and monitor and report upon their implementation; (2) reach out beyond the current boundaries of the sport sector to the global women's equality movement and develop closer partnerships between sport and women's organisations on the one side, and representatives from sectors such as education, youth, health, human rights and employment on the other. Develop strategies that help other sectors obtain their objectives through the medium of sport and at the same time further sport objectives; (3) promote and share information about the positive contribution that girls' and women's involvement in sport makes, including social, health and economic issues; (4) build the capacity of women as leaders and decision makers and ensure that women play meaningful and visible roles in sport at all levels. Create mechanisms that ensure that young women have a voice in the development of policies and programs that affect them; (5) avert the "world crisis in physical education" by establishing and strengthening quality physical education programs as key means for positive introduction to young girls of the skills and other benefits they can acquire through sport. Further, create policies and mechanisms that ensure progression from school to community-based activity; (6) encourage the media to portray positively and significantly cover the breadth, depth, quality and benefits of girls' and women's involvement in sport; (7) ensure a safe and supportive environment for girls and women participating in sport at all levels by taking steps to eliminate all forms of harassment and abuse, violence and exploitation, and gender testing; (8) ensure that policies and programs provide opportunities for all girls and women in full recognition of the differences and diversity among them – including such factors as race, ability, age, religion, sexual orientation, ethnicity, language, culture or their status as an indigenous person; (9) recognise the importance of governments to sport development and urge them to develop appropriate legislation, public policy and funding monitored through gender impact analysis to ensure gender equality in all aspects of sport; (10) ensure that Official Development Assistance programs provide equal opportunities for girls' and women's development and recognise the potential of sport to achieve development goals; and (11) encourage more women to become researchers in sport and more research to be undertaken on critical issues relating to women in sport.

In order for the women and sport movement to continue globally after 1994, a coordinated and strategic plan was developed by the International Working Group on Women and Sport (IWG). It was determined that the IWG, an informal, coordinating body of selected nongovernmental and governmental organisations whose objectives included the development of girls and women in sport and physical activity throughout the world, would monitor progress. The IWG network from 1994 – 1998 included representatives from the Namibian government, UK Sports Council, WomenSport International (WSI), Colombian Olympic Academy, Sport Canada, Australian Sports Commission, Hillary Commission, Commonwealth Games Federation, International Association of Physical Education and Sport for Girls and Women (IAPESGW), Fiji Sports Council, Finnish Sports Federation, Alexandria University and Smith College. The focus of the approach involved converting policy to practice. The approach included: (1) commitment to the implementation of the principles and actions contained within the Brighton Declaration and the Windhoek Call for Action; (2) maximising international coordination mechanisms, such as cooperation between the IWG, the International Olympic Committee (IOC) Working Group on Women and Sport, WSI, IAPESGW, United Nations Commission on the Status of Women and others; (3) regular conferences and opportunities for information exchange; (4) developing and maintaining strategic alliances with the global women's development movement; (5) the continuation of the IWG in a facilitating, supporting and monitoring role; and (6) the staging of the 3rd World Conference on Women and Sport in Canada in 2002.

In 2002, the 3rd World Conference on Women and Sport, *Investing in Change*, was hosted in Montreal, Canada. Each participant received a toolkit of materials appropriate for use in a variety of cultural settings, and each began work on an individual action plan that also included the publishing of two articles on women and sport issues upon returning home. Those at the conference committed to the creation of sustainable infrastructure for girls and women in sport globally which included: (1) safe and supportive spaces for play and physical development for girls and women; (2) quality physical education in schools for children, to develop fundamental motor skills and abilities, the basis of lifelong participation in physical activity; (3) equal opportunity for competition and training; (4) sport and physical activity as health promotion and to develop awareness of the power of sport in avoiding risk behaviours like early sexual activity and teen pregnancy, substance abuse, HIV/AIDS, inactivity and obesity and in promoting the ability to choose positive lifestyles; (5) strong links between practice, policy and research, including research to provide evidence for advocacy and improved delivery of sport and physical activity programs; (6) effective networks and communication between women working in all roles and levels in sport and physical activity; and (7) strategic approaches to gender equity in sport and physical activity.

The 4th World Conference on Women and Sport, hosted in Kumamoto, Japan in 2006, addressed the following objectives that were laid out in 2002: (1) to recommend to the United Nations and to conference participants that countries include sport and physical activity for women as a section in reports to the monitoring group for the Convention on the Elimination of All Forms of Discrimination against Women; (2) to present an official report of the 2002 World Conference on Women and Sport to the 2004 Meeting of the Ministers of Physical Education and Sport to be held under the auspices of UNESCO (MINEPS IV); (3) to monitor progress against the principles of the Brighton Declaration, the Windhoek Call for Action, and to collect evidence for the use of sport in developing women, communities and nations; (4) to develop and extend the availability of resources for people working in women and sport across the world; (5) to record and evaluate examples of gender mainstreaming in sport and physical activity policy and provision; (6) to continue to monitor the effects of the World Crisis in Physical Education and to promote evidence-based advocacy on the value of physical education for girls and young women; (7) to support and encourage the work of the international organisations of women's sport and physical education and to encourage sustainable activities and structures for development of women and sport, including working with government organisations (Gos), non-government organisations (NGOs), National Olympic Committees (NOCs) and physical education organisations at national and regional levels; (8) to encourage active cooperation between health, education and gender equity agencies at national and international levels; (9) to work cooperatively with International Federations and the Olympic Movement; and (10) to work proactively to encourage cultural and structural change among sports organisations towards gender equity.

Prior to the groundswell of activity after the Brighton Declaration, the International Association of Physical Education and Sport for Girls and Women (IAPESGW) was founded in 1949. It is the oldest, continuing international organisation in the world that has girls and women in physical education and sport as its prime focus. It is the only organisation that has held world congresses on physical and sport for girls and women quadrennially since its inception. A membership-driven organisation, it supports its members working in the areas of women and girls in physical education and sport and provides professional development and international cooperation. As a non-profit organisation and member of the International Council of Sport Science and Physical Education (ICSSPE) and the IWG, IAPESGW also helps to construct international policy and aid work in sport and physical education.

More recently, WomenSport International (WSI) was founded in 1994 to meet the challenge of ensuring that sport and physical activity receive the attention and priority they deserve in the lives of girls and women. A non-profit organisation, it is an evidence-based advocacy group committed to create and support actions that bring about increased opportunities and positive changes for girls and women in all levels of involvement in sport and physical activity/education. As a member of ICSSPE and IWG, it also helps in the construction of international policy in sport and physical activity.

At the 105th IOC Session in 1996, the IOC ratified its first Working Group on Women and Sport. By the end of that year, the International Olympic Committee (IOC) held its first IOC Women and Sport Conference. Since then, an international conference has been held quadrennially where the primary objective is to assess and analyze the progress made within the Olympic Movement and to determine actions to improve and increase participation of women in sport. The conferences have been held in Lausanne, Paris, Marrakech and Amman. The IOC institutionalised the Working Group to a Commission in 2004.

Despite policies, declarations and calls for action globally, there continues to be a lack of targeted affirmative action policies, gender-focused development initiatives, training and education, and motivational programs focused on women's participation in several areas of the world. Their absence has negatively impacted gender equality and equity in and through sport, globally. The needs of women must be addressed specifically in cultural contexts to challenge gender relations and provide access and opportunity so that women can assume their rightful place in sport, sport leadership and society.

References

IAPESGW (2008).
    www.iapesgw.org
IAPESGW (2007). *Strategic plan. Kennesaw*, GA: IAPESGW.
ICSSPE (2008).
    www.icsspe.org
IOC (2008).
    www.olympic.org
IWG (2007). *Embracing change: Annual meeting workbook.* Kuala Lumpur: IWG.
IWG (2008).
    www.iwg-gti.org/
UK Sport (2006). *Women in sport: The state of play 2006.* London: UK Sport:
    World class success.

1.2. Function

Those working in the area of women and sport are ultimately involved in the construct of gender mainstreaming through sport. The principles of the Brighton Declaration provide a basis for continued exploration of the area, with particular emphasis upon inclusion of all girls and women.

1.3. Body of Knowledge

The body of knowledge relative to women and sport can be divided into three basic categories: (1) women and sport participation; (2) women and sport leadership; and (3) women, sport and development. Each of the areas is emerging and being developed by researchers using interdisciplinary strategies and research models.

1.4. Methodology

The methods used for the assessment and evaluation of programs, projects and initiatives can include case studies, descriptive analysis, surveys and capacity analysis tools. They may include auditing of policies and practices that organisations undertake. Qualitative analysis has also been used.

1.5. Relationship to Practice

The area of women and sport connects easily with practice. Many of the above-mentioned strategies associate directly into the daily lives of people across the world.

1.6. Future Perspectives

Sensitisation programs at all levels can be geared to refute myths surrounding women's capabilities and dedication to work and physical activity outside the home, improving managerial understanding of gender and family issues, and endorsing the valuable contribution women can make to sport, its management and its productivity. In order to contribute productively and equally with men, it is essential that women have access to education, leadership and management training and experience, mentors and role models at the highest levels, as well as admittance to formal and informal networks and communication channels in the work place and outside the home.

Recruitment, job assignment, career planning, grading, wages, transfers and promotions must be transparent, objective and fair for all those involved with the participation in and management of sport. All organisations must find ways to monitor their own progress through the institutionalisation of documents such as the Brighton Declaration, in policies, procedures and practices.

Women and leadership development programs can be implemented, sustained and evaluated that can include mentor support, networking, development of personal development plans, effective communication, national and international sport governance structures and function, women's sport development, sports law, advocacy/influencing skills, conflict prevention/resolution, sponsorship and marketing, external partnership development, good management practice, and application and/or nomination skills within cultural frameworks.

2. Information Sources

2.1. Journals

Women in Sport and Physical Activity Journal
Journal of Women in Coaching

2.2. Reference Books, Encyclopaedias, etc.

Encyclopedia of Women and Sport in America
International Encyclopedia of Women and Sport
Cohen, G. L. (2001). *Women in sport: Issues and controversies.* Reston, VA: AAHPERD.
Dong, J. (2002). *Women, sport and sociology in modern China: Holding up more than half the sky.* London: Routledge.
Hargreaves, J. (1994). *Sporting females: Critical issues in the history and sociology of women's sports.* New York: Routledge.
Hartman, I. And Pfister, G. (2007). *Sport and women: Social issues in international perspective.* London: Taylor & Francis.
Hawkes, N.R. and Seggar, J.F. (2000). *Celebrating women coaches: A biographical dictionary.* Westport, CT: Greenwood Press.
Kluka, D., Scoretz, D. and Melling, C. (2004). *Women, sport and physical activity: Sharing good practice.* Berlin: ICSSPE.
O'Reilly, J. and Cahn, S. *Women and sports in the U. S.: A documentary reader.* Holliston, MA: Northeastern Publishing.

2.3. Book Series

Not applicable.

2.4. Conference/Workshop Proceedings

IAPESGW Congresses from 1953 – 2009 (quadrennially)
IOC Conferences on Women and Sport, 1996 – 2008 (quadrennially)

2.5. Data Banks

Main data banks used to locate works on women and sport are SPORT DISCUS
and INDEX MEDICUS and are accessible via the internet, for a price.

2.6. Internet Sources

International Association of Physical Education and Sport for Girls and Women
    www.iapesgw.org
International Council of Sport Science and Physical Education
    www.icsspe.org
Special Olympics International
    www.specialolympics.org
International Paralympic Committee
    www.paralympic.org
International Olympic Committee
    www.olympic.org
International Working Group on Women and Sport
    www.iwg-gti.org/
UK Sport
    www.uksport.org
Australian Sport Commission
    www.ausport.gov.au
Canadian Association for the Advancement of Women and Sport and Physical Activity
    www.caaws.ca
European Women and Sport
    www.ews-online.com
Feminist Majority Foundation
    http://feminist.org/
Harassment and abuse prevention in sport
    www.harassmentinsport.com

Hillary Commission
www.hillarysport.org.nz
Japanese Association for Women and Sport
www.jws.or.jp
Olympic women
www.olympicwomen.co.uk
UNESCO
www.unesco.org
Women in Sport Career Foundation
www.wiscfoundation.org
Women's Sport and Fitnesss Foundation UK
www.wsf.org.uk
Women's Sports Foundation USA
www.womenssportsfoundation.org
WomenSport International
www.sportsbiz.bz/womensportinternational/
German Olympic Sports Federation
www.dosb.de/de/
International Table Tennis Federation
www.ittf.com

3. Organisational Network

3.1. International Level

International Association of Physical Education and Sport for Girls and Women (IAPESGW)
International Council of Health, Physical Education, Recreation, Sport and Dance (ICHPERSD) Girls and Women in Sport Commission
International Olympic Committee Women and Sport Commission
International Working Group on Women and Sport (IWG)
WomenSport International (WSI)

3.2. Regional Level

African Women in Sport Association
Arab Women and Sport Association

Asia Women and Sport
European Women and Sport
Association of Women, Sport and Culture in the Mediterranean

3.3. National Level

Algerian Association for Women's Sport Development
Australian Sports Commission Women and Sport Unit
Association of Women in Sports Sierra Leone
Black Women in Sports Foundation USA
Cambodian Commission of Women and Sport
Canadian Association for the Advancement of Women and Sport and Physical Activity
Egyptian Women's Sports Association
Faculty of Physical Education for Girls, Alexandria University
Indonesian Association of Physical Education and Sports for Girls and Women
Israel's Unit for the Advancement of Women in Sport
Japan Association of Physical Education for Women
Japanese Association for Women in Sport
Madagascar Women and Sport
Myanmar Women Sports Federation
National Association of Women, Physical Activity and Sport Morocco
National Association for Girls and Women in Sport USA
National Commission on Women and Sport Mauritius
Nigerian Association of Women in Sport
National Olympic Committee – Women and Sport Commission Portugal
Philippine Women's Sports Foundation
Singapore Council of Women's Sport Organisations
Sport Society of Egyptian Women
Uganda Sports Women's Association
Women and Sport Association of Togo
Women's Sports Association of Ghana
Women in Sport Foundation Zimbabwe
Women's Sports Foundation South Africa
Women's Sports Foundation UK
Women's Sports Foundation USA
Womensport Australia

3.4. Specialised Centres

Melpomene Institute for Women, USA
Trinity Center for Girls and Women in Sport, USA
Tucker Research Center on Women in Sport, USA
Women's Sports Medicine Center, New York NY USA

3.5. Specialised International Degree Programmes

Presently, there is no international degree program in the field.

4. Appendix Material

4.1. Terminology

Gender mainstreaming, equality, equity, discrimination, inclusion, sexual orientation, sexual harassment, human rights, community development, leadership, sport-specific topics: coaching the female athlete, female athlete performance, sport participation, pregnancy and participation

4.2. Position Statement(s)

Bratislava Resolution on the Prevention of Sexual Harassment and Abuse of Women, Young People and Children in Sport (Council of Europe)
CASM/ACMS (Canadian Association of Sports Medicine) Masquerading as Females in Women's Sports
Doha Conference, 2nd Asian Women and Sport Conference
FEPSAC (European Federation of Sport Psychology) Gender and Sports Participation
FEPSAC Sexual Exploitation
FIMS Female Athlete Triad
ICSSPE Equality and Equity plan
Helsinki Spirit 2000
IOC Medical Commission on Girls and Women in Sport
IWG communique – Montreal Toolkit
MINEPS IV Commission III Women and Sport Recommendations

National Coalition on Women and Girls in Education Title IX at 30: Report card on gender equity

NAGWS Homophobia, Homonegativism and Heterosexism in Sport

NCAA Committee on Women's Athletics

US Lacrosse Eye Protection in Women's Lacrosse

US Lacrosse Helmet Protection in Women's Lacrosse

Women's Sports Foundation USA Addressing the issue of Verbal, Physical and Psychological Abuse of Athletes

Women's Sports Foundation USA Coaching: Do female athletes prefer male coaches?

Women's Sports Foundation USA Gender Discrimination and the Olympic Games

Women's Sports Foundation USA Girls and Contact sports

Women's Sports Foundation USA Homophobia and Sport Policy

Women's Sports Foundation USA Sexual Harassment and Sexual Relationships between Coaches, Other Athletic Personnel and Athletes

WSI Sexual Harassment and Abuse of Girls and Women in Sport

Declarations/Calls for Action:

Brighton Declaration on Women and Sport – 1994

Declaration of Athens MINEPS IV – 2005

Paris Call for Action European Women and Sport - 2005

Windhoek Call for Action – 1998

FIFA by the 4th Women's Football Symposium - 2006

IAPESGW Accept and Respect: Muslim women in sport - 2008

Manila Declaration on women and sport – 1996

Yemen Challenge – 2005

4.3. Varia

Not applicable.

4.4. Free Statement

None available.

Appendix

Sport Science Careers:
Search Strategies and Decision-Making for University Students;
Sources in English Speaking Countries

Gretchen Ghent

Introduction

The purpose of this guide is to outline sources of information to assist the college and university student in exploring the numerous options for finding meaningful work (or further education). It is assumed that students will have or, are in the process of, obtaining a Bachelor's or Master's degree in kinesiology, sport science, physical education or recreation.

This resource guide begins with information on the

1. personal exploration of interests, followed by a section on
2. researching possible jobs in the sport sciences, then
3. where to find information on performing an effective job search strategy with websites that list sports jobs in many different settings.
4. The last section is a listing of the possible occupations and job titles for various specializations in the sport sciences, medicine and allied health fields.

The job titles and positions have been gleaned from many publications and career planning websites found during the research for this paper. (Please note that the term sport sciences will be used throughout this resource guide to indicate the whole range of subjects including kinesiology, human performance, exercise science, physical education and recreation).

1. Exploration of Interests, Values, Skills and Workplace Possibilities

During the first or second year of university, a student may want to probe one's personal interests, talents, skills and workplace preferences. Staff members at the university's career guidance center are available for consultation. They can administer tests that will indicate a person's skills, talents and strengths and weaknesses. These staff members can help students decide which programs or faculties would develop a person's talents and skills. Also found in a university guidance center are pamphlets and other information material that describe professions and jobs plus educational requirements.

Many faculties and schools of sport sciences also have undergraduate or faculty advisors who also can provide career counseling for current or potential undergraduate students. These individuals can offer assistance in mapping out a course of instruction, discussing career paths of recent graduates and otherwise advising students in the more specific details of university programs and courses.

2. Researching Possible Occupations in the Various Sport Sciences Fields

Since the mid 1990s there is much information on the Internet about career possibilities in the sport sciences with some websites being devoted to actual job listings. Exploration of these websites can give the student an excellent idea of the wide range of job possibilities. In some cases where a province or state or professional association has legislated specific certification or qualification requirements, many of the websites will note these requirements.

There are three types of websites that can be consulted where information on potential or actual job titles/descriptions may be found. They include

- websites of professional associations,
- university sport sciences faculties/schools, and
- commercial "sports jobs" websites.

For each of these areas, a few websites are listed in the sections below.

2.1. Searching Professional Association Websites:

Some professional associations have a section on education and careers for their particular discipline (biomechanics, sport psychology, exercise science, sport medicine).

For a list of professional associations consult the following two websites:

Scholarly Sport Sites, Associations (Scholarly/Expert/Specialist)
www.ucalgary.ca/lib-old/ssportsite/assoc.html
SPORTQuest, Associations
www.sirc.ca/online_resources/sportquest_associations.cfm
Two examples of professional association career information include:

Professional Fields of Study in Sport and Movement Sciences (National Association for Sport and Physical Education
www.aahperd.org/NASPE/CUPEC_field_of_studies.html
Careers in Sports Medicine and Exercise Science (American College of Sports Medicine)
www.acsm.org/ (Click on Certification)
National Strength and Conditioning Association Career Resources
www.nsca-lift.org/careerresources/default.shtml
PE Central, Becoming a Physical Education Teacher
www.pecentral.org/professional/becomingapeteacher.html

2.2. Searching University Faculties/Schools of Kinesiology, Exercise Science, Human Performance

University and colleges that have faculties or schools of sport sciences often have a section on their website that describes the career possibilities for their faculty's graduates. In some cases, the websites include surveys of job titles and further education pursuits of recent graduates. In today's changing job market, these recent surveys are very valuable and provide excellent ideas for potential employment. Consult:

2.3. Career Advice – North America

University of Waterloo Applied Health Sciences
    www.ahs.uwaterloo.ca/prospective/kin/careers.html
Careers in Kinesiology (Kansas State University)
    www.k-state.edu/kines/careers.htm
Career Services (University of Michigan. Faculty of Kinesiology)
    www.kines.umich.edu/career/
What can I do with a Major in Kinesiology (University of Wisconsin – Eau Claire)
    www.uwec.edu/career/Students/Major/kinesiology.htm
Internships & Careers (Indiana University, HPER, Dept of Kinesiology
    www.indiana.edu/~kines/careers/

2.4. Career Advice – Oceania, South Africa, UK

Careers (Griffith University. School of Physiotherapy and Exercise Science)
    www.griffith.edu.au/school/pes/
Career Paths in Physical Education (University of Otago, School of Physical Education)
    http://physed.otago.ac.nz/prospective/careers.html
Stellenbosch University. Faculty of Education. Sport Science: General Information
    htttp://academic.sun.ac.za/education/faculty/sport/studentinfo.html
University of Edinburgh. Moray House School of Education. Careers Planning for
    Sports Students
    www.education.ed.ac.uk/careers/sports.html
Australian Sports Commission Topics: Careers in Sport
    www.ausport.gov.au/info/topics/careers.asp

3. The Job Search Strategy

Finding the appropriate job can be very time-consuming, but rewarding if the out-come results in a positive job placement. A number of university websites provide succinct advice on the strategy to employ to find that good and challenging job. These sources include sections on:

- Resume writing
- Job interview preparation

- Job search strategies
- Networking
- Direct contact
- Job fairs
- Responding to advertisements and public announcements in newspapers, periodicals

See:
- University of New Hampshire. University Advising and Career Center
  www.unh.edu/uacc/
- Career CyberGuide (York University Career Services)
  www.yorku.ca/careers/cyberguide/introduction.html

Some of the commercial sports jobs websites also give advice on the job search. See

How to Conduct an Effective Job Search (MonsterTRAK)
  http://static.jobtrak.com/job_search_tips/search.html

3.1. Searching the Internet for Positions Available

A search for jobs available in a wide range of sport sciences can be done on some of the meta search engines including:
- Google www.google.com/
- Yahoo, www.yahoo.com/

Use many search terms such as:

- Coaching jobs
- Physical education jobs
- Employment sports
- Employment recreation
- Employment exercise science

3.2. Positions Available – North America

Job Opportunities in Sport (Canadian Interuniversity Sport)
   www.universitysport.ca/e/jobs/index.cfm
ISB Job Market (International Society of Biomechanics)
   www.isbweb.org/jobs/index.html
Sporting Goods Manufacturers Association Job Postings
   http://sgma.affiniscape.com./jobbankdisplaylistings.cfm

3.3. Positions Available – Oceania, UK

Australian Sports Commission, Jobs – Current Vacancies
   www.ausport.gov.au/jobs/index.asp
Sport and Recreation New Zealand (SPARC) Job Seekers
   www.sparc.org.nz/dashboard/job-seekers
Leisure Jobs UK
   www.leisurejobs.com/
Sports Coaching Jobs (UK)
   www.jobs.ac.uk/categories/sportscoaching

3.4. Subscription-Based Job Websites

Each of the websites listed here require a weekly, monthly or yearly subscription fee to be paid by the job seeker.

Careers in Sports, (SBRnet: Sports Business Research Network)
   www.sbrnet.com/
Women'Sports Job Wire&trade
   www.womensportsjobs.com/

4. Possible Occupations for University/College Graduates in Sport Sciences

Sub-Categories	Job Titles/Positions
Administration/Management	Event Manager Facility Manager General Manager (selected sports) Pharmaceutical/medical sales position Player Marketing Representative Public Relations /Sports Info Manager Regional/National Coaching Manager Risk Management/Insurance Manager Sports Center Manager Sport Marketing Director/Manager Sports Statistician Sporting Goods Product/Creative Director Sales /Marketing Manager
Athletics/Coaching/Training	Athletic Director/Manager Athletic Trainer Coach (professional athletes & teams) Coach (school, after school, city recreation) Coach (specific sports, national teams)
Exercise/Science/Ergonomics/ Adapted/ Rehabilitation	Cardiology Technician Massage Therapist Occupational Therapist Orientation & Mobility Instructor Paramedic Physical Therapist* Physiotherapist * Rehabilitation Therapist/Manager Respiratory Therapist Sport Equipment Developer Ultrasound Sonographer Vocational Rehabilitation Specialist Worker Health and Productivity Coordinator

Fitness/ Health/Health Promotion	Aerobics/Group Exercise Instructor Corporate Fitness Center Manager Clinical Kinesiologist/Research Assistant Fitness /Health Club Manager Fitness/ Strength/Conditioning/Exercise Prescription Specialist Health Promotion/Lifestyle/Wellness Specialist Nutritionist/Dietitian * Occupational Health/Safety Manager Personal Trainer Physical Activity Projects Officer
Mass Media/Information Technology	Information Manager & Technology Officer Sports Journalist Sports Commentator Sports Website Designer Venue Technology Manager Medical Information Technologist Health/Fitness Newsletter Editor
Physical Education/Teaching	Physical Educator (K-12) Specialist Physical Educator (armed services, police) Aquatics Supervisor Sports Program Officer Athlete Career and Education Advisor Dance & Movement Educator Sports Official/Judge/Umpire
Sports Psychology/Sociology	Sports Psychologist* Counselor Career Advisor

Recreation/Leisure/Outdoor	Adventure Consultant (adventure tourism, eco-tourism) Camp Instructor/Manager Outdoor Education Manager/Supervisor Park Ranger Recreation Program Officer Recreation Supervisor Resort Activity Leader Sports & Tourism Program Manager Sports Tour Operator Wilderness Therapy Program Instructor
Sport Sciences (Biomechanics, Biomedical Engineering, Sports Medicine, Research	Biomechanist Clinical Kinesiologist Ergonomist/Human Factors Engineer Exercise Physiologist * Injury Management Advisor Kinanthropometrist Psychomotor Behaviorist

*Notes:

Many states and provinces have legislated educational or credential requirements for these occupations. Check with the faculty career advisor for course or program requirements.

For example, the American College of Sports Medicine certifies the following:

- ACSM Health/Fitness Instructor
- ACSM Exercise Specialist®
- ACSM Registered Clinical Exercise Physiologist®

Physical Therapists /Physiotherapists are also regulated provincial and state bodies. See

- Federation of State Boards of Physical Therapy (US)
  www.fsbpt.org/
- Canadian Alliance of Physiotherapy Regulators
  www.alliancept.org/

About the Editor

Jan Borms holds an MSc from the University of Oregon and a PhD from the Vrije Universiteit Brussel (VUB) in Belgium. He is now professor emeritus of human biometry and health promotion after an academic career spanning five decades. From 1996 to 2001 he was head of the department of human biometry and bio-mechanics at the faculty of physical education at the VUB.

He was the founding president of the International Society for the Advancement of Kinanthropometry (ISAK) and is a member of several other international scientific organizations. Dr. Borms served for many years as an officer of the International Council of Sport Science and Physical Education (ICSSPE) and chaired its editorial board from 2000 to 2004. He was the recipient of the Philip Noel Baker Research Award in 1989 and is now honorary member of ICSSPE. He published widely in the field of kinanthropometry, human growth, and health promotion and has spoken at many conferences around the world. Dr. Borms is also the editor of the Karger series *Medicine and Sport Science*.